Advances in Analysis, Probability and Mathematical Physics

Mathematics and Its Applications

Volume 314

Advances in Analysis, Probability and Mathematical Physics

Contributions of Nonstandard Analysis

edited by

Sergio A. Albeverio
Faculty of Mathematics,
Ruhr-Universität Bochum,
Bochum, Germany

Wilhelm A. J. Luxemburg
Department of Mathematics,
California Institute of Technology,
Pasadena, California, U.S.A.

and

Manfred P. H. Wolff
Mathematical Institute,
University of Tübingen,
Tübingen, Germany

KLUWER ACADEMIC PUBLISHERS
DORDRECHT / BOSTON / LONDON

Library of Congress Cataloging-in-Publication Data

Advances in analysis, probability, and mathematical physics : contributions of nonstandard analysis / edited by Sergio A. Albeverio, Wilhelm A.J. Luxemburg, and Manfred P.H. Wolff.
p. cm. -- (Mathematics and its applications ; v. 314)
Proceedings of an international conference held July 1992 at the Heinrich Fabri Institut of the University of Tübingen.
Includes index.
ISBN 0-7923-3191-5
1. Nonstandard mathematical analysis--Congresses. I. Albeverio, Sergio. II. Luxemburg, W. A. J., 1929- . III. Wolff, M. P. H. (Manfred P. H.) IV. Series: Mathematics and its applications (Kluwer Academic Publishers) ; v. 314.
QA299.82.A38 1995
515'.33--dc20 94-35635

ISBN 0-7923-3191-5

Published by Kluwer Academic Publishers,
P.O. Box 17, 3300 AA Dordrecht, The Netherlands.

Kluwer Academic Publishers incorporates
the publishing programmes of
D. Reidel, Martinus Nijhoff, Dr W. Junk and MTP Press.

Sold and distributed in the U.S.A. and Canada
by Kluwer Academic Publishers,
101 Philip Drive, Norwell, MA 02061, U.S.A.

In all other countries, sold and distributed
by Kluwer Academic Publishers Group,
P.O. Box 322, 3300 AH Dordrecht, The Netherlands.

Printed on acid-free paper

Printed in the Netherlands

Table of Contents

PREFACE

During the last week of July, 1992 an international conference was held on nonstandard analysis at the "Heinrich Fabri Institut" of the University of Tübingen in Blaubeuren, Germany. The conference was dedicated to Professor Dr. D. Laugwitz on the occasion of his sixtieth birthday. With this dedication the participants wished to honor Professor Laugwitz for the pivotal role he played in the development of the theory of infinitesimals and for his contributions to its history.

The year 1958 is an important year in the history of the theory of infinitesimals. In that year there appeared in the Mathematische Zeitschrift the paper "Eine Erweiterung der Infinitesimalrechnung", written jointly by C. Schmieden, a professor of aeronautics at the "Technische Hochschule" of Darmstadt, and D. Laugwitz. The contents of this important paper were inspired by an idea of the late Professor Schmieden to extend the real number system in a certain way to include infinitely small as well as infinitely large numbers in order to obtain a number system which would allow a representation of the Dirac δ-function by means of a "point" function with values in the new number system. This was accomplished by embedding the real number system as constant sequences in the ring of sequence of reals with the natural identification for two sequences to define the same number when they are equal from a certain point onwards, a definition that in a way is akin with Cantor's definition of the reals. In this ring null sequences may be viewed as infinitesimals and those that approach infinity as infinitely large numbers. By introducing the notion of a "normal" function, Laugwitz showed in 1959 how the real functions of the calculus can be extended to normal functions on the ring. The family of normal functions, however, contains far more functions to allow the Dirac δ-function to be represented by normal functions. This representation gave a satisfactory answer to the question raised by Schmieden in terms of his ideas concerning the new ring of numbers extending the reals.

It seems that at the end of the fifties the theory of infinitesimals, which had survived since the days of Leibniz and Newton only as a matter of speech, was to enjoy a survival. Around 1960, from a completely different point of view and independently of the ideas of Schmieden and Laugwitz, Abraham Robinson introduced, using the ideas of model theory, an entirely new version of the theory of infinitesimals. He showed that there exist non-archimedean totally ordered fields containing the reals that in a model theoretic sense have the same properties as the reals. The announcement of this approach to the theory of infinitesimals appeared in 1961 under the title "Nonstandard Analysis," where nonstandard referred to the nature of the new fields of numbers as defined as nonstandard models of the first order theory of the reals. It turned out that the new systems of numbers were closely related. The ring of Schmieden and Laugwitz may be viewed as a reduced power of the theory of the reals and so shares with the reals fewer properties than the hyperreal number systems introduced by Robinson, being nonstandard models of the theory of the real number system.

During the last thirty years the use of nonstandard models in mathematics has taken its rightful place among the various methods employed by mathematicians. This development was in an essential way spurred on and inspired by Laugwitz. He was one of the initiators and inspiring organizers of a series of very successful "Tagungen" at the Mathematical Research Institute in Oberwolfach of which the first one was held in 1970.

Because of the controversial nature of the notion of infinitesimals throughout the history of the calculus Laugwitz, inspired by Robinson's historical account of the history of the theory of infinitesmals with which he finished his by now classical treatise on nonstandard analysis, started to examine further from a historical point of view the works of Leibniz, Newton, Euler and particularly Cauchy to trace the rise and fall of the use of the infinitesimal arguments in the calculus. From these historical investigations a much clearer picture has emerged of the role infinitesimals has played in the development of the calculus[1].

The contributions in this book, dedicated to Professor D. Laugwitz, have been selected by the Editors to present a panoramic view of the various directions in which nonstandard analysis is advancing and we hope may be a source of inspiration for the future.

The first group of contributions consists of papers in the areas *"Analysis, Topology and Topological groups"*.
In the contribution by *S. Albeverio, D. Guido, A. Ponosov, S. Scarlatti*, on *"Singular traces and nonstandard analysis"*, singular traces on the space of compact operators in a separable Hilbert space (of importance e.g. in non commutative geometry) are characterized in terms of nonstandard analysis, extending results of Dixmier and Varga. Corresponding ergodic states are also discussed, relating to previous work on Banach-Mazur limits by methods of nonstandard analysis.
In their article *"Navier–Stokes equations" M. Capiński* and *N. Cutland* prove the existence of weak solutions of a very general equation of this type. Moreover they succeed also in solving the correspondent stochastic differential equation.
The contribution by *E.I. Gordon, "Hyperfinite approximations of commutative topological groups"*, presents a nonstandard approach by hyperfinite approximations to harmonic analysis on arbitrary locall compact abelian groups having a compact open subgroup.
T. Norberg in *"A note on the myope topology"* discusses the identy of two natural topologies, the myope and Lawson's one, on the collection of compact subsets of a locally compact Hausdorff space. A characterization of the monads of the myope

[1]See the companion volume, also dedicated to Professor D. Laugwitz, "Analysis and Geometry", edited by B. Fuchststeiner and W. A. J. Luxemburg, BI Wissenschaftsverlag, Mannheim 1992.

topology is given, opening the way to the study of capacities in this setting.
M. Oberguggenberger in "Nonlinear theories of generalized functions" presents an overview of Rosinger-Colombeau's theory, pointing out its relations with nonstandard analysis. One of the main advantages of this theory is its ability to handle some of the multiplication problems for distributions. Applications to non linear partial differential equations are also discussed.
The contribution by *H. Osswald, "A nonstandard approach to the Pettis integral"*, presents lifting theorems for stochastic processes with values in a nonseparable locally convex space. Various applications are given including the existence of solutions for nondeterministic Peano-Caratheodory differential equations.
H. Ploss in *"A counterexpample to the spectral mapping Theorem revisited from a nonstandard point of view"* uses non standard analysis methods to provide a natural explanation for the nonvalidity of the spectral mapping theorem for the exponential function.
H. Render in *"Non standard polynomials in several variables"* extends results of C. Impens on nonstandard polynomials to the case of several variables and relates polynomial approximations in function theory with complex and real micro continuity of non standard polynomials.
In his contribution *"An existence result for a class of partial differential equations with smooth coefficients" T. Todorov* proves the existence of a solution for a class of linear partial differential equations with C^∞–coefficients in a space of generalized functions larger than the space of Schwartz distributions.
B. Wietschorke in *"On the generation of topology by external equivalence relations"* extends Reeken's theory of external equivalence relations in Nelson's Internal Set Theory to vector spaces. He gives a necessary and sufficient condition for the smallest monad containing the external set of infinitesimals to be the monad of a topology. Applications to locally convex spaces are given. *B. Zimmer* in *"Nonstandard hull of Lebesgue-Bochner spaces"* discusses the structure of the nonstandard hull of $L_p(\mu, E)$, showing that between it and $L_p(\hat{\mu}, \hat{E})$ (with $\hat{E}$ resp. $\hat{\mu}$ the nonstandard hull of E resp. the Loeb measure associated with μ) there is a space isometric to the space $M_p(\hat{\mu}, \hat{E})$ of $\hat{E}$-valued functions that have an SL^p -integrable-lifting. She also shows in particular that the Bochner integral extends naturally to $M_p(\hat{\mu}, \hat{E})$.

The second group consists of papers in *Probability Theory.*
The contribution by *H. Akiyama "A nonstandard approach to diffusions on manifolds and nonstandard heat kernels"* uses internal methods to give nonstandard constructions for diffusions and heat kernels on manifolds.
N.J. Cutland and *Siu-Ah Ng* in *"A nonstandard approach to the Malliavin Calculus"* give an intuitive approach to the Malliavin calculus for the classical Wiener space, showing in particular that the basic operators of this calculus have natural descriptions as classical differential operators on a nonstandard space R for some infinite natural number N. This has applications to stochastic differential equations.
Ergodicity of set transformations within the Alternative Set Theory is studied in the contribution by *M. Kalina "Ergodic transformations in AST"*. In fact ergodic transformations are shown to be homomorphic to a cycle of some suitable length.
In their contribution *"Nonstandard characterization for a general invariance prin-*

ciple" *D. Landers* and *L. Rogge* show that every invariance principle of probability theory is equivalent to a nonstandard construction of internal S-continuous processes, which all represent, modulo an infinitesimal error, the limit process. The contribution by *T. Lindstrøm "Anderson's brownian motion and the infinite dimensional Ornstein-Uhlenbeck process"* presents a natural extension of Anderson's construction of Brownian motion as infinitesimal random walk to the infinite dimensional Ornstein- Uhlenbeck process. A Brownian sheet representation is also given, as well as two other hyperfinite constructions, which are discussed in their respective merits.
In *"Two applications of NSA in the theory of Stochastic dynamical systems" A. Ponosov* presents a new infinite dimensional multiplicative ergodic theorem for generalized cocycles. The proof shows how methods of nonstandard analysis permits to extend Oseledec's methods to the infinite dimensional case in a natural way. Another application of nonstandard analysis (Loeb measures) to a proof of a new result about standard objects is also given, namely a construction of stationary solutions for nonlinear infinite dimensional stochastic differential equations with a monotone principal part, extending results of E. Browder.
The contribution by *D.A. Ross "Nonstandard methods and the space of experiments"* gives nonstandard proofs of two results of the theory of statistical experiments. The first one shows that the space of experiments is compact with respect to a certain weak topology. The second proof gives the validity of a criterium for the equivalence of experiments.

The third group of papers concerns applications of nonstandard analysis to *Mathematical Physics.*
L. Arkeryd in *"Infinite range forces and strong L^1-asymptotics for the space-homogeneous Boltzmann equation"* illustrates the strength of the applications of nonstandard analysis in the kinetic gas theory by discussing the strong L^1-convergence in time to Maxwellian limits for standard solutions of the Boltzmann equation starting far from equilibrium. In particular the convergence is proven for the space homogeneous Boltzmann equation without angular cut off, in a situation for which no nonstandard proof is known as yet.
In *"A nonstandard analysis approach to the theory of quantum meanfield systems" M.P.H. Wolff* develops an approach to the theory of (quantum mechanical) mean field systems in which the relevant algebras of observables are quotients of subalgebras of a hyperfinite tensor product of the relevant one-particle algebra. In particular, this underlines the statistical character of meanfield phenomena. A detailed discussion of meanfield dynamics and of the relation between the quantum and classical case is also given. The present approach also serves as a basis for a theory of fluctuations, to be discussed in subsequent work.

We would like to thank Professor Michael Hazewinkel and the Publishing House Kluwer for their offer to publish this book in their Mathematical Sciences Series.

The Editors

August 10, 1994

Part I:

ANALYSIS

SINGULAR TRACES AND NONSTANDARD ANALYSIS

S. Albeverio
Fakultät für Mathematik, Ruhr-Universität Bochum
D-44780 Bochum,
SFB-237 Essen-Bochum-Düsseldorf, CERFIM, Locarno

D. Guido
Dip. di Matematica, Universitá di Roma Tor Vergata, Italy

A. Ponosov †
Fakultät für Mathematik, Ruhr-Universität Bochum
D-44780 Bochum

S.Scarlatti
Dip. di Matematica, Universitá di L'Aquila, Italy

Abstract

We discuss non trivial singular traces on the compact operators, extending some results by Dixmier and Varga. We also give an explicit description of these traces and associated ergodic states using tools of non standard analysis.

1 Introduction

Let H be a complex separable Hilbert space and let $B(H)$ be the Banach algebra of all bounded linear operators on H. It is well-known that every non trivial normal (i.e. continuous in some sense; see Section 2 for precise definitions) trace on $B(H)$ is proportional to the usual one (see e.g. [6]). On the other hand Dixmier proved in 1966 [7] that there exist non trivial traces on $B(H)$ which are not normal (moreover they are "singular" in the sense of definition 2.2 below).

The basic idea of Dixmier for constructing singular traces was to consider the compact operators for which the usual trace diverges at a given suitable rate and to associate to any such compact operator a suitable element of the Banach space

†Supported by the Deutsche Forschungsgemeinschaft

$\ell^\infty(\mathbb{N})$ of all bounded sequences. The singular trace is then obtained evaluating on such elements a state on $\ell^\infty(\mathbb{N})$ which is invariant under "2-dilations".

The importance of singular traces is well-known now due to their applications in non commutative geometry and quantum field theory (see e.g. [3]).

The general question which operator ideals in $B(H)$ support traces has been recently studied by Varga [16] who used different kinds of states on $\ell^\infty(\mathbb{N})$.

In our two preceding papers [2], [3] we gave explicit formulas for Dixmier–type traces and introduced a new class of singular traces.

The present paper is organized as follows. In the first part of Section 2 we deal with the problem of describing which kind of sequences can be "summed" by a non trivial singular trace. We shall give a complete answer to this problem, generalizing some results by Varga [16].

In the second part of Section 2 we generalize the construction of non-normal traces given by Dixmier in [7].

In Section 3 we describe ergodic states giving rise to both kinds of singular traces introduced in Section 2. The basic technique we use is related to non standard analysis (NSA). The Section 3 also involves the representation of Banach-Mazur limits by NSA. Such representation have been discussed before - e.g. in [12], [13].

In Section 4 we work out explicitly the computation of the Dixmier traces of an operator again using the NSA framework in an essential way.

2 Singular traces and generalized eccentric operators

The content of this section is purely standard. So we will omit all the proofs which can be found in [4]. We start recalling some basic definitions and results.

Let $\mathcal{R}$ a von Neumann algebra and $\mathcal{R}^+$ the cone of its positive elements. A *weight* on $\mathcal{R}$ is a linear map

$$\phi : \mathcal{R}^+ \to [\prime,+\infty]$$

Any weight can be extended by linearity on the natural domain given by the linear span of $\{T \in \mathcal{R}^+ | \phi(\mathcal{T}) < +\infty\}$

A weight τ which has the property:

$$\tau(T^*T) = \tau(TT^*) \qquad \forall T \in \mathcal{R}$$

is called a *trace* on $\mathcal{R}$.

The natural domain of a trace is a two–sided ideal denoted by $\mathcal{I}_\tau$. For instance the natural domains of the trivial traces on $\mathcal{R}$ given by $\tau \equiv 0$ and $\tau \equiv +\infty$ are respectively the ideals $\mathcal{R}$ and $\{0\}$ while the usual trace on $B(H)$, the bounded linear operators on a complex, separable Hilbert space H, is associated with the ideal $L^1(H)$ of the trace class operators.

A weight ϕ on $\mathcal{R}$ is called *normal* if for every monotonically increasing generalized sequence $\{T_\alpha, \alpha \in I\}$ of elements of $\mathcal{R}^+$ such that $T = \sup_\alpha T_\alpha$ one has

$$\phi(T) = \lim_\alpha \phi(T_\alpha)$$

From now on the von Neumann algebra $\mathcal{R}$ will be fixed to be $B(H)$.

A classical result [6] concerning normal traces on $B(H)$ is the following:

Theorem 2.1 Dixmier *Every non trivial normal trace on $B(H)$ is proportional to the usual trace.*

By a theorem of Calkin (see [10]), each proper two-sided ideal in $B(H)$ contains the finite rank operators and is contained in the ideal $K(H)$ of the compact linear operators on H. Therefore all traces on $B(H)$ live on the compact operators, and the following definition makes sense:

Definition 2.2 *A trace τ on $B(H)$ will be call* singular *if it vanishes on the set $F(H)$ of finite rank operators.*

Proposition 2.3 *Any trace τ on $K(H)$ can be uniquely decomposed as $\tau = \tau_1 + \tau_2$, where τ_1 is a normal trace and τ_2 is a singular trace.*

In view of this proposition, in the rest of the paper we shall restrict our attention to the singular traces.

For T a compact operator on H, $\{\mu_n(T)\}_{n=1}^\infty$ will denote the non increasing sequence of the eigenvalues of $|T|$ with multiplicity.

We shall also set $\sigma_n(T) \equiv \sum_{k=1}^{n} \mu_k(T)$.

Definition 2.4 *Let T be a compact operator. We call* integral sequence *of T the sequence $\{S_n(T)\}_{n=0}^\infty$ which is an indefinite integral (w.r.t. the counting measure) of $\{\mu_n(T)\}_{n=1}^\infty$, i.e. $S_n(T) - S_{n-1}(T) = \mu_n(T), n \geq 1$, and such that*

$$S_0(T) \equiv \begin{cases} 0 & T \notin L^1(H) \\ -tr(T) & T \in L^1(H) \end{cases}$$

Notice that if $T \notin L^1(H)$, $S_n(T) = \sigma_n(T)$, $n \geq 1$, while if $T \in L^1(H)$, then $S_n(T) = \sigma_n(T) - tr(T) \to 0$. We also give the following

Remark 2.5 *If T does not belong to $L^1(H)$ and τ is a trace which is finite and non-zero on $|T|$ then τ is necessarily singular, that is, the existence of traces which are non trivial on T is equivalent to the existence of non trivial singular traces on T. Since for $T \in L^1(H)$ the existence of a non trivial trace is obvious, it follows that the relevant question is not the mere "traceability" of a compact operator T, but the existence of a singular trace which is non trivial on $|T|$.*

Let us also notice that a trace τ is finite on $|T|$ if and only if the principal ideal $\mathcal{I}(T)$, i.e. the (two–sided) ideal generated by T in $B(H)$, is contained in $\mathcal{I}_\tau$.

Lemma 2.6 *Let T be a compact operator. The following are equivalent*

[1] 1 is a limit point of the sequence $\left\{\frac{S_{2n}(T)}{S_n(T)}\right\}_{n=0}^{\infty}$

[2] There exists an increasing sequence of natural numbers $\{p_k\}$ such that $\lim_{k\to\infty} \frac{S_{kp_k}(T)}{S_{p_k}(T)} = 1$

Definition 2.7 *A compact operator T which satisfies one of the equivalent properties of Lemma 2.6 will be called generalized eccentric.*

Remark 2.8 *The class of generalized eccentric operators which are not in $L^1(H)$ coincides with the class of eccentric operator considered in [16].*

We can now state the main result of this section.

Theorem 2.9 *Let T be a compact operator. Then the following are equivalent:*

[1] There exists a singular trace τ such that $0 < \tau(|T|) < +\infty$.

[2] T is generalized eccentric.

As we have already mentioned the proof of the theorem can be found in [4]. Nevertheless we are going to discuss here some possible procedures to construct singular traces on $K(H)$ because it is important for the non standard part of this work. Basically, all singular traces in question can be represented as follows:

$$\tau_\psi(A) \equiv \psi\left(\left\{\frac{S_n(A)}{S_n(T)}\right\}\right) \quad , \quad A \in \mathcal{I}(\mathcal{T}) \tag{1}$$

where ψ is a generic non normal state on $\ell^\infty(\mathbb{N}) \equiv \ell^\infty$.

Recall that the state is just a linear bounded positive functional with the property $\psi(\{1\}) = \|\psi\| = 1$. It is easy to see that for such a ψ the function τ_ψ has all properties of traces with the only exception of additivity which requires some more conditions to be imposed upon. In the literature there are two ways to do this. Although both of them were introduced in more particular situations than we are dealing with they can be adapted to our situation too. The first way is due to Dixmier [D] and consists in choosing ψ to be 2-dilation invariant (therefore ψ vanishes on the set c_0 of the infinitesimal sequences, hence it is not normal):

$$\psi(\{a_n\}) = \psi(\{a_{2n}\}) \qquad \forall a \in \ell^\infty.$$

In this case the generalized eccentric operator T must satisfy the stronger condition:

$$\lim_{n\in\mathbb{N}} \frac{S_{2n}(T)}{S_n(T)} = 1,$$

(compare with the conditions given in Lemma 2.6).The second way of getting additivity was suggested by Varga [16] who considered non normal states supported by "fast" sequences, i.e. the states of the form:

$$\psi(\{a_n\}) = \varphi(\{a_{n_k}\}) \qquad \forall a \in \ell^\infty. \tag{2}$$

where φ is an arbitrary non normal state on ℓ^∞ and the sequence n_k is given by $n_k \equiv kp_k$ with p_k defined in Lemma 2.6. We now formulate briefly two results on the existence of singular traces.

Theorem 2.10 *Let T be a generalized eccentric operator. The functional τ_ψ defined in (1) and (2) linearly extends to a singular trace on the ideal $\mathcal{I}(\mathcal{T})$.*

Theorem 2.11 *If ψ is a two-dilation invariant state and $\lim_{n\in\mathbb{N}} \frac{S_{2n}(T)}{S_n(T)} = 1$, then τ_ψ is a trace on $\mathcal{I}(\mathcal{T})$. Moreover, in this case formula (1) gives rise to a singular trace (which will be denoted by τ_ψ as well) even on the (larger) ideal*

$$\mathcal{I}_m(T) \equiv \left\{A \in K(H) | \left\{\frac{S_n(A)}{S_n(T)}\right\} \in \ell^\infty\right\}$$

We would like to point out that, when $T \notin L^1(H)$, the ideal $\mathcal{I}_m(T)$ is a maximal norm ideal in the sense of Shatten [15] (see also [10]).

Remark 2.12 *We should stress that we did get new types of singular traces. Of course, if the operator $T \notin L^1(H)$ then the traces described in Theorems 2.10 and 2.11 are exactly the traces discussed by Dixmier in [7] and by Varga in [16] respectively, and they vanish on $L^1(H)$.. In the case $T \in L^1(H)$ our Theorems 2.10 and 2.11 produce new classes of singular traces, which in a sense are the inverse images inside $L^1(H)$ of the class of Varga-type traces and of the class of Dixmier-type traces respectively. We notice also that if the trace τ_2 in the decomposition of proposition 2.3 does not vanish, it should be of this new type, i.e. it should not vanish on $L^1(H)$.*

The existence of new Dixmier-type traces was announced in [AGPS 2].

3 Two-dilation invariant states and ergodicity

The main problem we are going to discuss in this section concerns extremal (ergodic) states which give rise to singular traces. In our opinion it is non standard analysis which supplies us the most convenient tools for this purpose. We first briefly discuss extremal states corresponding to the Varga-type traces described in Theorem 2.11. Let us denote by $\Delta_{\{n_k\}}$ the set of all extremal points in the set of non normal states of the form (2).

Proposition 3.1 *The set $\Delta_{\{n_k\}}$ consists of the states*

$$\psi(\{a_n\}) \equiv {}^{\circ}({}^*a_{n_m})$$

for some $m \in {}^*\mathbb{N}_\infty = {}^*\mathbb{N} - \mathbb{N}$.

Proof: It immediately follows from the fact that extremal states on ℓ^∞ are all of the form

$$\varphi(\{a_n\}) = {}^{\circ}({}^*a_m), m \in {}^*\mathbb{N}. \tag{3}$$

Since additionally φ must vanish on the set c_0 then m becomes infinitely large. ∎

Remark 3.2 *Of course, instead of infinitely large numbers one can equivalently use the Stone-Čech compactification $\overline{\mathbb{N}}$ of $\mathbb{N}$ and the isomorphism $\ell^\infty(\mathbb{N}) \simeq C(\overline{\mathbb{N}})$ given by the Gelfand transform in order to describe extremal states in Proposition 3.1. Namely, they will be given by Dirac measures supported by the set $\overline{\mathbb{N}} \setminus \mathbb{N}$. On the contrary, the classification of ergodic 2-dilation invariant states does require NSA (see e.g. the next remark).*

Now we come to the much more difficult problem of classification of two dilation invariant states. First, we remark that in order to prove the existence of such states, Dixmier invoked the amenability of the affine group. As promised, here we shall adopt an alternative point of view which relies on the use of NSA and related methods (see e.g. [11], [1]).

Theorem 3.3 *The map $\omega \to \varphi_\omega$, $\omega \in {}^*\mathbb{N}_\infty$, defined by*

$$\varphi_\omega(a) \equiv {}^\circ\left(\frac{1}{\omega}\sum_{k=1}^{\omega} {}^*a_{2^k}\right) \tag{4}$$

takes values in the convex set of 2–dilation invariant states over ℓ^∞.

Proof: Let $b_n \equiv \frac{1}{n}\sum_{k=1}^{n} a_{2^n}$. Since $\{a_n\}$ is bounded $\{b_n\}$ is also bounded so that $\varphi_\omega(a) = {}^\circ({}^*b_\omega)$ is well defined for all ω. Obviously, φ_ω is a state. It is also 2–dilation invariant since:

$$\varphi_\omega(\{a_{2n}\}) - \varphi_\omega(\{a_n\}) = {}^\circ\left(\frac{1}{\omega}({}^*a_{2^{\omega+1}} - {}^*a_2)\right) = 0. \blacksquare$$

A consequence of this theorem is that an explicit formula for the previously introduced traces can easily be given.

Corollary 3.4 *If T is an operator verifying $\lim_n \frac{S_{2n}}{S_n} = 1$ and ω is an infinite hypernatural then*

$$\tau_\omega(A) \equiv {}^\circ\left(\frac{1}{\omega}\sum_{k=1}^{\omega}\frac{{}^*S_{2^k}(A)}{{}^*S_{2^k}(T)}\right) \qquad A \in \mathcal{I}_\Updownarrow(\mathcal{T}) \tag{5}$$

is one of the singular traces described in Theorem 2.12.

The proof of this corollary follows immediately from Theorems 2.12 and 3.3.

There is a simple generalization of the formula (4) which describes 2-dilation invariant states. If $j \in {}^*\mathbb{N}$ and $n \in {}^*\mathbb{N}_\infty$ the map

$$\{a_k\} \to {}^\circ\left(\frac{1}{n}\sum_{i=1}^{n} {}^*a_{j2^i}\right) \tag{6}$$

is a 2–dilation invariant state over ℓ^∞ and therefore gives rise to a singular trace. Since any hypernatural j can be written in a unique way as a product of an odd number and a power of $2, j = (2m-1)2^{k-1}$, we may rewrite the previous states as

$$\varphi_{k,m,n}(a) = {}^\circ\left(\frac{1}{n}\sum_{i=k+1}^{k+n} {}^*a_{(2m-1)2^{i-1}}\right) \quad k,m \in {}^*\mathbb{N}, n \in {}^*\mathbb{N}_\infty \tag{7}$$

In the rest of the section we shall study states of the form (7) in relation to the problem of ergodicity.

Let $\Delta : \mathbb{N} \to \mathbb{N}$ the multiplication by 2, Δ_* the corresponding morphism on $\ell^\infty(\mathbb{N})$, $\Delta_*(\{a_n\}) = \{a_{2n}\}$, we shall say that the state φ is Δ-invariant if $\varphi \circ \Delta_* = \varphi$. We shall give necessary conditions for extremality in the (convex compact) set of Δ-invariant states in terms of NSA.

It is known (see e.g. [8], p.113) that the states on $\ell^\infty(\mathbb{N})$ can be identified with the finitely additive probability measures on $\mathbb{N}$, therefore we shall denote any such a state by μ, and the notation $\mu(A)$ with $A \subset \mathbb{N}$ makes sense.

Moreover extremality of a Δ-invariant state μ can be expressed in terms of ergodicity of μ seen as a measure, i.e. for each $A \subset \mathbb{N}$ such that $\Delta A = A$, $\mu(A) = 0$ or 1.

Remark 3.5 *Using the Stone-Čech compactification $\overline{\mathbb{N}}$ of $\mathbb{N}$ and the isomorphism $\ell^\infty(\mathbb{N}) \simeq C(\overline{\mathbb{N}})$ given by the Gelfand transform once again we get an identification of the states on $\ell^\infty(\mathbb{N})$ with the σ-additive probability Radon measures on $\overline{\mathbb{N}}$. On the other hand a transformation on $\mathbb{N}$ extends to a continuous transformation on $\overline{\mathbb{N}}$. We shall denote with $\overline{\mu}$, $\overline{\Delta}$ the measure and the transformation on $\overline{\mathbb{N}}$ induced by μ and Δ respectively. It turns out that ergodicity of $\overline{\mu}$ is equivalent to ergodicity of the finitely additive measure μ. This equivalence can be shown using well known criteria for ergodicity (see e.g. [10]).*

Remark 3.6 *Let us consider the correspondence $\eta : \mathbb{N} \times \mathbb{N} \to \mathbb{N}$ defined by $(m,n) \to (2m-1)2^{n-1}$, which is a bijection. It induces an isomorphism: $\eta_* : \ell^\infty(\mathbb{N}) \to \ell^\infty(\mathbb{N} \times \mathbb{N})$ given by*

$$(\eta_* a)_{m,n} \equiv a_{\eta(m,n)},$$

$\Delta_* \equiv \eta_*^{-1}\Delta\eta_*$ becoming the translation T in the second variable: $\Delta_*(m,n) = (m, n+1) \equiv (m, Tn)$. It might be thought that the isomorphism η_* gives rise to the splitting of the dynamical system $(\overline{\mathbb{N}}, \overline{\Delta})$ in a product of two dynamical systems $(\overline{\mathbb{N}}, id)$ and $(\overline{\mathbb{N}}, \overline{T})$, thus furnishing a standard approach to the considerations we shall make below. Unfortunately this is not true since the spaces $\overline{\mathbb{N}} \times \overline{\mathbb{N}}$ and $\overline{\mathbb{N} \times \mathbb{N}}$ are not homeomorphic (see e.g. [9]). Our idea is to exploit the features of non-standard analysis, in particular the nice functorial property ${}^*\mathbb{N} \times {}^*\mathbb{N} = {}^*(\mathbb{N} \times \mathbb{N})$.

Let M_Δ denote the set of extremal Δ-invariant (i.e. ergodic) states on $\ell^\infty(\mathbb{N})$.

Proposition 3.7 *Any $\mu \in M_\Delta$ coincides with one of the states $\varphi_{k,m,n}$ for some $k, m \in {}^*\mathbb{N}$, $n \in {}^*\mathbb{N}_\infty$.*

Proof: Evidently, for each state μ and each finite dimensional $E \in \ell^\infty(\mathbb{N}) \equiv \ell^\infty$ there exist numbers $s(E)$ and $\lambda_j(E)$ such that

[1] $\mu(a) = \sum_{j=1}^{s(E)} \lambda_j(E) a_j \quad (\forall a \in E)$,

[2] $\lambda_j(E) \geq 0$,

[3] $\sum_{j=1}^{s(E)} \lambda_j(E) = 1$.

For every hyperfinite dimensional space E (see e.g. [1], p.55), $\ell^\infty \subset E \subset {}^*\ell^\infty$, there is therefore an internal set $\{\lambda_1, \dots, \lambda_s\}$ satisfying (ii) and (iii) and such that

$$\mu(a) = {}^*\mu({}^*a) = \sum_{j=1}^{s} \lambda_j {}^*a_j \qquad (\forall a \in \ell^\infty)$$

If μ is 2-dilation invariant, then for all finite n

$$\mu(a) = \sum_{j=1}^{s} \lambda_j \frac{1}{n} \sum_{i=1}^{n} {}^*a_{j2^{i-1}} \qquad (\forall a \in \ell^\infty).$$

As it can be easily checked, for each finite dimensional space $F \subset \ell^\infty$, and each $n \in \mathbb{N}$ we have

$$\alpha_{n,{}^*F} \equiv \sup_{a \in {}^*F, \|a\| \leq 1} \sup_{1 \leq j \leq s} \left\{ |{}^*\mu(a) - \sum_{j=1}^{s} \lambda_j \frac{1}{n} \sum_{i=1}^{n} a_{j2^{i-1}}| + \right.$$

$$\left. + |{}^*\varphi_{k,m,n}(a) - \frac{1}{n} \sum_{i=1}^{n} a_{j2^{i-1}}| \right\} \approx 0 \tag{8}$$

where $j = (2m-1)2^{k-1}$. By saturation, there are a hyperfinite dimensional space F , $\ell^\infty \subset F \subset {}^*\ell^\infty$ and a number $n \in^* \mathbb{N}_\infty$ such that (8) remains valid. This means that

$$^*\mu(a) \approx \sum_{j=1}^{s} \lambda_j {}^*\varphi_{m,k,n}(a) \quad (j = (2m-1)2^{k-1}) \tag{9}$$

for all $a \in \ell^\infty$ and immediately implies that the set of all $T-$invariant states coincides with the closed convex hull $\overline{co}M$ of the set $M \equiv \{\mu_{k,n} \mid k \in^* \mathbb{N},\, n \in^* \mathbb{N}_\infty\}$.

Finally, we show that M is closed which, according to [8], p.708, would imply the inclusion $M_\Delta \subset M$. Consider a directed set $\Gamma \subset^* \mathbb{N}^3$ and assume that $\varphi_{k,m,n} \xrightarrow{\Gamma} \mu$. Clearly, for every finite dimensional $E \subset \ell^\infty$ there exists a subsequence $\{(k_i, m_i, n_i)\} \in \Gamma$ such that

$$^*\varphi_{k_i,m_i,n_i}(a) \approx {}^*\mu(a) \quad \text{for some} \quad i \in^* \mathbb{N}_\infty \quad \text{and for all} \quad a \in {}^*E,\ \|a\| \le 1.$$

If μ is Δ invariant then by saturation the latter relation holds for a certain hyperfinite dimensional $E \supset \ell^\infty$ with $k \in^* \mathbb{N}$ and $n \in^* \mathbb{N}_\infty$. This concludes the proof. ∎

Corollary 3.8 *The problems of description of extremal 2-dilation invariant and extremal translation invariant states are equivalent.*

Proof: Recalling the map η_* given by $(\eta_* a)_{m,n} = a_{\eta(m,n)}$ where $\eta_*(m,n) = (2m-1)2^{n-1}$ and applying the proposition just proved we conclude that for each fixed m the states $\varphi_{k,m,n} \circ \eta_*$ coincide with the translation invariant states $\mu_{k,n}$ on ℓ^∞ defined by

$$\mu_{k,n}(a) \equiv^\circ \left(\frac{1}{n} \sum_{i=k+1}^{k+n} {}^*a_i \right),$$

so that any $\mu \in \eta_*^{-1}(M_\Delta)$ is contained in one of the sets $\{\delta_m \otimes \nu | \nu \in M_T \cap M\}$ where δ_m is given by $\delta_m(a) \equiv {}^\circ({}^*a_m), m \in^* \mathbb{N}$, M_T stands for the set of extremal translation invariant states on ℓ^∞ and $M \equiv \{\mu_{k,n} | k \in^* \mathbb{N}, n \in^* \mathbb{N}_\infty\}$. On the other hand it is known (see [12], [13], or also [2] where the result was proved independently) that $M_T \subset M$ which completes the proof. ∎

Remark 3.9 *Of course, the representation given in Proposition 3.7 is not unique due to Corollary 3.8 and the following trivial*

Proposition 3.10 *If* $\frac{k-l}{n} \approx 0$ *and* $\frac{p-n}{n} \approx 0$, *then* $\mu_{k,n} = \mu_{l,p}$.

Now we formulate the main result in this section more precisely.

Theorem 3.11 *If* μ *is an extremal 2-dilation invariant state on* ℓ^∞ *then* $\mu = \varphi_{k,m,n}$ *(with* $\varphi_{k,m,n}$ *defined by (7)) for some* $m \in^* \mathbb{N}$ *and infinitely large hypernaturals* k *and* n *such that* $\frac{n}{k} \approx 0$.

Proof: By virtue of Corollary 3.8 it suffices to show that if $\nu \in M_T$ then $\nu = \mu_{k,n}$ for some $k, n \in^* \mathbb{N}_\infty$. We first prove that $k \in^* \mathbb{N}_\infty$. Suppose it is not the case and k is finite. Without loss of generality we can assume $k = 1$, and, due to proposition 3.10, $n = 2m$. If we show that $\mu_{1,n} \ne \mu_{1,m}$ then the representation $\mu_{1,n} = \frac{1}{2}(\mu_{1,m} + \mu_{m,m})$ implies $\mu_{1,n}$ is not extremal.

For $b \equiv m2^{-p}$, choose $p \in^* \mathbb{N}_\infty$ such that $0 < {}^\circ b < \infty$ and consider a non decreasing sequence $\{b_j\}$ such that ${}^*b_{2p} \approx b$. Define a sequence $\{c_j\}$ of natural numbers by putting $c_j \equiv [2^j b_j]$ ($[\cdot]$ denotes the integer part). Since

$$c_j \leq \frac{1}{2}[2^{j+1}b_j] + 1 \leq \frac{1}{2}[2^{j+1}b_{j+1}] + 1 = \frac{1}{2}c_{j+1} + 1,$$

one gets $c_{j+1} \geq 2c_j - 2$. At the same time,

$$\begin{aligned} {}^o\left(\frac{{}^*c_{2p} - m}{m}\right) &= {}^o\left(\frac{{}^*c_{2p} - m}{2^{2p}}\right){}^o\left(\frac{2^{2p}}{m}\right) = {}^o\left(\frac{[2^{2p*}b_{2p}]}{2^{2p}} - b\right)\frac{1}{{}^ob} \\ &\leq \frac{1}{{}^ob}{}^o({}^*b_{2p} - b) = 0. \end{aligned}$$

By the same reason, $\frac{{}^*c_{2p+1} - n}{n} \approx 0$. Applying proposition 3.10 once again we obtain

$$\frac{1}{m}\sum_{i=1}^{m}{}^*a_i \approx \frac{1}{{}^*c_{2p}}\sum_{i=1}^{{}^*c_{2p}}{}^*a_i\,; \quad \frac{1}{n}\sum_{i=1}^{n}{}^*a_i \approx \frac{1}{c_{2p+1}}\sum_{i=1}^{{}^*c_{2p+1}}{}^*a_i \qquad (\forall a \in \ell^\infty) \tag{10}$$

Now we introduce a set $B = \cup_{q=1}^\infty [c_{2q-1}, c_{2q}] \subset \mathbb{N}$ and a sequence $\chi = \{\chi_i\}$ where $\chi_i = 1$ for $i \in B$ and 0 otherwise. By (10),

$$\begin{aligned} \mu_{1,n}(B) &= {}^o\left(\frac{1}{{}^*c_{2p+1}}\sum_{i=1}^{{}^*c_{2p+1}}{}^*\chi_i\right) \\ &= {}^o\left(\frac{{}^*c_{2p}}{{}^*c_{2p+1}}\right){}^o\left(\frac{1}{{}^*c_{2p}}\sum_{i=1}^{{}^*c_{2p}}{}^*\chi_i + \frac{1}{{}^*c_{2p}}\sum_{i={}^*c_{2p}+1}^{{}^*c_{2p+1}}{}^*\chi_i\right) \\ &= {}^o\left(\frac{[2^{p*}c_p]}{[2^{p+1*}c_{p+1}]}\right)\mu_{1,m}(B) \leq \frac{1}{2}\mu_{1,n}(B). \end{aligned}$$

It remains to prove that $\mu_{1,m}(B) \neq 0$. In order to see this, let us observe that ${}^*\chi_i = 1$ for $i \in [{}^*c_{2p-1}, {}^*c_{2p}]$ and that ${}^*c_{2p} - {}^*c_{2p-1} \geq {}^*c_{2p-1} - 2$. Hence, $\sharp\{i \mid {}^*\chi_i = 1\} \geq \frac{1}{2}{}^*c_{2p} - 1$ and

$$\mu_{1,m}(B) = {}^o\left(\frac{1}{{}^*c_{2p}}\sum_{i=1}^{{}^*c_{2p}}{}^*\chi_i\right) \geq {}^o\left(\frac{1}{{}^*c_{2p}}\left(\frac{1}{2}{}^*c_{2p} - 1\right)\right) = \frac{1}{2},$$

so that $\mu_{1,n}$ is not extremal. We continue the proof assuming $k \in^* \mathbb{N}_\infty$ and $\frac{n}{k} \not\approx 0$ or, equivalently, ${}^o\left(\frac{k}{n}\right) < \infty$. We have to show that again $\mu_{k,n}$ is not extremal. First we notice that

$$0 \leq \frac{-k + [\sqrt{k} + 1]^2}{n} \leq \frac{2\sqrt{k} + 1}{n} \approx \frac{2\sqrt{k}}{n} = \frac{2k}{n\sqrt{k}} \approx 0$$

and analogously,

$$\frac{-k+[\sqrt{k}]^2}{n}\approx 0,\quad \frac{k+n-[\sqrt{k+n}]^2}{n}\approx 0,\quad \frac{k+n-[\sqrt{k+n}+1]^2}{n}\approx 0.$$

Applying now proposition 3.10, one can assume that $k=(2r)^2$, $k+n=(2s)^2$. Let us introduce a set $C\equiv\cup_{i=1}^{\infty}[(2i-1)^2,(2i)^2]$ and observe that

$$\mu_{k,n}(C\Delta TC)\leq {}^{o}\left(\frac{\sharp\{i\mid i^2\in[k,k+n]\}}{n}\right)={}^{o}\left(\frac{\sharp\{i\mid 2r\leq i\leq 2s\}}{(2s)^2-(2r)^2}\right)=0.$$

Extremality of μ would imply, therefore, that $\mu_{k,n}(C)$ should have been equal to 0 or 1. On the other hand,

$$\begin{aligned}\mu_{k,n}(C) &= {}^{o}\left(\frac{1}{n}\sum_{i=r+1}^{s}\left((2i)^2-(2i-1)^2\right)\right)\\ &= {}^{o}\left(\frac{(4s+1+4r+3)(r-s)}{8r^2-8s^2}\right)=\frac{1}{2}.\end{aligned}$$

This contradiction implies the result. ∎

Remark 3.12 *The necessary conditions in theorem 3.11 are surely not sufficient. To see this, one can consider a state* $\mu\equiv\mu_{4^p-n,n}$ *for arbitrary* $p, n\in{}^*\mathbb{N}_\infty$. *If* $n\geq 2^{2p-1}$, *then non extremality of* μ *follows from the theorem. Otherwise, we may introduce a set* $A\equiv\cup_{q=1}^{\infty}(2^{2q-1},2^{2q}]$ *which is easily seen to be* $\mu-$*a.e.* $T-$*invariant, but* $\mu(A)=\frac{1}{2}$.

Now we give a Corollary which relates the results of this section with the description of the Dixmier-type traces (for the proof cf. [2]).

Corollary 3.13 *Let* τ *be a Dixmier-type trace on the ideal* $\mathcal{I}_m(T)$ *(see section 2). Then* τ *is in the closure of the convex hull of the family*

$$\{\tau_{k,m,n}\mid m\in{}^*\mathbb{N},\ k,n\in{}^*\mathbb{N}_\infty,\ (\frac{n}{k})\approx 0\},$$

where $\tau_{k,m,n}$ *is the trace associated with the state* $\tau_{\varphi_{k,m,n}}$ *via formula (1) on the same domain* $\mathcal{I}_m(T)$.

Remark 3.14 *The states* $\mu_{k,n}$ *can be seen intuitively as averages on intervals of the set* ${}^*\mathbb{N}$. *This suggests to call ergodic all the intervals associated with ergodic states.*

Then it is easy to show that if the interval I is ergodic and a subinterval J is such that $\frac{|I|}{|J|} \not\approx 0$ (where $|I|$ denotes the length of I), then $\mu_I = \mu_J$. A sketch of the proof is the following: let $I = I_0 \cup J \cup I_1$ be a partition of I into subintervals. It turns out that

$$^o\left(\frac{|I_0|}{|I|}\right)\mu_{I_0} + {}^o\left(\frac{|J|}{|I|}\right)\mu_J + {}^o\left(\frac{|I_1|}{|I|}\right)\mu_{I_1} = \mu_I,$$

hence, by the ergodicity of I, $\mu_I = \mu_J$.

4 A computational example

We shall now discuss some advantages of representing singular traces by means of NSA.

A remarkable advantage lies, in our opinion, in the increased computability of the value of a singular trace on a given operator when such a trace is parameterized by some infinite number.

In what follows, we shall work out an example in which we explicitly calculate the value of the Dixmier trace of an operator, even though it depends on the non-standard parameter.

To this aim we shall make use of formula (5) choosing a compact operator T such that $S_n(T) = \log n$. The choice of "summing" logarithmic divergences has extensively been used by Connes in some applications to non-commutative geometry [3].

Let $q \geq 1$ be a fixed natural number, we consider a positive compact operator A_q whose sequence of eigenvalues $(\lambda_n \mid n = 3, 4. \ldots)$ is defined in the following way: let $(n_k \mid k = 0, 1, \ldots)$ be an unbounded increasing sequence of natural numbers (with $n_0 \equiv 1$) whose explicit dependence on q will be given below. For $n \in (2^{n_k}, 2^{n_{k+1}}]$, we define

$$\lambda_n := \frac{n_{k+1} - n_k}{2^{n_{k+1}} - 2^{n_k}} \tag{11}$$

For $m \geq 2$ we consider the sum $\sigma_{2^m} := \sum_{j=3}^{2^m} \lambda_j$. Let $n_k < m \leq n_{k+1}$, then we have

$$\sigma_{2^m} = n_k + \frac{2^m - 2^{n_k}}{2^{n_{k+1}} - 2^{n_k}} \cdot (n_{k+1} - n_k) - 1 \tag{12}$$

since

$$\sigma_{2^m} = \sum_{r=0}^{k-1} \sum_{j=2^{n_r}+1}^{2^{n_{r+1}}} \lambda_j + \sum_{j=2^{n_k}+1}^{2^m} \lambda_j.$$

Now let $p > 1$ and hence $n_s < p \leq n_{s+1}$ for some s, we have

$$\frac{1}{p}\sum_{m=1}^{p}\frac{\sigma_{2^m}}{\log 2^m} = \frac{1}{\log 2}\cdot\frac{1}{p}\left(\sum_{k=0}^{s-1}\sum_{m=n_k+1}^{n_{k+1}}\frac{\sigma_{2^m}}{m} + \sum_{m=n_s+1}^{p}\frac{\sigma_{2^m}}{m}\right) \tag{13}$$

We now proceed to estimate the sums appearing on the r.h s. of (13). By means of (12) we have

$$\begin{aligned}\sum_{m=n_k+1}^{n_{k+1}}\frac{\sigma_{2^m}}{m} &= \left[n_k - 1 - \frac{2^{n_k}}{2^{n_{k+1}}-2^{n_k}}(n_{k+1}-n_k)\right]\sum_{m=n_k+1}^{n_{k+1}}\frac{1}{m} \\ &+ \frac{n_{k+1}-n_k}{2^{n_{k+1}}-2^{n_k}}\cdot\sum_{m=n_k+1}^{n_{k+1}}\frac{2^m}{m}\end{aligned} \tag{14}$$

We notice that the following equalities hold

$$\sum_{m=n_k+1}^{n_{k+1}}\frac{1}{m} = \log\frac{n_{k+1}}{n_k} + O\left(\frac{1}{n_k} - \frac{1}{n_{k+1}}\right) \tag{15}$$

$$\sum_{m=n_k+1}^{n_{k+1}}\frac{2^m}{m} = \frac{1}{\log 2}\left(\frac{2^{n_{k+1}}}{n_{k+1}} - \frac{2^{n_k}}{n_k}\right)\left(1 + O\left(\frac{1}{n_k}\right)\right) \tag{16}$$

from which it follows

$$\sum_{m=n_k+1}^{n_{k+1}}\frac{\sigma_{2^m}}{m} = \left[n_k\log\frac{n_{k+1}}{n_k} + O\left(1 - \frac{n_k}{n_{k+1}}\right)\right]\left[1 + O\left(\frac{1}{n_k}\right)\right] \tag{17}$$

under the assumption $O\left(\frac{1}{n_k}\right) \geq O\left(\frac{2^{n_k}}{2^{n_{k+1}}}\right)$.

To verify such a condition we fix the initial sequence $(n_k|k = 0, 1, \ldots)$ to be of the form $n_k := 2^{kq}$, where $q \in \mathbb{N}$.

Formula (17) takes then the form:

$$\sum_{m=n_k+1}^{n_{k+1}}\frac{\sigma_{2^m}}{m} = \left[2^{kq}q\log 2 + O(1)\right]\left[1 + O(2^{-kq})\right] \tag{18}$$

Therefore we obtain

$$\frac{1}{p\log 2}\sum_{k=0}^{s-1}\sum_{m=n_k+1}^{n_{k+1}}\frac{\sigma_{2^m}}{m} = \frac{1}{q\log 2}\left(p\log 2\frac{2^{sq}-1}{2^q-1} + O(s)\right) \tag{19}$$

Now, by definition, taking $p \in {}^*\mathbb{N}_\infty$, we have

$$\begin{aligned}
\tau_p^{\mathrm{Dix}}(A_q) &= {}^\circ\left(\frac{1}{p}\sum_{m=1}^{p}{}^*\left(\frac{\sigma_{2^m}}{\log 2^m}\right)\right) \\
&= {}^\circ\left(\frac{1}{p\log 2}\sum_{k=0}^{s-1}\sum_{m=n_k+1}^{n_{k+1}}{}^*\left(\frac{\sigma_{2^m}}{m}\right)\right) + \\
&+ {}^\circ\left(\frac{1}{p\log 2}\sum_{m=n_s+1}^{p}{}^*\left(\frac{\sigma_{2^m}}{m}\right)\right) \qquad (20)\\
&= {}^\circ\left(\frac{2^{sq}}{p}\right)\frac{q}{2^q-1} + {}^\circ\left(\frac{1}{p\log 2}\sum_{m=n_s+1}^{p}{}^*\left(\frac{\sigma_{2^m}}{m}\right)\right)
\end{aligned}$$

where, in the last equality, we have used (19) and the fact that ${}^\circ\left(\frac{s}{p}\right) = 0$.

To end the computation of the Dixmier trace of A_q we need to evaluate the second term in the r.h.s. of (20).

We have, for $p \in \mathbb{N}$,

$$\frac{1}{p\log 2}\sum_{m=n_s+1}^{p}\frac{\sigma_{2^m}}{m} = \qquad (21)$$

$$\begin{aligned}
&= \frac{1}{\log 2}\left(\frac{n_s}{p}\right)\left(1+O\left(\frac{1}{n_s}\right)\right)\left(\log\frac{p}{n_s}+O\left(\frac{1}{n_s}\right)\right) + \\
&+ \frac{1}{p(\log 2)^2}\left(\frac{n_{s+1}-n_s}{2^{n_{s+1}}-2^{n_s}}\right)\left(\frac{2^p}{p}-\frac{2^{n_s}}{n_s}\right)\left(1+O\left(\frac{1}{n_s}\right)\right)
\end{aligned}$$

Hence, by estimates similar to the previous ones, and taking $p \in {}^*\mathbb{N}_\infty$ we obtain

$${}^\circ\left(\frac{1}{p\log 2}\sum_{m=n_s+1}^{p}{}^*\left(\frac{\sigma_{2^m}}{m}\right)\right) = \frac{1}{\log 2}{}^\circ\left(\frac{2^{sq}}{p}\right)\log{}^\circ\left(\frac{p}{2^{sq}}\right) \qquad (22)$$

From (20) and (22) it follows

$$\tau_p^{Dix}(A_q) = t\left(\frac{q}{2^q-1} - \log_2(t)\right) \qquad (23)$$

where $t := {}^\circ\left(\frac{2^{sq}}{p}\right)$.

In general t can take any value in the interval $[2^{-q},1]$. In particular, in the case $p = 2^{sq+r}$, $1 \le r \le q$, formula (23) becomes

$$\tau_p^{Dix}(A_q) = 2^{-r}\left(\frac{q}{2^q-1} + r\right) \qquad (24)$$

References

[1] Albeverio, S., Fenstad, J.E., Høegh-Krohn, R., Lindstrøm, T.: *Non standard methods in stochastic analysis and mathematical physics.*, Acad. Press, Orlando (1986).

[2] Albeverio S., Guido D., Ponosov A., Scarlatti S., "*Dixmier traces and non standard analysis*", Proceedings of the III International Conference "Stochastic Processes: Physics and Geometry", Locarno June 1991, Edts. S. Albeverio et al., World Scientific Publ. (1993).

[3] Albeverio S., Guido D., Ponosov A., Scarlatti S., "*Non standard representation of non normal traces*", Proceedings of the German-French ZiF-Meeting "Dynamics of the Complex and Irregular Systems" Bielefeld, Germany, December 1991, Edts. Ph. Blanchard et al. World Scientific Publ. (1993).

[4] Albeverio S., Guido D., Ponosov A., Scarlatti S. "*Singular traces and compact operators.I.*", Preprint, Ruhr-Universität Bochum, April 1993.

[5] Connes, A.: *Géometrie non commutative*, Intereditions, Paris (1990).

[6] Dixmier, J.: *Von Neumann algebras*, North Holland (1981)

[7] Dixmier, J.: *Existence de traces non normales*, C.R. Acad. Sci. Paris **262**, (1966).

[8] Edwards, R.E.: *Functional analysis, theory and applications*, New York (1965).

[9] Glicksberg, I.: *Stone-Čech compactifications of products*, Trans. AMS **90** (1959).

[10] Gohberg, I., Krein, M.G.: *Introduction to the theory of non-selfadjoint operators*, Mos- cow (1985).

[11] Hurd, A.E, Loeb, P.A.: *An introduction to non-standard real analysis*, Acad. Press, Orlando (1985).

[12] Keller, G., Moore, L.C., Jr.: *Invariant means on the group of integers*, in: Analysis and Geometry, Ed. B. Fuchssteiner and W.A.J. Luxemburg, Wissenschaftsverlag, (1992) 1.

[13] Luxemburg, W.A.J.: *Nonstandard hulls, generalized limits and almost convergence*, in: Analysis and Geometry, Ed. B. Fuchssteiner and W.A.J. Luxemburg, Wissenschaftsverlag, (1992) 19.

[14] Mané, R.: *Ergodic theory and differentiable dynamics*, Springer, Berlin-Heidelberg-N.Y. (1987).

[15] Shatten, R.: *Norm ideals of completely continuous operators*, Springer, Berlin - Heidelberg - N.Y. (1970).

[16] Varga, J.V.: *Traces on irregular ideals*, Proc. of Am. Math. Soc. **107**, (1989) 715.

Navier-Stokes equations

M.Capiński
Instytut Matematiyki Uniwersytet Jagiellonski
Ul. Reymonta 1, 30–059 Krakow, Poland

N.J.Cutland
Dep. Pure Mathematics, Univ. of Hull,
Cottingham Road, Hull HU 6 7RX, England

Abstract

The methods of nonstandard analysis are applied to the study of Navier–Stokes equations. We give a construction of weak solutions, solve general stochastic Navier–Stokes equations, and show how to obtain statistical solutions in the general stochastic case.

1 Introduction

The Navier–Stokes equations

$$\begin{cases} u_t - \nu \Delta u + \langle u, \nabla \rangle u = f + \nabla p \\ \operatorname{div} u = 0 \\ u|_{\partial D} = 0 \end{cases} \tag{1}$$

describe the time evolution of the velocity field $u = (u_1, \ldots, u_n)$ of an incompressible viscous fluid. If $n = 3$, which is most interesting from the point of view of physical interpretation, this is a system of four equations. The unknown functions are u_1, u_2, u_3, where $u_i : D \times [0, \infty) \to \mathbb{R}$ ($D \subset \mathbb{R}^3$ is a bounded domain with C^2 boundary) that represent the components of the velocity vector, and $p : D \times [0, \infty) \to \mathbb{R}$ representing the pressure. The equation $\operatorname{div} u = 0$ is the incompressibility condition. It allows us to eliminate the pressure and to reduce the problem to the study of the time evolution of u. We equip this system with the Dirichlet boundary condition.

Nonstandard analysis has proved to be a powerful tool in tackling a wide range of problems concerned with Navier–Stokes equations. The first step was the proof of existence of weak solutions given in [5]. This proof is substantially simpler than the standard ones and pointed the way to further results. The most important ([4]) was the existence of solutions to general stochastic Navier–Stokes equation of the form

$$du = (\nu\Delta u - \langle u, \nabla\rangle u + f(t,u))dt + g(t,u)dW_t.$$

This solved a problem which was open for 18 years: in 1973 Bensoussan and Temam wrote the first paper [2] on stochastic Navier–Stokes equation with constant g. This equation has drawn the attention of many authors (see [13], [14], [9], for example) but the general case of feedback in g was first treated by Bensoussan [1] for dimension 2, simultaneously with the above mentioned paper covering the 3-dimensional case. Another nonstandard existence result [6] allows some special unbounded operators to appear in the noise term.

The nonstandard construction of solutions is based on a certain hyperfinite-dimensional representation $\mathbb{H}_N$ of the function spaces involved. Using the transfer principle we solve a system of ODE's for the deterministic case (SODE's in the stochastic case). The solution is then shown to be nearstandard in a suitable topology. The final step is to show that the standard part satisfies the appropriate Navier-Stokes equation.

We give a complete proof for the deterministic case in Sect.3 and outline the main steps of proof in the stochastic case in Sect.4. In both cases the theorems given are stronger than those in [5], [4]. We conclude with a proof of existence of so-called statistical solutions for the stochastic equation (hence in particular we have existence of statistical solutions for the deterministic one).

2 Preliminaries

For a fixed domain $D \subset \mathbb{R}^n$, $n \leq 4$, with boundary of class C^2 let $\mathbb{H}$ be the closure of the set

$$\mathcal{D} = \{u \in C_0^\infty(D, \mathbb{R}^n) : \operatorname{div} u = 0\}$$

in the L^2 norm $|u| = (u,u)^{1/2}$, where

$$(u,v) = \sum_{j=1}^{n} \int_D u^j(x)v^j(x)dx.$$

The letters u, v, w will be used for elements of $\mathbb{H}$.

The symbol $\mathbf{V}$ denotes the closure of $\mathcal{D}$ in the norm $|u| + \|u\|$ where $\|u\| = ((u,u))^{1/2}$ and

$$((u,v)) = \sum_{j=1}^{n} \Big(\frac{\partial u}{\partial x_j}, \frac{\partial v}{\partial x_j}\Big).$$

$\mathbb{H}$ and $\mathbf{V}$ are Hilbert spaces with scalar products $(\cdot,\cdot)$ and $((\cdot,\cdot))$ respectively, and $|\cdot| \leq c\|\cdot\|$ for some constant c.

By A we denote the self-adjoint extension of the operator $-\Delta$ in $\mathbb{H}$ and by $\{e_k\}$ an orthonormal basis of its eigenfunctions with the corresponding eigenvalues λ_k, $\lambda_k \geq 0$, $\lambda_k \nearrow \infty$. For $u \in \mathbb{H}$ we write $u_k = (u, e_k)$. We write Pr_m for the projection of $\mathbb{H}$ on the subspace $\mathbb{H}_m$ spanned by $\{e_1, \ldots, e_m\}$.

By $\mathbf{V}'$ we denote the space dual to $\mathbf{V}$, with the duality extending the scalar product in $\mathbb{H}$.

In general, we define the spaces $\mathbb{H}^r$ for $r > 0$ by

$$\mathbb{H}^r = \{u \in \mathbb{H} : \sum_{k=1}^{\infty} \lambda_k^r u_k^2 < \infty\}$$

(so $\mathbb{H} = \mathbb{H}^0$, $\mathbf{V} = \mathbb{H}^1$). The elements of the dual spaces $\mathbb{H}^{-r}$ ($\mathbf{V}' = \mathbb{H}^{-1}$) are characterized by

$$|u|_{-r} := \sum_{k=1}^{\infty} \lambda_k^{-r} u_k^2 < \infty$$

where $u_k = u[e_k]$, so that

$$u = \sum_{k=1}^{\infty} u_k e_k,$$

the limit being taken in $\mathbb{H}^{-r}$.

The operator A can be extended to $A : \mathbf{V} \to \mathbf{V}'$ by $Au(v) = ((u, v))$ for $v \in \mathbf{V}$.

We put

$$b(u, v, z) = \sum_{i,j=1}^{n} \int_D u^j(x) \frac{\partial v^i}{\partial x_j}(x) z^i(x) dx = (\langle u, \nabla \rangle v, z)$$

whenever the integrals make sense. Note that for $u, v, z \in \mathbf{V}$ we have

$$b(u, v, z) = -b(u, z, v), \tag{2}$$

hence

$$b(u, v, v) = 0.$$

We shall need some well-known inequalities for b (see [11] for example) and we list them here for reference:

$$|b(u, v, z)| \leq c\,\|u\|\,\|v\|\,\|z\|, \tag{3}$$

$$|b(u, v, z)| \leq c\,|u|\,\|v\|\,|Az|, \tag{4}$$

$$|b(u, v, z)| \leq c\,\|u\|\,|v|\,|Az|, \tag{5}$$

$$|b(u, v, z)| \leq c\,|u|\,\|v\|\,|z|_r \quad \text{if } r > \tfrac{n}{2}, \tag{6}$$

for suitable u, v, and z.

The inequality (3) allows us to define a $\mathbf{V}'$-valued bilinear form $B(u,v)$ by

$$B(u,v)(z) = b(u,v,z).$$

We now introduce a subspace of ${}^*\mathbb{H}$ which will play a crucial role in all our considerations. Fix an infinite integer $N \in {}^*\mathbb{N}\backslash\mathbb{N}$ and let $\mathbb{H}_N$ be the space spanned by the vectors $\{E_1, \ldots, E_N\}$ where $E_k = {}^*e_k$.

By Pr_N we denote the orthogonal projection of ${}^*\mathbb{H}$ onto $\mathbb{H}_N$. We use U, V for elements of $\mathbb{H}_N$. We write $U_k = (U, E_k)$ for $k \in {}^*\mathbb{N}$ and we have

$$(U,V) = \sum_{k=1}^{N} U_k V_k, \qquad |U|^2 = \sum_{k=1}^{N} U_k^2,$$
$$((U,V)) = \sum_{k=1}^{N} \lambda_k U_k V_k, \qquad \|U\|^2 = \sum_{k=1}^{N} \lambda_k U_k^2.$$

We shall also use the $\mathbb{H}^r$ norm of elements of $\mathbb{H}_N$ for $r \in \mathbb{R}$ (we shall mainly need $r = 1, 0, -1$) :

$$|U|_r^2 = \sum_{k=1}^{N} \lambda_k^r U_k^2.$$

In the following proposition we sum up some basic facts from nonstandard functional analysis that we shall need in what follows.

Proposition 2.1 *For $U \in \mathbb{H}_N$ and $u \in \mathbb{H}^r$, $r \in \mathbb{R}$, we have*

(a) *$U \approx u$ in the weak topology of $\mathbb{H}^r$ iff $(U, {}^*v) \approx (u,v)$ for all $v \in \mathbb{H}^{-r}$ (and then $u_k = {}^\circ U_k$ for all finite k);*

(b) *if U is nearstandard in $\mathbb{H}^r$, then U is weakly nearstandard in $\mathbb{H}^r$ and the standard parts coincide.*

We use $\mathrm{st}_{\mathbb{H}^r} U$ ($\text{w-st}_{\mathbb{H}^r} U$) to denote the standard part (weak standard part) of U in $\mathbb{H}^r$. For $U \in \mathbb{H}_N$ we write

$${}^\circ U = \sum_{k=1}^{\infty} {}^\circ U_k e_k,$$

where we allow ${}^\circ U_k = \pm\infty$. In general this is a formal expression but if

$$\sum_{k=1}^{\infty} \lambda_k^{r\,\circ} U_k^2 < \infty$$

for some r, then it makes sense and in fact ${}^\circ U = \text{w-st}_{\mathbb{H}^r} U$:

The next two lemmas are concerned with weak and strong nearstandardness, respectively.

Lemma 2.2 *Let $r \in \mathbb{R}$ and $U \in \mathbb{H}_N$. Then*

(a) $|{}^\circ U|_r \leq {}^\circ|U|_r$ *(we allow infinity on either side);*

(b) *if U is weakly $\mathbb{H}^r$-nearstandard, then ${}^\circ U = \text{w-st}_{\mathbb{H}^r}(U)$;*

(c) *if $|U|_r < \infty$, then U is weakly $\mathbb{H}^r$-nearstandard.*

Lemma 2.3 *For $U \in \mathbb{H}_N$ and $r \in \mathbb{R}$,*

(a) *U is (strongly) nearstandard in $\mathbb{H}^r$ iff*

$$\sum_{k=1}^{\infty} \lambda_k^r\, {}^\circ U_k^2 \approx \sum_{k=1}^{N} \lambda_k^r U_k^2 < \infty$$

(we use $x < \infty$ to mean 'x is finite');

(b) *if $|U|_r < \infty$, then U is $\mathbb{H}^q$-nearstandard for all $q < r$.*

Remark Lemma 2.3 (b) is a nonstandard expression of the compactness of the embedding of $\mathbb{H}^r$ in $\mathbb{H}^q$ for $r > q$.

The following crucial lemma has many applications.

Lemma 2.4 *If $\|U\|$ and $\|V\|$ are finite with $u = {}^\circ U$, $v = {}^\circ V$, and $z \in \text{dom}A$ then*

$${}^*b(U, V, {}^*z) \approx b(u, v, z).$$

Proof By transfer of the properties (4), (5) concerning b we have

$$|{}^*b({}^*u - U, {}^*z, v)| \leq c\,|{}^*u - U|\,|Az|\,\|{}^*v\|,$$

$$|{}^*b(U, {}^*z, {}^*v - V)| \leq c\,|{}^*v - V|\,|Az|\,\|U\|.$$

Since $|{}^*u - U| \approx 0 \approx |{}^*v - V|$ and $|Az|$, $\|U\|$, and $\|v\|$ are finite, we readily obtain the result. ∎

3 Weak Solutions

We first give the definition of weak solution for the case of forces f depending on u. Such a general case is relevant since it allows us to deduce results about other equations of hydromechanics such as equations of thermohydraulics or magneto-hydro-dynamics (cf. [12], where the linear dependence of f on u in two-dimensional case is dealt with).

Definition 3.1 Given $u_0 \in I\!H$ and $f : [0,T) \times I\!H \to \mathbf{V}'$ the function $u : [0,\infty) \to I\!H$ is a *weak solution* of the Navier-Stokes equation if

(i) $u \in L^2(0,T;\mathbf{V}) \cap L^\infty(0,T;\mathbb{H})$ *for all* $T < \infty$,

(ii) *for all* $v \in \mathbf{V}$, *for all* $t \geq 0$

$$(u(t),v) - (u_0,v) = \int_0^t [-\nu((u(s),v)) - b(u(s),u(s),v) + (f(s,u(s)),v)]ds \qquad (7)$$

where $(f(s,u(s)),v)$ denotes the duality between $\mathbf{V}'$ and $\mathbf{V}$ extending the scalar product in $I\!H$

Theorem 3.2 *Suppose that* f *is continuous and satisfies*

$$|f(t,u)|_{-1} \leq c(t)\|u\|^\alpha |u|^{1-\alpha}$$

for some $\alpha \in [0,1)$, *where* $c \in L^p(0,T)$ *for all* T, *where* $p \geq \frac{2}{1-\alpha}$. *Then there is a weak solution to Navier–Stokes equation.*

Proof Consider an $I\!H_N$ valued function

$$U(\tau) = \sum_{k=1}^N U_k(\tau)E_k$$

and the following nonstandard equation in $I\!H_N$:

$$\dot{U}_k(\tau) = -\nu\lambda_k U_k(\tau) - {}^*b(U(\tau),U(\tau),E_k) + F_k(\tau,U(\tau)) \qquad (8)$$

$k = 1,\ldots,N$, where $\tau \in {}^*(0,\infty)$,

$$F_k(\tau,V) = ({}^*f(\tau,V),E_k)$$

and the initial condition is

$$U_0 = \sum_{k=1}^N ({}^*u_0,E_k)E_k.$$

By transfer we have local existence of a solution to (8). We shall show now that this solution exists for all τ and satisfies a certain inequality. Multiplying both sides of (8) by $U_k(\tau)$ and summing over k we find that

$$|U(\sigma)|^2 + \nu\int_0^\sigma \|U(\tau)\|^2 d\tau = |U_0|^2 + \int_0^\sigma (F(\tau,U(\tau)),U(\tau))d\tau$$

(we have used the property $b(u,v,v)=0$). Now, the right hand side can be estimated using the Young inequality ($ab \le \varepsilon a^p + c_\varepsilon b^q$ for $p>1$, $\varepsilon>0$, where $q=\frac{p}{p-1}$, $c_\varepsilon = \frac{p-1}{p^q \varepsilon^{1/p-1}}$) and the hypothesis:

$$\begin{aligned}\int_0^\sigma (F(\tau,U(\tau)),U(\tau))d\tau &\le \int_0^\sigma |F(\tau,U(\tau))|_{-1}\|U(\tau)\|d\tau \\ &\le \int_0^\sigma c(\tau)\|U(\tau)\|^{1+\alpha}|U(\tau))|^{1-\alpha}d\tau \\ &\le \varepsilon\int_0^\sigma \|U(\tau)\|^2 d\tau + c_\varepsilon \int_0^\sigma c^{\frac{2}{1-\alpha}}(\tau)\,|U(\tau)|^2 d\tau\end{aligned}$$

where we require $\varepsilon < \nu$. This inequality gives

$$|U(\sigma)|^2 + (\nu-\varepsilon)\int_0^\sigma \|U(\tau)\|^2 d\tau \le |U_0|^2 + c_\varepsilon \int_0^\sigma c^{\frac{2}{1-\alpha}}(\tau)\,|U(\tau)|^2 d\tau$$

which by Gronwall's lemma implies

$$|U(\sigma)|^2 + (\nu-\varepsilon)\int_0^\sigma \|U(\tau)\|^2 d\tau \le |U_0|^2 \exp\left\{\int_0^\sigma c^{\frac{2}{1-\alpha}}(\tau)d\tau\right\} \tag{9}$$

where the right hand side is finite for finite σ. The aim now is to show that $U(\tau)$ has a standard part (in an appropriate sense) that is a weak solution of the Navier-Stokes equations.

By (9), for finite τ, $|U(\tau)| < \infty$ so we have the weak standard part

$$\text{w-st}_{I\!H} U(\tau) = {}^\circ U(\tau) \in I\!H$$

with

$$({}^\circ U(\tau))_k = {}^\circ\,(U_k(\tau))$$

for finite k.

Note that for finite T and each finite k, ${}^*b(U(\tau),U(\tau),E_k)$ is S-integrable on ${}^*[0,T]$ because using (4), and (9) we have

$$\int_0^T {}^*b(U(\tau),U(\tau),E_k)^2 d\tau \le c\int_0^T |U(\tau)|^2\|U(\tau)\|^2\lambda_k^2 d\tau < \infty.$$

Clearly, the other terms on the right hand side of (8) are S-integrable. Hence if $\sigma \approx \tau < \infty$, ${}^\circ U_k(\sigma) = {}^\circ U_k(\tau)$ for all finite k and so $U(\sigma) \approx U(\tau)$ in the weak topology of $I\!H$ (i.e. U is *S-continuous* with respect to the weak topology of $I\!H$). Consequently, we can define a standard function

$$u : [0,\infty) \to I\!H$$

by $u({}^\circ\tau) = {}^\circ U(\tau)$ and it is continuous with respect to the weak topology of $I\!H$. We claim that u is a weak solution to the Navier-Stokes equations.

Let $T < \infty$. By Lemma 2.2 (a) for $r = 0$, and (9) we have $u \in L^\infty(0,T;I\!H)$. By Lemma 2.2 (a), $r = 1$, and basic Loeb theory we have

$$\int_0^T \|u(t)\|^2 dt = \int_0^T \|u(^\circ\tau)\|^2 d_L\tau = \int_0^T \|^\circ U(\tau)\|^2 d_L\tau \leq {}^\circ\!\int_0^T \|U(\tau)\|^2 d\tau < \infty;$$

hence $u \in L^2(0,T;\mathbf{V})$.

Finally it is sufficient to verify (7) for $v = e_k$. For any $t \geq 0$ we have

$$\begin{aligned}(u(t), e_k) &= {}^\circ U_k(t)\\ &= (u_0, e_k) - {}^\circ\!\int_0^t (\nu\lambda_k U_k(\tau) + {}^*b(U(\tau), U(\tau), {}^*e_k) - ({}^*f(\tau), {}^*e_k))\, d\tau.\end{aligned}$$

Clearly

$${}^\circ\!\int_0^t ({}^*f(\tau), {}^*e_k) d\tau = \int_0^t (f(s), e_k) ds;$$

and since $U_k(\tau)$ is S-continuous we have

$${}^\circ\!\int_0^t \lambda_k U_k(\tau) d\tau = \int_0^t \lambda_k u_k(^\circ\tau) d_L\tau = \int_0^t \lambda_k u_k(s) ds = \int_0^t ((u(s), e_k)) ds.$$

Finally, ${}^*b(U(\tau), U(\tau), {}^*e_k)$ is S-integrable as noted earlier. Moreover, from (9) $\|U(\tau)\| < \infty$ for a.a. τ and so by Lemma 2.4

$${}^*b(U(\tau), U(\tau), {}^*e_k) \approx b(u(^\circ\tau), u(^\circ\tau), e_k).$$

Thus

$$\begin{aligned}&{}^\circ\!\int_0^t {}^*b(U(\tau), U(\tau), {}^*e_k) d\tau = \int_0^t {}^{\circ *}b(U(\tau), U(\tau), {}^*e_k) d_L\tau\\ &= \int_0^t b(u(^\circ\tau), u(^\circ\tau), e_k) d_L\tau = \int_0^t b(u(s), u(s), e_k) ds\end{aligned}$$

which completes the proof. ∎

4 The Stochastic Case

We now turn to a general system of stochastic Navier-Stokes equations in dimensions $n \leq 4$

$$\begin{cases} du = [\nu\Delta u - \langle u, \nabla \rangle u + f(t,u) + \nabla p]dt + g(t,u)dw_t \\ \operatorname{div} u = 0 \end{cases} \tag{10}$$

where w is an $I\!H$-valued Wiener process of prescribed covariance Q (so $Q \in \mathcal{L}(I\!H, I\!H)$ is nonnegative, trace class), and the stochastic integral is a natural extension of the finite dimensional Itô integral.

Let us now introduce the following notations for the path spaces

$$\mathcal{Z}_T = L^\infty(0,T;I\!H) \cap L^2(0,T;\mathbf{V}) \cap C(0,T;I\!H_{\text{weak}})$$

$$\mathcal{Z} = \bigcap_{T<\infty} \mathcal{Z}_T$$

We can explain now what we mean by a solution of (10).

Definition 4.1 Let $u_0 \in I\!H$, $f : [0,\infty) \times \mathbf{V} \to \mathbf{V}'$, and $g : [0,\infty) \times \mathbf{V} \to \mathcal{L}(I\!H, \mathbf{V}')$. A stochastic process $u(t,\omega)$ with a.a. paths in $\mathcal{Z}$ is a *weak solution of the stochastic Navier-Stokes equations* (10) if

$$E\left(\sup_{t\in[0,T]} |u(t)|^2 + \int_0^T \|u(t)\|^2 dt\right) \le c(T) < \infty \text{ for all } T \tag{11}$$

and

$$u(t) = u_0 + \int_0^t [-\nu Au(s) - B(u(s),u(s)) + f(s,u(s))]ds + \int_0^t g(s,u(s))dw(s) \tag{12}$$

holds as an identity in $\mathbf{V}'$ (the first integral is understood in the sense of Bochner).

Note that ∇p as an element of $\mathbf{V}'$ is equal to 0: $\nabla p[v] = (p, \text{div} v) = 0$.

The properties of the two integrals involved mean that an equivalent formulation of (12) is that for each $v \in \mathbf{V}$

$$\begin{aligned}(u(t),v) &= (u_0,v) + \int_0^t [-\nu((u(s),v)) - b(u(s),u(s),v) + (f(s,u(s)),v)]ds \\ &\quad + \int_0^t (v, g(s,u(s))dw(s)).\end{aligned}$$

We first prove an existence theorem and then we discuss the possibility of relaxing some conditions on the coefficients. We prefer to give first the proof in a particular, though fairly general, situation to avoid some technical complications which might obscure the simple underlying idea.

The Main Theorem

Let $K_m = \{v : \|v\| \le m\} \subseteq \mathbf{V}$, which we consider with the strong topology of $I\!H$. In the theorem below, continuity on each K_m turns out to be the appropriate condition for the coefficients f, g. Note that this is weaker than continuity on $\mathbf{V}$ in either the $I\!H$-norm or the weak topology of $\mathbf{V}$.

Theorem 4.2 *Suppose that* $u_0 \in \mathbb{H}$ *and*

$$f : [0,\infty) \times \mathbf{V} \to \mathbf{V}', \qquad g : [0,\infty) \times \mathbf{V} \to \mathcal{L}(\mathbb{H},\mathbb{H})$$

are jointly measurable functions with the following properties

(i) $f(t,\cdot) \in C(K_m, \mathbf{V}'_{\mathrm{weak}})$ *for all* m,
(ii) $g(t,\cdot) \in C(K_m, \mathcal{L}(\mathbb{H},\mathbb{H})_{\mathrm{weak}})$ *for all* m,
(iii) $|f(t,u)|_{-1} + |g(t,u)|_{\mathbb{H},\mathbb{H}} \leq a(t)(1+|u|)$ *where* $a \in L^2(0,T)$ *for all* T.

Then equation (12) has a solution u *on a filtered Loeb space, with (11) (hence* $u(\cdot,\omega) \in \mathcal{Z}$ *a.s.).*

Proof We will sketch the main thread of the proof, with reference to [4] for full details. Consider the following internal N-dimensional SDE for a stochastic process $U(\tau,\omega) \in \mathbb{H}_N$

$$\begin{cases} dU(\tau) = [-\nu^* AU(\tau) - \overline{B}(U(\tau),U(\tau)) + F(\tau,U(\tau))]d\tau + G(\tau,U(\tau))dW(\tau) \\ U(0) = \mathrm{Pr}_N{}^* u_0 \end{cases} \tag{13}$$

where, putting $B_N = \mathrm{Pr}_N{}^* B$, the function $\overline{B}$ is given by

$$\overline{B}(U,V) = \begin{cases} {}^*B_N(U,V) & \text{if } U^2+V^2 \leq K \\ K^2(U^2+V^2)^{-1}B_N(U,V) & \text{otherwise} \end{cases}$$

for some infinite K, and F, G are given by

$$F(\tau,U) = \mathrm{Pr}_N{}^* f(\tau,U),$$

$$G(\tau,U)V = \mathrm{Pr}_N{}^* g(\tau,U)V.$$

The growth and continuity conditions on f, g and the boundedness of $\overline{B}$ ensure (using the transfer of the standard theory of SDE's) that (13) has an internal solution $U(\tau,\omega)$ for all $\tau \in {}^*[0,\infty)$ on some internal filtered space $(\Omega, \mathbf{A}, (\mathbf{A}_\tau)_{\tau\geq 0}, \Pi)$ carrying an internal Wiener process W with covariance Q_N; and U is adapted to $(\mathbf{A}_\tau)_{\tau\geq 0}$.

The transfer of Itô's formula gives

$$\begin{aligned} &|U(\tau)|^2 + 2\nu \int_0^\tau \|U(\sigma)\|^2 d\sigma \\ &= |U(0)|^2 + 2\int_0^\tau (F(\sigma,U(\sigma)),U(\sigma))d\sigma \\ &\quad + \int_0^\tau \mathrm{tr}[G(\sigma,U(\sigma))Q_N G(\sigma,U(\sigma))']d\sigma \\ &\quad + 2\int_0^\tau (U(\sigma),G(\sigma,U(\sigma))dW(\sigma)) \end{aligned} \tag{14}$$

Now

$$2|F(\tau,U),U)| \leq 2\|U\|^*a(\tau)(1+|U|) \leq \nu\|U\|^2 + \frac{1}{\nu}{}^*a^2(\tau)(1+|U|)^2 \tag{15}$$

and

$$\mathrm{tr}(G(\tau,U)Q_N G(\tau,U)') \leq \mathrm{tr}Q_N|G(\tau,U)|^2_{I\!H_N,I\!H_N} \leq \mathrm{tr}Q_N{}^*a^2(\tau)(1+|U|)^2. \tag{16}$$

So, substituting (16), (15) in (14) we have, for any T

$$\begin{aligned}\sup_{\tau\leq T}|U(\tau)|^2 + \nu\int_0^T &\|U(\tau)\|^2 d\tau \\ \leq\ &|U(0)|^2 + c\int_0^T {}^*a^2(\tau)(1+|U(\tau)|^2)d\tau + 2\sup_{\tau\leq T}|M(\tau)|\end{aligned} \tag{17}$$

where $M(\tau)$ is the internal martingale

$$M(\tau) = \int_0^\tau (U(\sigma), G(\sigma, U(\sigma)))dW(\sigma).$$

Now M has quadratic variation

$$[M](\tau) = \int_0^\tau \mathrm{tr}[V(\sigma)Q_N V(\sigma)']d\sigma,$$

where

$$V(\sigma) = (U(\sigma), G(\sigma, U(\sigma))) \in I\!H_N.$$

Note that

$$\mathrm{tr}[V(\sigma)Q_N V(\sigma)'] \leq \mathrm{tr}(Q_N)|V(\sigma)|^2 \leq \mathrm{tr}(Q_N)^*a^2(\sigma)|U(\sigma)|^2(1+|U(\sigma)|)^2$$

So

$$[M](T) \leq c\sup_{\tau\leq T}|U(\tau)|^2\int_0^T {}^*a^2(\tau)(1+|U(\tau)|^2)d\tau \tag{18}$$

Now the Burkholder–Davis–Gundy inequality gives

$$E\left(\sup_{\tau\leq T}|M(\tau)|\right) \leq \kappa E\left([M](T)^{\frac{1}{2}}\right)$$

for some fixed finite constant κ. From (18) we have

$$[M](T)^{\frac{1}{2}} \leq \frac{1}{3\kappa}\sup_{\tau\leq T}|U(\tau)|^2 + c_1\int_0^T {}^*a^2(\tau)(1+|U(\tau)|^2)d\tau$$

and so

$$E\left(\sup_{\tau\leq T}|M(\tau)|\right) \leq \frac{1}{3}E\sup_{\tau\leq T}|U(\tau)|^2 + c_2 E\int_0^T {}^*a^2(\tau)(1+|U(\tau)|^2)d\tau.$$

Substituting in (17) we obtain

$$E\left(\sup_{\tau\leq T}|U(\tau)|^2+\int_0^T\|U(\tau)\|^2d\tau\right)\leq c+c_3E\int_0^T {}^*a^2(\tau)|U(\tau)|^2d\tau.$$

and we can apply Gronwall's lemma with $y(\tau)=|U(\tau)|^2$, $g(\tau)=c_3{}^*a^2(\tau)$, and $h(\tau)=c(T))$ to give $E|U(\tau)|^2\leq c_4$ for $\tau\leq T$. This implies

$$E\left(\sup_{\tau\leq T}|U(\tau)|^2+\int_0^T\|U(\tau)\|^2d\tau\right)\leq c_1(T) \tag{19}$$

with $c_1(T)$ finite for $T<\infty$.

Now put $\Psi(\tau,\omega)=G(\tau,U(\tau,\omega))$. We have

$$|G(\tau,U(\tau,\omega))|_{I\!H_N,I\!H_N}\leq {}^*a(\tau)(1+|U(\tau,\omega)|)$$

which together with (19) shows that

$$\tau\mapsto|\Psi(\tau,\omega)|^2_{I\!H_N,I\!H_N}$$

is S-integrable for a.a. ω, and

$$\begin{aligned}E\left(\int_0^T|\Psi(\tau)|^2_{I\!H_N,I\!H_N}d\tau\right) &\leq E\left(\int_0^T {}^*a^2(\tau)(1+|U(\tau)|)^2d\tau\right)\\ &\leq E\left(\sup_{\tau\leq T}(1+|U(\tau)|)^2\int_0^T {}^*a^2(\tau)d\tau\right)\\ &\leq c_2(T)\end{aligned}$$

by (19). So by the Hilbert space extension of the theory of Itô integration on an adapted Loeb space (see [4], Theorem 5.1, for details)

$$\int_0^\tau G(\sigma,U(\sigma))dW(\sigma)$$

is S-continuous in $|\cdot|$ for a.a. ω.

By (19) again, $\overline{B}(U(\tau),U(\tau))=B_N(U(\tau),U(\tau))$ for all finite τ (for a.a. ω), and so by Theorem 5.1 of [4]

$$\int_0^\tau[-\nu{}^*AU(\sigma)-\overline{B}(U(\sigma),U(\sigma))+F(\sigma,U(\sigma))]d\sigma$$

is a.s. S-continuous in $\mathbf{V}'$. Hence $U(\tau)$ is a.s. S-continuous in $\mathbf{V}'$ for finite τ and we may define

$$u:[0,\infty)\times\Omega\to I\!H$$

by

$$u(t,\omega) = {}^\circ U(\tau,\omega)$$

for any $\tau \approx t$. We will see that this u is the required solution. The estimate (19) gives condition (11) of the theorem.

To see that u satisfies equation (12), we have from Theorem 3.6 of [4]

$$\begin{aligned} &{}^\circ \int_0^t [-\nu\, {}^*AU(\sigma) - \overline{B}(U(\sigma),U(\sigma)) + F(\sigma,U(\sigma))]d\sigma \\ &= \int_0^t [-\nu Au(s) - B(u(s),u(s)) + f(s,u(s))]ds. \end{aligned} \tag{20}$$

Finally, condition (iii) together with (19) shows that $G(\tau,U(\tau))$ is a.s. weakly nearstandard. Anderson's Luzin theorem shows that for a.a. finite τ we have

$${}^\circ G_{ij}(\tau,U) = g_{ij}({}^\circ\tau, {}^\circ U),$$

all finite i, j, whenever $\|U\| < \infty$ and so for a.a. ω, for a.a. finite τ

$${}^\circ G(\tau,U(\tau,\omega)) = g({}^\circ\tau, u({}^\circ\tau,\omega))$$

i.e. $G(\tau,U(\tau,\omega))$ lifts $g(t,u(t,\omega))$. Now by Theorem 5.4 of [4] have for a.a. ω

$${}^\circ \int_0^t G(\sigma,U(\sigma,\omega))dW(\sigma) = \int_0^t g(s,u(s,\omega))dw(s) \tag{21}$$

for all $t < \infty$. Putting (20) and (21) together we see that u solves equation (12). ∎

Remark We can easily adapt the proof to cover the case when u_0 is random and independent of w. If u_0 is given by a probability measure μ on $I\!H$ then we take $\Omega = I\!H_N \times {}^*C(0,\infty; I\!H_N)$ and $P = (\mu_N \times \Lambda_N)_L$, where $\mu_N = {}^*\mu \circ \mathrm{Pr}_N^{-1}$ and Λ_N is the internal Wiener measure.

Further Generalization

We replace condition (iii) of Theorem 4.2 by two separate conditions for f and g:

C1) for a certain $\theta \in [0,1)$, and for $a_1 \in L^q(0,T)$ with $q = \frac{2}{1-\theta}$ we have

$$|f(t,u)|_{-1} \le a_1(t)\|u\|^\theta |u|^{1-\theta}$$

for all $u \in \mathbf{V}$.

C2) for a certain $\kappa \in [0,1)$, and for $a_2 \in L^q(0,T)$ with $q = \frac{1}{1-\kappa}$ we have

$$|g(t,u)|_{I\!H,I\!H} \le a_2(t)\|u\|^\kappa |u|^{1-\kappa}$$

for all $u \in \mathbf{V}$.

We claim that Theorem 4.2 remains valid if we replace (iii) by C1–2. The proof remains the same and we only present the necessary modifications.

First, using the Young inequality ($a = \|U\|^{1+\theta}\nu^{\frac{1+\theta}{2}}$, $b = a_1(\tau)|U|^{1-\theta}\nu^{-\frac{1+\theta}{2}}$, $p = \frac{2}{1+\theta}$) we have

$$2|(F(\tau,U),U)| \leq 2^*a_1(\tau)\|U\|^{1+\theta}|U|^{1-\theta} \leq \nu\|U\|^2 + 2^{\frac{2}{1-\theta}}\nu^{-\frac{1+\theta}{1-\theta}}{}^*a_1^{\frac{2}{1-\theta}}(\tau)|U|^2$$

to replace (15). Next, again by the Young inequality, we have

$$\begin{aligned}\mathrm{tr}(G(\tau,U)Q_N G(\tau,U)') &\leq \mathrm{tr}Q_N|G(\tau,U)|^2_{I\!H_N,I\!H_N} \\ &\leq \mathrm{tr}Q_N{}^*a_2^2(\tau)\|U\|^{2\kappa}|U|^{2-2\kappa} \\ &\leq \frac{\nu}{2}\|U\|^2 + ca_2^{\frac{1}{1-\kappa}}(\tau)|U|^2\end{aligned}$$

which replaces (16). The resulting estimate (17) remains the same with a suitably chosen a.

Now, in the same fashion as above, we arrive at

$$[M](T) \leq \mathrm{tr}(Q_N)|V(\sigma)|^2 \leq c\sup_{\tau\leq T}|U(\tau)|^2\int_0^T {}^*a_2^2(\tau)\|U(\tau)\|^{2\kappa}|U(\tau)|^{2-2\kappa}d\tau$$

which corresponds to (18). We can apply the Burkholder–Davis–Gundy inequality and we get

$$\begin{aligned}E\left(\sup_{\tau\leq T}|M(\tau)|\right) &\leq \kappa E\left(\sup_{\tau\leq T}|U(\tau)|\left(\int_0^T {}^*a_2^{\frac{1}{1-\kappa}}(\tau)\|U(\tau)\|^{2\kappa}|U(\tau)|^{2-2\kappa}d\tau\right)^{\frac{1}{2}}\right) \\ &\leq \frac{1}{3}E\sup_{\tau\leq T}|U(\tau)|^2 + \int_0^T {}^*a_2^{\frac{1}{1-\kappa}}(\tau)\|U(\tau)\|^{2\kappa}|U(\tau)|^{2-2\kappa}d\tau \\ &\leq \frac{1}{3}E\sup_{\tau\leq T}|U(\tau)|^2 + \frac{\nu}{2}E\int_0^T \|U(\tau)\|^2 d\tau \\ &\quad + cE\int_0^T {}^*a_2^{\frac{1}{1-\kappa}}(\tau)|U(\tau)|^2 d\tau.\end{aligned}$$

Substituting in (17) we obtain

$$E\left(\sup_{\tau\leq T}|U(\tau)|^2 + \int_0^T \|U(\tau)\|^2 d\tau\right) \leq c_5(T) + c_6E\int_0^T {}^*a^2(\tau)|U(\tau)|^2 d\tau$$

and Gronwall's lemma gives (19) as before. In the proof of Theorem 4.2 we have not employed the full power of (14) as on the right of (16) and (18) we only had $|U(\tau)|^2$. Despite weaker estimates on F and G, (19) allows us now to draw the same conclusions about S-continuity of internal integrals and to complete the proof in the same way. At each step we use the Young inequality so that the terms involving $\|U(\tau)\|^2$ on the right are accompanied by sufficiently small constants.

Statistical Solutions

First we give the definition of statistical solution for the stochastic system (10).

Definition 4.3 A family μ_t of Borel probability measures on $I\!H$ is called a *statistical solution* of the stochastic Navier–Stokes equations if

(i) the function $t \mapsto \int_{I\!H} |u|^2 d\mu_t(u)$ is $L^\infty(0,T)$ for all $T < \infty$

(ii) $\int_0^T \int_{I\!H} \|u\|^2 d\mu_t(u)dt < \infty$ for all $T < \infty$

(iii) for any φ of the form $\varphi(u) = e^{i(u,v)}, v \in V,$ and for any $t > 0$

$$\begin{aligned}\int_{I\!H} \varphi(u)d\mu_t(u) &= \int_{I\!H} \varphi(u)d\mu_0(u) + \int_0^t \int_{I\!H} \Big(-\nu((u,\varphi'(u))) - b(u,u,\varphi'(u)) \\ &\quad +(f(s,u),\varphi'(u)) \\ &\quad +\frac{1}{2}\mathrm{tr}(Qg^T(s,u)\varphi''(u)g(s,u))\Big) d\mu_s(u)ds \qquad (22)\end{aligned}$$

where Q is the covariance of w (g^T denotes the transposed operator and φ', φ'' are the functional derivatives of φ).

We shall call the equation (22) the *Foias equation* for the system (10). Comparing it with that derived by Foias [8] for the deterministic case it has an additional term involving the second order derivative $\varphi''(u)$. The equation (22) is similar to that considered in [14] where $g(t,u) = \mathrm{Id}$.

We now show that a solution of (22) is provided by the measures given by any solution to (10).

Theorem 4.4 *Assume the hypotheses of Theorem 4.2. Suppose that the initial function $u_0 \in \mathbb{H}$ is random and independent from the Wiener process. Let μ be the distribution of u_0 and suppose that*

$$\int_{\mathbb{H}} |u|^2 d\mu(u) < \infty.$$

Suppose that $u(t,\omega)$ is an adapted solution to (10), satisfying (11), on a probability space $(\Omega, \mathcal{F}, P)$ with filtration (see Remark following Theorem 4.2).

Then the family of measures

$$\mu_t(A) = P(u(t) \in A)$$

is a solution to the Foias equation (22), with $\mu_0 = \mu$.

Proof Taking $\varphi(u) = e^{i(u,v)}$ with $v \in V$, the infinite dimensional Itô lemmma in [10] gives

$$\begin{aligned} d\varphi(u(t)) &= \Big[-\nu((u(t),\varphi'(u(t)))) - b(u(t),u(t),\varphi'(u(t))) \\ &\quad +(f(t,u(t)),\varphi'(u(t))) + \frac{1}{2}\mathrm{tr}(Qg^T(t,u(t))\varphi''(u(t))g(t,u(t)))\Big]dt \\ &\quad +(\varphi'(u(t)),g(t,u(t))dw(t)) \\ &= \eta(t,u(t))dt + \theta(t,u(t))dw(t) \end{aligned}$$

say. We shall prove below that

$$\text{(a)} \quad E\left(\int_0^T |\eta(t,u(t))|dt\right) < \infty \text{ for all } T < \infty$$

$$\text{(b)} \quad E\left(\int_0^T |\theta(t,u(t))|^2 dt\right) < \infty \text{ for all } T < \infty.$$

Taking (a) and (b) for granted, we can quickly complete the proof of the theorem:

$$\begin{aligned} \int_{I\!H}\varphi(u)d\mu_t(u) - \int_{I\!H}\varphi(u)d\mu_0(u) &= E(\varphi(u(t)) - \varphi(u(0))) \\ &= E\left(\int_0^t \eta(s,u(s))ds + \int_0^t \theta(s,u(s))dw(s)\right) \\ &= \int_0^t E\eta(s,u(s))ds \end{aligned}$$

using (a) and Fubini for the first term and the zero mean property of the stochastic integral given condition (b) ([10], Prop.1.4). Finally, note that

$$E\eta(s,u(s)) = \int_{I\!H}\eta(s,u)d\mu_s(u)$$

to see that (22) is satisfied. The conditions (i) and (ii) of Definition 4.3 follow from (11) automatically.

It remains to check (a) and (b) above. For (a) note first that $\|\varphi'(u)\| = \|v\|$ and $|\varphi''(u)|_{I\!H,I\!H} = |v|^2$ so for any $u \in I\!H$

$$|\eta(t,u)| \le \nu\|u\|\,\|v\| + c\|u\|^2\,\|v\| + |f(t,u)|_{V'}\,\|v\| + \frac{1}{2}\mathrm{tr}Q.|g(t,u)|^2_{I\!H,I\!H}|v|^2.$$

The growth conditions (iii) on f and g, and (11) on $|u(t)|$ ensure that (a) holds.

For (b) it is enough to observe that

$$|\theta(t,u)| \le |\varphi'(u)|\,|g(t,u)|_{I\!H,I\!H} = |v|\,|g(t,u)|_{I\!H,I\!H}$$

and use (iii) together with (11) again to see that

$$E\left(\int_0^T |g(t,u(t))|^2_{I\!H,I\!H}dt\right) < \infty. \quad \blacksquare$$

References

[1] A. Bensoussan, A model of stochastic differential equation in Hilbert space applicable to Navier–Stokes equation in dimension 2, in: Stochastic Analysis, Liber Amicorum for Moshe Zakai, eds. E.Mayer–Wolf, E.Merzbach, and A.Schwartz, Academic Press 1991, pp.51–73.

[2] A.Bensoussan and R.Temam, Equations stochastiques du type Navier–Stokes, *J. Functional Analysis* **13** (1973), 195–222.

[3] M.Capiński and N.J.Cutland, Statistical solutions of Navier–Stokes equations by nonstandard densities, *Mathematical Models and Methods in Applied Sciences* **1**:4 (1991), 447–460.

[4] M.Capiński and N.J.Cutland, Stochastic Navier–Stokes equations, *Acta Applicanda Mathematicae* **25** (1991), 59–85.

[5] M.Capiński and N.J.Cutland, A simple proof of existence of weak and statistical solutions of Navier–Stokes equations, *Proceedings of the Royal Society, London, Ser.A*, **436** (1992), 1–11.

[6] M.Capiński and N.J.Cutland, Navier–Stokes equations with multiplicative noise, *Nonlinearity* **6** (1993), 71–77.

[7] M.Capiński and N.J.Cutland, Foias and Hopf statistical solutions for stochastic Navier-Stokes equations, to appear in *Stochastics.*

[8] C.Foias, Statistical study of Navier-Stokes equations I, *Rend. Sem. Mat. Univ. Padova* **48** (1973), 219–348

[9] H. Fujita Yashima, *Equations de Navier–Stokes stochastiques non homogenes et applications*, Thesis, Scuola Normale Superiore, Pisa 1992.

[10] A.Ichikawa, Stability of semilinear stochastic evolution equations, *Journal of Mathematical Analysis and Applications* **90** (1982), 12–44.

[11] R.Temam, *Navier-Stokes equations*, North-Holland, Amsterdam 1979.

[12] R.Temam, *Infinite-Dimensional Dynamical Systems in Mechanics and Physics*, Springer–Verlag, New York–Berlin–Heidelberg 1988.

[13] M. Viot, *Solutions Faibles d'Equations aux Derivees Partielles non Lineaires*, Thesis, Universite Paris VI (1976).

[14] M.I.Vishik and A.V.Fursikov, *Mathematical Problems of Statistical Hydromechanics*, Kluwer, Dordrecht – London 1988.

Hyperfinite Approximations of Commutative Topological Groups

E. I. Gordon
Department of Mathematics
University of Nizhny Novgorod
Prospect Gagarina, 23
603600 Nizhny Novgorod, Russia

Abstract

In the paper [1] a nonstandard approach to the theory of Fourier series was based on the approximation of the group $\mathbf{S} = \{z \in \mathbb{C} : |z| = 1\}$ by the hyperfinite group $\mathbf{G} = \{z \in \mathbb{C} : z^N = 1\}$ where N is an infinite large natural number. Here we prove the existence of such a hyperfinite approximation for an arbitrary LCA group with a compact open subgroup.

Introduction

In the paper [1] a nonstandard approach to the theory of Fourier series was developed. This approach was based on the approximation of the group $\mathbf{S} = \{z \in \mathbb{C} | \, |z| = 1\}$ by the hyperfinite group $G = \{z \in \mathbb{C} | z^N = 1\}$, where N is an infinitely large natural number. In the paper [2-6] we considered the notion of a hyperfinite approximation (h.a.) for an arbitrary locally compact abelian (LCA) group, which made it possible to extend the ideas of the paper [1] to harmonic analysis on arbitrary separable LCA groups. The additive groups of the fields $\mathbb{R}$ and $\mathbb{Q}_p$ were considered in [2-6] as the examples. The existence of a h.a. for an arbitrary compact abelian group was proved in [5]. Here we prove the existence of a h.a. for an arbitrary LCA group with a compact open subgroup. This theorem was announced in [4]. We also present here some new examples of the hyperfinite approximations.

The nonstandard universe is assumed to be ω_1 – saturated. We use the following abbreviations:

$$\begin{array}{llll} \forall^{st} a & \ldots = & \forall a & (a \text{ standard } \Rightarrow \ldots) \\ \forall^{int} A & \ldots = & \forall A & (A \text{ internal } \Rightarrow \ldots) \end{array}$$

(Similar for $\exists$).

1 The main theorem

In this paper $\mathbf{g}$ is a standard LCA group, $^*\mathbf{g}$ its nonstandard extension, $\mathcal{M}_{\mathbf{g}}$ the monad of zero in $^*\mathbf{g}$, $\mathcal{M}_{\mathbf{g}} = \cap\{\ ^*U | 0 \in U \subseteq \mathbf{g},\ U \text{ is open }\}$; $\xi \underset{\mathbf{g}}{\approx} \eta := \xi - \eta \in \mathcal{M}_{\mathbf{g}}\quad \forall \xi, \eta \in {}^*\mathbf{g}$ (the indices $\mathbf{g}$ in the last two notations usually will be omitted); $ns(\ ^*\mathbf{g}) = \{\xi \in {}^*\mathbf{g} | \exists^{st}\eta \in \mathbf{g}(\xi \approx \eta)\}$: the homomorphism $st : ns(\ ^*\mathbf{g}) \to \mathbf{g}$ is defined by the condition $st(\xi) \approx \xi$.

Definition 1. Let G be a hyperfinite group, and $j : G \to\ ^*\mathbf{g}$ an internal map, $G_f = j^{-1}(ns(\ ^*\mathbf{g}))$. The pair $< G, j >$ is called **a hyperfinite approximation** (h.a.) of the group $\mathbf{g}$ if it satisfies the following conditions:

1) $j(O_G) = O_{\mathbf{g}}$;
2) $\forall^{st}\xi \in \mathbf{g}\exists g \in G(j(g) \approx \xi)$;
3) $\forall g_1, g_2 \in G_f(j(g_1 \pm g_2 \approx j(g_1) \pm j(g_2))$.

Definition 2. Let $\hat{\mathbf{g}}$ be the dual group to $\mathbf{g}$, and $\hat{G}$ the internal dual group to G. A h.a. $< \hat{G}, \hat{j} >$ of $\hat{\mathbf{g}}$ is called **dual** to a h.a. $< G, j >$ of $\mathbf{g}$ if

1) $\forall \chi \in \hat{G}(\chi | G_f \approx 1 \Rightarrow \hat{j}(\chi) \approx \hat{O})$
2) $\forall \chi \in \hat{G}_f \forall g \in G_f(\hat{j}(\chi)(j(g)) \approx \chi(g))$.

Here $\hat{G}_f = \hat{j}^{-1}(ns(*\mathbf{g}))$, $\hat{O} \in \hat{\mathbf{g}}$ is the zero of $\hat{\mathbf{g}}$, so $\forall \xi \in \mathbf{g}\, \hat{O}(\xi) = 1$.

Let now $\mathbf{g}$ be separable, $< G, j >$ – a h.a. of $\mathbf{g}$ and $U \subseteq \mathbf{g}$ a relatively compact neighbourhood of zero in $\mathbf{g}$. Let

$$\Delta = \frac{|j^{-1}(^*U)|}{|G|}, \qquad \hat{\Delta} = \frac{1}{\Delta |G|} = \frac{1}{|j^{-1}(^*U)|} \tag{1}$$

Define an internal measure ν_Δ on G such that $\forall A \subset G(\nu_\Delta(A) = \Delta \cdot |A|)$, and let λ_Δ be the corresponding Loeb measure. The definition of $\nu_{\hat{\Delta}}$ and $\lambda_{\hat{\Delta}}$ on $\hat{G}$ is analogous. We consider G and $\mathbf{g}$ ($\hat{G}$ and $\hat{\mathbf{g}}$) as the measurable spaces with σ–algebras of Loeb measurable sets and Haar measurable sets respectively.

Theorem 1. i) *The map $st \circ j : G_f \to \mathbf{g}(st \circ \hat{j} : \hat{G}_f \to \hat{\mathbf{g}})$ is measurable and the Loeb measure $\lambda_\Delta\,(\lambda_{\hat{\Delta}})$ on $G_f\,(\hat{G}_f)$ induces the Haar measure $\mu_\Delta\,(\mu_{\hat{\Delta}})$ on $\mathbf{g}\,(\hat{\mathbf{g}})$.*

ii) *If $f : \mathbf{g} \to \mathbb{C}$ is standard, then $f \in L_1(\mathbf{g})$ iff there exist an S–integrable lifting $F : G \to^* \mathbb{C}$ satisfying the following condition:*

$$\forall^{int} A \subset G \backslash G_f\, (\Delta \cdot \sum_{g \in A} |F(g)| \approx 0). \tag{2}$$

In this case $\int_{\mathbf{g}} f d\mu_\Delta = {}^\circ(\Delta \circ \sum_{\mathbf{g} \in G} F(g))$. Certainly the same theorem is true for $f \in L_1(\hat{\mathbf{g}})$.

If standard $p \geq 1$, $f \in L_p(\mathbf{g})$ and $F : G \to^* \mathbb{C}$ is a lifting of $f \cdot st \cdot j$ such that $|F|^p$ is S–integrable and satisfies (2) then F is called an S_p–integrable lifting of f.

Theorem 1 was proved in [4], where we also investigated the conditions for ${}^*f \circ j$ to be an S_p-interable lifting of $f \in L_p(\mathbf{g})$. In the case of groups $\mathbb{R}$ and $\mathbb{Q}_p$ the condition (2) is equivalent to some standard condition of fast enough decreasing of f at infinity [2.4]. This theorem has a standard interpretation, which deals with the approximations of Haar inegral on LCA groups by Riemann sums.

Let $\mathcal{F} : L_2(\mathbf{g}) \to L_2(\hat{\mathbf{g}})$ be the Fourier transform on $\mathbf{g}$:

$$\mathcal{F}(f)(\kappa) = \int_{\mathbf{g}} f(\xi)\overline{\kappa}(\xi)d\mu_\Delta(\xi),$$

and $\Phi : G^{*\mathbf{C}} \to \hat{G}^{*\mathbf{C}}$ the internal Fourier transform on G:

$$\Phi(\phi)(\xi) = \Delta \sum_{\mathbf{g} \in G} \phi(g)\overline{\xi}(g)$$

In this case $\phi^{-1}(\psi)(g) = \hat{\Delta} \sum_{\xi \in \hat{G}} \psi(\xi)(g)$, where $\hat{\Delta}$ is defined by formula (1).

Theorem 2. *If* $\mathbf{g}$ *has an open compact subgroup, and if* $\{< G, j >, < \hat{G}, \hat{j} >\}$ *is the pair of dual h.a. of the groups* $\mathbf{g}$ *and* $\hat{\mathbf{g}}$ *respectively then*

i) the Haar measure $\mu_{\hat{\Delta}}$ *on* $\hat{\mathbf{g}}$ *is associated with the Haar measure* μ_Δ *on* $\mathbf{g}$ *(this means that the Fourier transform* $\mathcal{F}$ *preserves normes, defined by these measures);*

ii) if $f \in L_2(\mathbf{g})$ *and* $F : G \to^* \mathbb{C}$ *is an* S_2*-integrable lifting of* f*, then* $\phi(f)$ *is an* S_2*-integrable lifting of* $\mathcal{F}(f)$.

This theorem was proved in [6]. In [2,4] we had constructed the pair of dual h.a. of $\mathbb{R}$ and $\hat{\mathbb{R}} \cong \mathbb{R}$, which satisfies i) and ii) of theorem 2. Here we are going to prove.

Theorem 3. *If* $\mathbf{g}$ *has an open compact subgroup, then* $\mathbf{g}$ *has a h.a.* $< G, j >$.

Now, using theorem 4.1 of [4], it is easy to show that in this case $\hat{\mathbf{g}}$ has a h.a., dual to $< G, j >$.

It is well known that every LCA group $\mathbf{g}$ is isomorphic to $\mathbb{R}^n \times H$, where H has an open compact subgroup. So, the following statement is true.

Corollary *If* $\mathbf{g}$ *is a separable LCA group, then a pair of dual h.a. of* $\mathbf{g}$ *and* $\hat{\mathbf{g}}$ *exists, which satisfies i) and ii) of theorem 2.*

It is easy to present the standard version of this corollary, which establishes that the Fourier transform on arbitrary separable LCA group may be approximated by the Fourier transform on some finite group.

2 The proof of the main theorem

We shall use the following well known van Kampen's theorem:

If $\mathbf{g}$ *is a compact abelian group then in every open neighbourhood of zero it can be found such a closed subgroup* H *of* $\mathbf{g}$, *that* $\mathbf{g}/H$ *is isomorphic to* $\mathbf{S}^m \times K$ *for some natural* m, *and some finite group* K.

Let U be a compact open subgroup of $\mathbf{g}, D = \mathbf{g}/U$. D is a countable discrete group (we assume that $\mathbf{g}$ is separable). Using ω_1–saturation of the nonstandard universe we can find a hyperfinite set $T \subseteq {}^*D$ such that $D \subseteq T$. Let ${}^*D(T)$ be the internal subgroup of ${}^*D$, generated by T. If $\pi : \mathbf{g} \to D$ is the canonical projection, $H = {}^*\pi^{-1}({}^*D(T))$, $\varepsilon = {}^*\pi|H$, then we have the following commutative diagram

$$\begin{array}{ccccc} U & \subseteq & \mathbf{g} & \xrightarrow{\pi} & D \\ {\scriptstyle id}\downarrow & & {\scriptstyle id}\downarrow & & {\scriptstyle id}\downarrow \\ {}^*U & \subseteq & H & \xrightarrow{\varepsilon} & {}^*D(T), \end{array} \tag{3}$$

the lines of which are short exact sequences.

Using the structural theorem about the finitely generated abelian groups and the transfer principle we find that ${}^*D(T) = D_1 \oplus D_2$ where D_1 is a hyperfinite abelian group and D_2 an internal free abelian group with a hyperfinite set of generators. Let $H_i = \varepsilon^{-1}(D_i)\,(i = 1,2)$. Then $H = H_1 + H_2$, $H_1 \cap H_2 = {}^*U$ and the following sequences are exact

$${}^*U \subseteq H_1 \xrightarrow{\varepsilon_1} {}^*D_1(T) \tag{4}$$

$${}^*U \subseteq H_2 \xrightarrow{\varepsilon_2} {}^*D_2(T) \tag{5}$$

Here $\varepsilon_i = \varepsilon|H_i\,(i = 1,2)$.

Consider the infinitesimal neighbourhood of zero $V \subseteq {}^*U$ (so $V \subseteq \mathcal{M}$). Using van Kampen's theorem and the transfer principle we find a hypernatural number k, a hyperfinite group R and a surjective continuous homomorphism $\phi : {}^*U \to \mathbf{S}^k \oplus R$, such that $Ker\varphi \subseteq V$. Using the extension theory for abelian groups we find such an internal group L, and such homomorphisms $\gamma : H_1 \to L$, $\kappa :{}^* \mathbf{S}^k \oplus R \to L$, $\delta : L \to D_1$, that the diagram

$$\begin{array}{ccccc} {}^*U & \subseteq & H_1 & & \\ \phi\downarrow & & \gamma\downarrow & \searrow \varepsilon_1 & \\ {}^*\mathbf{S}^k \oplus R & \xrightarrow{\kappa} & L_1 \xrightarrow{\delta} & & D_1 \end{array} \tag{6}$$

is commutative, its lower line is exact and γ is surjective.

By the equality $Ext(D_1,\ \mathbf{S}^k \oplus R) = Ext(D_1, R)$ there exist two short exact sequences

$${}^*\mathbf{S}^k \xrightarrow{\kappa_1} L_1 \xrightarrow{\delta_1} D_1, \tag{7}$$

$$R \xrightarrow{\kappa_2} L_2 \xrightarrow{\delta_2} D_2, \tag{8}$$

which are connected with the lower line of diagram (6) in following way:

$$L = \{< 1_1, 1_1 >\in L_1 \oplus L_2 | \delta_1(1_1) = \delta_2(1_2),\ \kappa = < \kappa_1, \kappa_2 >,\ \delta(1_1, 1_2) = \delta_1(1_1) = \delta_2(1_2)\}.$$

The group ${}^*\mathbf{S}^k$ is injective. So there exist a monomorphism $\chi : D_1 \to L_1$, such that $\delta_1 \cdot \chi = id$. The group L_2 is hyperfinite, because the groups R and D_1 are hyperfinite. Note that ϕ is open, because it is a surjective continuous homomorphism of compact groups. So $\phi(V)$ is an open neighbourhood of zero in $\mathbf{S}^k \oplus R$, and $\phi(V) \cap^* \mathbf{S}^k$ is a neighbourhood of zero in ${}^*\mathbf{S}^k$. It is easy to see that there is a hyperfinite subgroup $F \subseteq^* \mathbf{S}^k$ such that

$$F + (\phi(V) \cap^* \mathbf{S}^k) = {}^*\mathbf{S}^k \tag{9}$$

Consider now the hyperfinite group $M \subseteq L$ such that

$$M = \{< \kappa_1(f) + \chi(d),\ 1 > | f \in F,\ d \in D_1,\ l \in L_2,\ \delta_2(1) = d\}. \tag{10}$$

The homomorphism γ is surjective. So we can find an internal set $G_1 = \{g_m | m \in M\}$ such that $\forall m \in M\ \gamma(g_m) = m$.

Define the binary operation $+_1$ on G_1 by the formula $g_{m_1} +_1 g_{m_2} = g_{m_1+m_2}$, so that $\gamma(g_{m_1} + g_{m_2}) = \gamma(g_{m_1}) +_1 \gamma(g_{m_2})$. It is easy to see that $< G_1, +_1 >$ is a hyperfinite abelian group.

Lemma 1.

i) $\forall m_1,\ m_2 \in M \quad g_{m_1} +_1 g_{m_2} \approx g_{m_1} + g_{m_2}$

ii) $\forall m \in M\,(-_1 g_m) \approx -g_m$

The proof follows immediately from the diagram (6).

Lemma 2. $\forall h \in H_1 \exists g \in G_1 (g \approx h)$.

Proof Let $h \in H_1$. By the previous considerations $\gamma(h) = < \kappa_1(s) + \chi(d), 1 >$ where $s \in {}^*\mathbf{S}^k$, $l \in L_2$, $\delta_2(1) = d$. By (9) find $f \in F$ such that $s - f \in \phi(V) \cap^* \mathbf{S}^k$. Then $m = < \kappa_1(f) + \chi(d),\ 1 > \in M$. It is easy to see that $g_m \approx h$.

Let $G_u =^* U \cap G_1$.

Lemma 3. *G_U is a subgroup of G_1. The pair $< G_U, c >$ is a h.a. of U.*

The proof follows immediately from the lemmas 1, 2 and the fact that U is a compact and open subgroup of $\mathbf{g}$.

Let us consider now the sequence (5). let $\nu : {}^*D(T) \to D_i$ $(i = 1, 2)$ be the canonical projection, and let m be hypernatural such that $(\nu_2(T) - \nu_2(T)) \cap mD_2 = \{0\}$. To prove the existence of such m it is necessary to apply the transfer principle to the following proposition which is evidently true:

If P is a finite subset of a free finitely generated abelian group H then there is a natural m such that $P \cap mH = \{0\}$.

Let $Q = D_2/mD_2$ and $\lambda : D_2 \to Q$ the canonical projection. Then Q is a hyperfinite abelian group and λ is injective on $\nu_2(T)$. Let $\tau : Q \to D_2$ be such a right inverse map to λ that $\nu_2(T) \subseteq \tau(Q)$. Let $G_3 = \tau(Q)$ and $\forall q \in Q\ \tau(q) = d_q$. Define the binary operation $+_3$ on G_3 by the formula $d_{q_1} +_3 d_{q_2} = d_{q_1+q_2}$ (note that τ is not a homomorphism!) so that $\lambda(d_{q_1} +_3 d_{q_2}) = \lambda(d_{q_1}) + \lambda(d_{q_2})$.

Lemma 4.

$$\text{i)}\quad \forall^{st} d \in D\ \exists q \in Q(\nu_2(d) = d_q);$$

$$\text{ii)}\quad \textit{if } d_{q_1}, d_{q_2} \in \nu_2(D) \textit{ then } d_{q_1} +_3 d_{q_2} = d_{q_1} + d_{q_2}$$

The operation + at the right side of the last equality is the group operation on D_2.

We omit the simple proof of this lemma. Note only that by our construction $D \subset T \subset {}^*D$, so $\nu_2(D) \subset \nu_2(T)$ and $\nu_2(D)$ is an external subgroup of D_2.

As D_2 is a free abelian group it is projective. So there is a monomorphism $\mu : D_2 \to H_2$ (5), which is right inverse to ε_2. Consider the set $G_2 = \{g + \mu(d_q) | g \in G_U,\ q \in Q\}$ and define the operation $+_2$ on G_2 by the formular $(g_1 + \mu(d_{q_1})) +_2 (g_2 + \mu(d_{q_2})) = g_1 +_U g_2 + \mu(d_{q_1} +_3 d_{q_2})$.

Using the facts that $H_2 = {}^*U \oplus \mu_2(U_2)$ and $G_u \subset {}^*U$ it is easy to see that $< G_2, +_2 >$ is a hyperfinite abelian group and $G_2 \cap^* U = G_U$.

Let now $G = G_1 \times G_2$ and define $j : G \to {}^*\mathbf{g}$ by the equality $j(< g_1, g_2 >) = g_1 + g_2$. Let us show that $< G, j >$ is a h.a. of $\mathbf{g}$. The first condition of the definition 1 is obvious. Let $\xi \in \mathbf{g}$. As $\mathbf{g} \subset H = H_1 + H_2$ (3), so $\xi = h_1 + h_2$, where $h_i \in H_i$. Then $d = \pi(\xi) = \varepsilon_1(h_1) + \varepsilon_2(h_2)$ by (3), (4), (5) and $\varepsilon_2(h_2) = \nu_2(d) \in \nu_2(D)$ because $d \in D$. By lemma 4 $\nu_2(d) = d_q$ for some $q \in Q$ and there is an element $g \in {}^*U$ such that $h_2 = g + \mu(d_q)$. By lemma 3 G_u approximates U and U is compact. So there is such $g_0 \in G_u$ that $g_0 \approx g$. Then $g_2 = g_0 + \mu(d_q) \approx g + \mu(d_q) = h_2$ and $g_2 \in G_2$. By lemma 2 there is $g_1 \in G_1$ such that $g_1 \approx h_1$. Then $g_1 + g_2 \approx h_1 + h_2 = \xi$. We proved the second condition of the definition 1.

Let $g_1 + g_2 \approx \xi \in \mathbf{g}$, $g_1' + g_2' \approx' \xi' \in \mathbf{g}$, $g_i, g_i' \in G_i (i = 1, 2)$. We must show that $(g_1 +_1 g_1') + (g_2 +_2 g_2') \approx \xi + \xi'$. By lemma 1 $(g_1 +_1 g_1') \approx g_1 + g_1'$. So it is enough to show that $g_2 +_2 g_2' \approx g_2 + g_2'$. Let $\pi(\xi) = d$. As $g_1 + g_2 \approx \xi$ so $g_1 + g_2 - \xi \in {}^*U$ and by (3) $\varepsilon(g_1 + g_2) = d$. By the definitions $g_i \in H_i$, so $\varepsilon_2(g_2) = \nu_2(d) \in \nu_2(D)$. Similar $\varepsilon_2(g_2') = \nu_2(d') \in \nu_2(D)$, where $d' = \pi(\xi')$. Now by the definition of G_2 $g_2 = \tilde{g} + \mu(\nu_2(d))$, $g_2' = \tilde{g}' + \mu(\nu_2(d'))$. By lemma 4 $g_2 +_2 g_2' = \tilde{g} +_u \tilde{g}' + \mu(\nu_2(d) + \nu_2(d'))$ $\tilde{g} +_u \tilde{g}' + \mu(\nu_2(d)) + \mu(\nu_2(d'))$. By lemma 3 $\tilde{g} +_u \tilde{g}' \approx \tilde{g} + \tilde{g}'$ and so $g_2 +_2 g_2' \approx g_2 + g_2'$.

The considerations for $g_1 - g_2$ are similar.

3 Some examples

Now we are going to present some examples of h.a. Let $\{< K_n, \phi_n > \mid n \in \mathbb{N}\}$ be a standard sequence of finite commutative abelian groups (rings) K_n and surjective homomorphisms $\phi_n : K_{n+1} \to K_n$, $K = \lim_{\leftarrow} < K_n, \phi_n >$ its projective limit.

By the definition of projective limits there is a sequence of surjective homomorphisms $\Phi_n : K \to K_n$ such that $\phi_n \circ \Phi_{n+1} = \phi_n$. Let $N \in {}^*\mathbb{N} \setminus \mathbb{N}$, then ${}^*K_N$ is a hyperfinite abelian group (ring). Let $j : {}^*K_N \to {}^*K$ be a right inverse map to ${}^*\Phi_N$. Then it is easy to prove the following

Proposition 1. *$< {}^*K_N, j >$ is a h.a. of the group (ring) K. Of course in the case of the rings we have to add to the definition 1 the condition 3) $\forall g_1, g_2 \in G_f \; j(a \cdot b) \approx j(a) \cdot j(b)$.*

For example, if $\tau = \{a_n | n \in N\}$ is a standard sequence of natural numbers, such that $\forall n \in \mathbb{N} \, a_n | a_{n+1}$, $K_n = \mathbb{Z}/a_n\mathbb{Z}$, $\phi_n : K_{n+1} \to K_n$ is defined by formula $\phi_n(a) = rem(a, a_n)$ – the remainder of division a by $a_n, a \in K_{n+1}$ (we present K_n as $\{0, 1, 2, \dots, a_n - 1\}$), then $K = \Delta_\tau$ is the ring of τ-adic integer numbers. The elements of K are the sequences $\{b_n | n \in \mathbb{N}\}$ of natural numbers, such that $0 \le b_n < a_n - 1$, $b_n = rem(b_{n+1}, a_n)$. We can define the injection $\nu : \mathbb{Z} \to K$ such that $\forall b \in \mathbb{Z} \quad \nu(b) = \{rem(b, a_n) | \nu \in \mathbb{N}\}$. It is easy to see that $\mathbb{Z}$ is dense in K. The following proposition follows immediately from Proposition 1.

Proposition 2. *If $N \in {}^*\mathbb{N} \setminus \mathbb{N}$, ${}^*K_N = \{0, 1, 2, \dots, {}^*a_N - 1\}$, $j = \nu | \, {}^*K_N$ then $< {}^*K_N, j >$ is a h.a. of the ring Δ_τ.*

Let us construct now a dual h.a. to the h.a. of Proposition 2.
Let $\mathbb{Q}^{(\tau)} = \left\{ \frac{m}{a_n} | m \in \mathbb{Z} \right\}$. Then $\mathbb{Q}^{(\tau)}$ is a subgroup of $\mathbb{Q}$ because $a_n | a_m$ for $n \le m$ and $\mathbb{Z} \subset \mathbb{Q}^{(\tau)}$. It is well known (see for example [7]) that $\mathbb{Q}^{(\tau)}/\mathbb{Z} \cong \hat{\Delta}_\tau$. The character $\chi_{\overline{m,n}}$ of Δ_τ, corresponding to the class $\overline{m,n}$ of the element $\frac{m}{a_n} \in \mathbb{Q}^{(\tau)}$ satisfies the formula

$$\chi_{\overline{m,n}}(\xi) = \exp 2\pi i \, \tfrac{m\psi_n(\xi)}{a_n}, \; \xi \in \Delta_\tau. \tag{11}$$

By transfer principle ${}^*\hat{K}_N \cong {}^*K_N$.

Proposition 3. *If $\hat{j} : {}^*K_N \to \hat{\Delta}_\tau$ is such that $\forall m \in {}^*K_N \hat{j}(m) = \chi_{\overline{m,n}}$, then $< {}^*K_N, \hat{j} >$ is a dual h.a. to the h.a. of Proposition 2.*

To prove the proposition 3 note that Δ_τ is compact and in the case of compact groups the first condition of the definition 2 is always true [6]. The proof of the second condition is trivial by (11).

In the case of a compact group **g** if **g** $\to \mathbb{C}$ is bounded and almost everywhere continuous then ${}^*f \circ j$ is an S_p-integrable lifting of f for every $p \ge 1$ (see the remarks after theorem 1). Then theorem 2 gives us the following approximation theorem for the Fourier transform on Δ_τ.

Proposition 4. *If $f : \Delta_\tau \to \mathbb{C}$ is almost everywhere continuous then*

$$\lim_{n\to\infty} \sum_{k=0}^{a_n-1} | \int_{\Delta_\tau} f(\xi) \exp\left(-2\pi i \frac{k\psi_n(\xi)}{a_n} d\mu_\tau(\xi) - a_n^{-1} \sum_{m=0}^{n-1} f(m) \exp(-\frac{2\pi ikm}{a_n} \right) |^2 = 0$$

Here μ_τ is the normed Haar measure on $\Delta_\tau(\mu_\tau(\Delta_\tau) = 1)$.

The following example is the τ-adic solenoid $\sum_\tau$. The set of this group is $[0,1]\times\Delta_\tau$. The operation $+_\tau$ is defined by formula

$$< x,\xi > +_\tau < y,\eta >=< \{x+y\},\ \xi+\eta+[x+y] >,$$

where $\{x\}$ is the fractional part and $[x]$–integer part of x. The definition of the compact topology on $\sum_\tau$ can be found in [7]. We describe it only in the nonstandard language:

$$< x,\xi >\approx 0 \Leftrightarrow (x\approx 0\wedge \xi\approx 0)\vee(x\approx 1\wedge \xi+1\approx 0)$$

Propositon 5. *Let $N\in^*\mathbb{N}\backslash\mathbb{N}$, and let G be the additive group of the ring*

$$^*\mathbb{Z}/^*a_N^2\,{}^*\mathbb{Z}=\{0,1,\ldots,a_N^2-1\}.$$

If $j:G\to\ {}^*\Sigma_\tau$ is such that $\forall a\in G\ j(a)=<\left\{\frac{a}{a_N}\right\},\left[\frac{a}{a_N}\right]>$, then $< G,j >$ is a h.a. of $\sum_\tau$.

We omit an elementary proof of this proposition. Theorem 1 gives now the following approximation theorem for the Haar integral on $\sum_\tau$.

Proposition 6. *If μ is the normed Haar measure on $\sum_\tau$ and $f:\sum_\tau\to\mathbb{C}$ is a bounded and almost everywhere continuous function on $\sum_\tau$ then*

$$\int_{\sum_\tau} f\,d\mu=\lim_{n\to\infty}\left(a_n^2\right)^{-1}\cdot\sum_{k=0}^{a_n^2-1} f\left(\left\{\frac{k}{a_n}\right\},\left[\frac{k}{a_n}\right]\right).$$

It is well known [7] that $\widehat{\Sigma}_\tau\cong\mathbb{Q}^{(\tau)}$. If $\alpha=\frac{m}{a_n}\in\mathbb{Q}^{(\tau)}$ then

$$\forall x\in[0,1]\,\forall\xi\in\Delta_\tau\,\chi_\alpha(x,\xi)=\exp\frac{2\pi i(x+\psi_n(\xi))m}{a_n}.$$

Now it is easy to construct the dual h.a. As above $\hat{G}\cong G$.

Proposition 7. *If N,G,j are such as in Proposition 5 and $\hat{j}:G\to\widehat{\Sigma}_\tau$ is defined by the formula $\hat{j}(b)=\chi_\beta$, where $\beta=\frac{b}{a_n}$, $b\in G$, then $< G,\hat{j} >$ is a h.a. of $\widehat{\Sigma}_\tau$, dual to $< G,j >$.*

Now it is quite easy to write the approximation theorem for the Fourier transformation on $\sum_\tau$ similar to Proposition 4.

If $\tau=\{p^n|n\in\mathbb{N}\}$ for a standard prime p, then $\Delta_\tau=\mathbb{Z}_p$ – the ring of integer p–adic numbers. If $\tau=\{n!|n\in\mathbb{N}\}$ then $\Delta_\tau\cong\Pi_p\mathbb{Z}_p$ and $\widehat{\Sigma}_\tau\cong\mathbb{Q}$ – the additive group of rationals in the discrete topology. So the Proposition 7 gives a h.a. for $\mathbb{Q}$.

References

[1] Luxemburg, W.A.J.: "'A nonstandard analysis approach to Fourier analysis, contributions to nonstandards analysis"' , Amsterdam, North Holland 1972, p. 16 – 39.

[2] Gordon, E.I.: "'On Fourier transform in nonstandard analysis"', Izv. Vyssh. Uchebn. Zaved. Math. 1989, N 2 p. 17 – 25 (in Russian).

[3] Gordon, E.I.: "'Hyperfinite approximations of locally compact abelian groups"', Soviet Math. Dokl. 1991, Vol. 42, N 2 p. 567 - 571.

[4] Gordon, E.I.: "'Nonstandard analysis and compact abelian groups"', Siberian Math. J. 1991, Vol 32. N 2 p. 26 – 40.

[5] Gordon E.J.: "'Nonstandard analysis and locally compact abelian groups"', Acta Applicandae Mathematicae 1991, Vol. 25 p. 221 – 239.

[6] Gordon, E.I.: "'Hyperfinite approximations of locally compact groups and some of their applications"', to appear in the Proceedings of the 5-th Siberian School "'Algebra and Analysis"', Irkutsk, 1991.

[7] Hewitt, E.; Ross, K.: "'Abstract Harmonic Analysis"', 1963, Vol. 1, Springer – Verlag, Berlin–Göttingen–Heidelberg.

A NOTE ON THE MYOPE TOPOLOGY

Tommy Norberg [†]
Dep. of Mathematics, CTH & GU, S–412 96 Göteborg, Sweden

Abstract

The note shows first that the myope and the Lawson topologies on the collection $\mathcal{K}$ of compact subsets of some locally compact Hausdorff space S are the same. Then, generalizing S to be some sober locally compact Q_δ space and topologizing $\hat{S} = S \cup \{\Delta\}$, where $\Delta \notin S$, with the topology having closed sets $\widehat{\mathcal{F}} = \mathcal{K} \cup \{\hat{S}\}$, where $\mathcal{K}$ now denotes the collection of compact saturated subsets of S, it is shown that Fell's topology on $\widehat{\mathcal{F}}$ relativizes to the myope (which by definition is the Lawson) topology on $\mathcal{K}$. In other words, $\widehat{\mathcal{F}}$ is a compactification of $\mathcal{K}$. A characterization of the (nonstandard) monad of $K \in \mathcal{K}$ w.r.t. the myope topology is also given. Let $\mathrm{m_{cc}}(s)$ denote the monad of $s \in S$ w.r.t. the cocompact topology. Then $L \in {}^*\mathcal{K}$ is near standard with standard part

$$\mathrm{st}(L) = \{s \in S : \mathrm{m_{cc}}(s) \cap L \neq \emptyset\}$$

if, and only if, $L \subseteq {}^*K$ for some $K \in \mathcal{K}$

1 Introduction

The *myope* topology is a topology on the collection $\mathcal{K}$ of compact subsets of some locally compact Hausdorff space S. It is generated by the two families

$$\mathcal{K}^F := \{K \in \mathcal{K} : K \cap F = \emptyset\}, \quad F \in \mathcal{F}$$

$$\mathcal{K}_G := \{K \in \mathcal{K} : K \cap G \neq \emptyset\}, \quad G \in \mathcal{G}$$

($\mathcal{F}$ and $\mathcal{G}$ denote resp. the collections of closed and open subsets of S). It is the relativization to $\mathcal{K}$ of the Vietoris topology topology on $\mathcal{F}$ (see, e.g., [14]). Unless S is compact, the myope topology is strictly finer than the trace of Fell's 'hit-or-miss' topology on $\mathcal{K}$, the latter being generated by the two collections $\mathcal{K}^K$, $K \in \mathcal{K}$, and $\mathcal{K}_G$, $G \in \mathcal{G}$. See Matheron [6], who shows Fell's topology to be compact Hausdorff and the myope topology to be locally compact Hausdorff.

[†]Supported by the Swedish Natural Science Research Council

Our main aim with this short note is to give a nonstandard characterization of the monads of the myope topology. This has been done already for Hausdorff-S. Refer to Wicks [14, Propositions 2.1 and 3.1] and the references therein. We are, however, also interested in certain non-Hausdorff settings. Moreover, the manner in which we arrive to the characterization is interesting also for those who wish to restrict their horizon to Hausdorff spaces.

We begin by showing that the second subbase family above consisting of sets of the form $\mathcal{K}_G$, can be thinned to G:s of the form $G = S \setminus K$ with $K \in \mathcal{K}$. This makes it possible to achieve our first goal, which is to extend the notion of the myope topology to certain non-Hausdorff settings in such a way that it is still being locally compact Hausdorff.

Its one-point compactification turns out to be homeomorphic to the collection $\widehat{\mathcal{F}}$ of all closed subsets of a sober locally compact space $\hat{S}$. The topology on $\widehat{\mathcal{F}}$ is Fell's [1], known to be compact Hausdorff. Moreover, its monads are known [8], so seeing the myope topology as the trace on $\mathcal{K}$ of Fell's topology tells us at once how its monads look like.

We proceed with a somewhat more detailed description of our results. The first, which is Theorem 1, states that the myope topology coincides with the Lawson topology on $\mathcal{K}$ (cf. [2, 3]). This is not at all surprising, since also the latter is known to be a locally compact Hausdorff topology on $\mathcal{K}$ and intuitively there cannot exist many topologies on $\mathcal{K}$ with such strong regularity properties.

We then increase the generality of our setting by allowing S to be an arbitrary sober locally compact Q_δ space, and redefining $\mathcal{K}$ to be the collection of all compact saturated subsets of S. (For our readers convenience we recall in Section 2 some notions from (non-Hausdorff) topology that we believe are not widely known; let us just note at this point that Hausdorff spaces are sober and Q_δ, and that all subsets of a Hausdorff space are saturated.) Relying on Theorem 1, we may (and will below) refer to the Lawson topology on $\mathcal{K}$ as the myope. It is well known that this is a locally compact Hausdorff topology on $\mathcal{K}$, see, e.g., [2, 3].

Next a new space $\hat{S}$ is formed by adding to S a point $\Delta \notin S$. The topology we provide $\hat{S}$ with is not the usual one used in the one-point compactification of locally compact Hausdorff spaces. Instead the weaker one with non-empty open sets $\hat{S} \setminus K$, $K \in \mathcal{K}$, is used. Its trace on S is the cocompact topology, cf. [3]. The topology we provided $\hat{S}$ with is sober and locally compact (Proposition 2 and Corollary 3). Thus Fell's topology on its collection $\widehat{\mathcal{F}}$ of closed sets is compact Hausdorff [1]. Our second characterization of the myope topology is Theorem 4, which states that the relativization of Fell's topology to $\mathcal{K}$ coincides with the myope topology.

By seeing the myope topology as the trace of Fell's topology we obtain with hardly any work Theorem 5, saying that $L \in {}^\star\mathcal{K}$ is near standard if, and only if, $L \subseteq {}^\star K$ for some $K \in \mathcal{K}$, and that in this case its standard part is

$$\mathrm{st}(L) = \{s \in S : \mathrm{m}_{\mathrm{cc}}(s) \cap L \neq \emptyset\}$$

($m_{cc}(s)$ is the monad of $s \in S$ w.r.t. the cocompact topology). By specializing to the case of a Hausdorff-S, we obtain in Theorem 7 the not surprising fact that a near standard $L \in {}^*\mathcal{K}$ has the standard part

$$\mathrm{st}(L) = \{s \in S : \mathrm{m}(s) \cap L \neq \emptyset\} = \{{}^\circ t : t \in L\}$$

(here $\mathrm{m}(s)$ and ${}^\circ t$ denote, resp., the monad of $s \in S$ and the standard part of $t \in L \subseteq {}^*S$ w.r.t. the original topology on S; note that it follows from the fact that L is near standard that all points of L are near standard).

Besides giving a more thorough understanding of the myope topology, this note lays a ground for a forthcoming study of capacities using nonstandard analysis. In particular the kind of capacities that are associated with probability distributions or more general measures on the collection $\mathcal{F}$ of closed subsets of S, cf. [6, 8].

Any such capacity may be viewed as an increasing upper semicontinuous function from $\mathcal{K}$ into the non-negative extended reals that maps $\emptyset$ on 0 [6]. Writing C for the collection of capacities, let us just note here that any (internal capacity) $c \in {}^*C$ has a standard part ${}^\circ c \in C$ w.r.t. Vervaat's sup vague topology, since this is a compact Hausdorff topology [12], and that it follows by [9] that

$$({}^\circ c)(K) = \mathrm{S}{-}\lim_{L \downarrow K} c(L)$$

for $K \in \mathcal{K}$. (Refer, e.g., to [11] for a definition of S−right limits of internal functions on the hyperreals. Its extension to our setting is straightforward.) In order to calculate the S−limit on the right hand side above, we need Theorems 5 and 7 giving necessary and sufficient conditions (in the general and the Hausdorff case, resp.) for $L \in {}^*\mathcal{K}$ to be in the monad of $K \in \mathcal{K}$.

The results are presented in detail in Section 2. Proofs that do not contribute significantly to the discussion are postponed to Section 3.

2 Results

We begin by assuming S locally compact Hausdorff. The object of study is the myope topology on its collection $\mathcal{K}$ of compact subsets. It is well known that $\mathcal{K}$ is an upper continuous lattice, which among other things means that it carries the Lawson topology (refer to Gierz, Hofmann, Keimel, Lawson, Mislove & Scott [2] or Hofmann & Mislove [3]; note that these references treat the dual notion of lower continuous lattices; also: their notion of a continuous lattice is slightly more restrictive than ours, because they require of a continuous lattice that it is also a complete lattice, while we call a lattice *continuous* if it is continuous as a poset, cf. [5, 2, 3]).

Theorem 1 *Let S be a locally compact Hausdorff space. On its collection $\mathcal{K}$ of compact subsets, the myope and the Lawson topologies coincide.*

The proof is postponed to Section 3. Let us only note at this point that the Lawson topology is generated by the two collections $\mathcal{K}^F$, $F \in \mathcal{F}$, and $\mathcal{K}_{K^c}$, $K \in \mathcal{K}$, cf. [3] (we write $B^c := S \setminus B$). Thus, trivially, the myope topology is finer than the Lawson topology.

We now interupt the general development and recall some general topological notions. Let X be a topological space, and take $B \subseteq X$. We write B^- and B^o for its closure and interior, resp. Moreover, B is called *saturated* if B coincides with the intersection of its open neigborhoods, i.e.,

$$B = \bigcap \{G \subseteq X : G \text{ is open and contains } B\} =: \text{sat}B$$

A simple exercise yields that $x \in \text{sat}B$ if, and only if, $\{x\}^- \cap B \neq \emptyset$. It follows, e.g., that all subsets of a T_1 space are saturated.

Note that we call X *locally compact* if every point $x \in X$ has a neighborhood basis of compact sets. This is not the standard definition of local compactness. It is, however, the definition used by Fell [1]. Note that $K \subseteq X$ is compact if, and only if, sat$K$ is so. Thus, in studying local compactness, restricting attention to compact saturated sets is no severe loss of generality. Hofmann & Mislove [3] claim that 'it is in fact advantageous to do so.' We do not disagree.

A closed set $F \subseteq X$ is said to be *irreducible* if, whenever $F \subseteq F_1 \cup F_2$ for any closed sets $F_1, F_2 \subseteq X$, then either $F \subseteq F_1$ or $F \subseteq F_2$. It is clear that each singleton closure is irreducible. Now X is called *sober* if every non-empty irreducible closed subset is the singleton closure of a unique point in X, i.e., the mapping $x \mapsto \{x\}^-$ is a bijection between X and its collection of non-empty irreducible closed sets. See [2, 3]. Sober spaces are T_0: given any pair of distinct points, one has a neighborhood which does not contain the other; and a T_0 space is called *super-sober* if the set of limit points of each ultrafilter is either empty or a singleton closure. All super-sober spaces are sober [2].

Note now that if both $K, L \subseteq X$ are compact saturated, then so is $K \cup L$ but not necessarily $K \cap L$. For an example of this phenomenon, see [10]. Next, let $\{K_i\}_i$ be a collection of compact saturated sets in X. If X is sober, and the collection $\{K_i\}_i$ filtered, then $\bigcap_i K_i$ is compact saturated [3] ($\{K_i\}_i$ is called *filtered* if for each pair K_{i_1}, K_{i_2} there is some $K_{i_3} \subseteq K_{i_1} \cap K_{i_2}$). By a Q_δ *space* we will understand a space in which the intersection of any non-empty collection of compact saturated sets is again compact saturated. If X is super-sober, then X is Q_δ. A proof due to van Rooij of this fact may be found in [10] (see also [3]). Let us also note here that a sober locally compact Q_δ space is super-sober [3]. So, for sober locally compact spaces, Q_δ (actually the fact that $K \cap L$ is compact, if both K, L so are) is equivalent to super-sobriety.

Returning to the general development, let S be a sober locally compact Q_δ (equivalently, a super-sober locally compact) space, and write $\mathcal{K}$ for its collection of compact saturated subsets. In the following we will refer to the topology on $\mathcal{K}$

which is generated by the two families

$$\mathcal{K}^F,\ F \in \mathcal{F}, \quad \text{and} \quad \mathcal{K}_{K^c},\ K \in \mathcal{K},$$

of sets in $\mathcal{K}$, as the *myope* topology. It is also known as the Lawson topology (cf. [2, 3]), where it is shown that it is locally compact Hausdorff.

Let Δ be an ideal point not in S and consider the disjoint union

$$\hat{S} := S \cup \{\Delta\}$$

We topologize $\hat{S}$ by declaring the sets $\hat{S} \setminus K,\ K \in \mathcal{K}$, open. Relativized to S, this topology is known as the *cocompact* topology [2, 3]. The set S endowed with the cocompact topology is denoted S_{cc}. Note that S_{cc} has a continuous topology. Its collection of closed subsets is the upper continuous lattice $\mathcal{F}_{cc} := \mathcal{K} \cup \{S\}$.

Everything would be fine if we could prove that S_{cc} is sober, since then [3, Proposition 2.11] would allow us to conclude that S_{cc} is locally compact. However, whether S_{cc} is sober or not depends critically on whether $S \in \mathcal{K}$ or not. If $S \in \mathcal{K}$ then $\mathcal{F}_{cc} = \mathcal{K}$ and we may conclude by [3, Lemma 2.23], that S_{cc} is sober (to see how, refer to the proof below of Proposition 2). If $S \notin \mathcal{K}$ then a simple argument yields that S itself is irreducible. This time [3, Lemma 2.23] tells us that S is not a singleton closure. Hence S_{cc} is not sober in this case. In order to obtain a sober space provided with a continuous topology, we must form the larger $\hat{S}$, the closed sets of which are the sets in $\widehat{\mathcal{F}} = \mathcal{K} \cup \{\hat{S}\}$.

Proposition 2 *Let S be a sober locally compact Q_δ space, and let $\hat{S} = S \cup \{\Delta\}$, where $\Delta \notin S$, be topologized by the sets in the collection $\{\hat{S} \setminus K : K \in \mathcal{K}\} \cup \{\emptyset\}$. Then $\hat{S}$ is sober.*

For a proof, see Section 3. As noted already the topology of $\hat{S}$ is continuous, so the next result follows by [3, Proposition 2.11].

Corollary 3 *$\hat{S}$ is locally compact.*

We now provide $\widehat{\mathcal{F}}$ with Fell's topology, which is known to be compact Hausdorff [1]. Recall that it is generated by the two families

$$\widehat{\mathcal{F}}^K := \{F \in \widehat{\mathcal{F}} : F \cap K = \emptyset\}, \quad K \in \widehat{\mathcal{K}}$$

$$\widehat{\mathcal{F}}_G := \{F \in \widehat{\mathcal{F}} : F \cap G \neq \emptyset\}, \quad G \in \widehat{\mathcal{G}}$$

of basic open sets ($\widehat{\mathcal{G}}$ and $\widehat{\mathcal{K}}$ denote resp. the collections of open and compact saturated subsets of $\hat{S}$). Being upper continuous, $\widehat{\mathcal{F}}$ also carries the Lawson topology, which is known to coincide with Fell's topology [2].

Also the proof of the next result is deferred (see Section 3).

Theorem 4 *Let S be a sober locally compact Q_δ space and write $\mathcal{K}$ for its collection of compact saturated sets. Then the trace on $\mathcal{K}$ of Fell's topology on $\widehat{\mathcal{F}} = \mathcal{K} \cup \{\hat{S}\}$ is the myope topology. In other words, $\widehat{\mathcal{F}}$ is an one-point compactification of $\mathcal{K}$.*

Next, we turn to a characterization of the monad $\mathrm{m}(K)$ of $K \in \mathcal{K}$ w.r.t. the myope topology. We work in a nonstandard enlargement of sufficiently high saturation (refer, e.g., to Hurd & Loeb [4]; our notations and terminology are standard). Recall that, for an arbitrary topological space X, the set

$$\mathrm{m}(x) = \bigcap \{{}^\star G : x \in G,\ G \text{ open in } X\}$$

is called the *monad* of $x \in X$. Moreover, $y \in {}^\star X$ is *near standard* if $y \in \mathrm{m}(x)$ for some $x \in X$. If distinct monads are disjoint, which they are, if, and only if, X is Hausdorff, then $x \in X$ is called the *standard part* of $y \in {}^\star X$ whenever $y \in \mathrm{m}(x)$. This is written $x = {}^\circ y$ or $x = \mathrm{st}(y)$.

Norberg [9] (cf. also [7, 13, 14]; however, these references treat the case when S is Hausdorff only) characterizes the monad $\widehat{\mathrm{m}}(F)$ of $F \in \widehat{\mathcal{F}}$ w.r.t. Fell's topology: Let $H \in {}^\star\widehat{\mathcal{F}}$, then H is near standard with standard part

$$\mathrm{st}(H) = \hat{H} := \{s \in \hat{S} : \widehat{\mathrm{m}}(s) \cap H \neq \emptyset\} \in \widehat{\mathcal{F}}$$

($\widehat{\mathrm{m}}(s)$ is the monad of $s \in \hat{S}$). Hence, for $F \in \widehat{\mathcal{F}}$, $H \in \widehat{\mathrm{m}}(F)$ if, and only if, $\hat{H} = F$. By Theorem 4,

$$\mathrm{m}(K) = \widehat{\mathrm{m}}(K) \cap {}^\star\mathcal{K}$$

showing that if $L \in {}^\star\mathcal{K}$, then L is near standard if, and only if, $\hat{L} \in \mathcal{K}$ which of course holds true if, and only if, $\widehat{\mathrm{m}}(\Delta) \cap L = \emptyset$.

Almost all statements of the following theorem are now proved.

Theorem 5 *Let S be a sober locally compact Q_δ space, and consider its collection $\mathcal{K}$ of compact saturated sets provided with the myope topology. Let $L \in {}^\star\mathcal{K}$. Then L is near standard if, and only if, $L \subseteq {}^\star K$ for some $K \in \mathcal{K}$. In this case, the standard part of L is*

$$\mathrm{st}(L) = \{s \in S : \mathrm{m}_{\mathrm{cc}}(s) \cap L \neq \emptyset\}$$

Proof We first prove the assertion that $\widehat{\mathrm{m}}(\Delta) \cap L = \emptyset$ if, and only if, $L \subseteq {}^\star K$ for some $K \in \mathcal{K}$. To see it, note first that

$$\widehat{\mathrm{m}}(\Delta) = \bigcap_{K \in \mathcal{K}} {}^\star\hat{S} \setminus {}^\star K$$

It follows that $\widehat{\mathrm{m}}(\Delta) \cap L = \emptyset$ if, and only if,

$$L \subseteq \bigcup_{K \in \mathcal{K}} {}^\star K$$

By saturation, this holds true if, and only if, $L \subseteq {}^\star K$ for some $K \in \mathcal{K}$.

Next note that if $L \in {}^\star\mathcal{K}$, then $L \subseteq {}^\star S$. And noting that

$$\mathrm{m}_{\mathrm{cc}}(s) = \widehat{\mathrm{m}}(s) \cap {}^\star S, \quad s \in S$$

we finally see that $\widehat{\mathrm{m}}(s) \cap L \neq \emptyset$ if, and only if, $\mathrm{m}_{\mathrm{cc}}(s) \cap L \neq \emptyset$. ∎

Note that, for $s \in S$,

$$\widehat{\mathrm{m}}(s) = \bigcap_{s \notin K \in \mathcal{K}} {}^\star\hat{S} \setminus {}^\star K$$

Hence $\widehat{\mathrm{m}}(\Delta) \subseteq \widehat{\mathrm{m}}(s)$. Moreover,

$$\mathrm{m}(\Delta) := \widehat{\mathrm{m}}(\Delta) \cap {}^\star S$$

may be regarded as the monad of infinity. So a member of ${}^\star\mathcal{K}$ is near standard if, and only if, it contains no infinite points. The next result cannot hold unless S is Hausdorff.

Proposition 6 *Suppose S is a locally compact Hausdorff space, and let S_{cc} be S endowed with the cocompact topology. Write* $\mathrm{m}(s)$ *and* $\mathrm{m}_{\mathrm{cc}}(s)$, *resp., for the monads of $s \in S$ w.r.t. the original and the cocompact topology. Then*

$$\mathrm{m}_{\mathrm{cc}}(s) = \mathrm{m}(s) \cup \mathrm{m}(\Delta)$$

Moreover, the two sets on the right hand side are disjoint.

Our final theorem is not surprising in the light of Matheron's result saying that the myope topology coincides with the relativization of Fell's topology on subsets of $\mathcal{K}$ that are compact w.r.t. the myope topology [6]. Being an obvious consequence of Proposition 6, it needs no further comment apart, perhaps, from the fact that if $L \in {}^\star\mathcal{K}$ then, by Robinson's characterization of compactness [4, Proposition III.1.12], every point of L is near a standard point in S. It can moreover be seen as a straightforward consequence of [14, Propositions 1.4 and 2.1].

Theorem 7 *Let S be a locally compact Hausdorff space, and consider its collection $\mathcal{K}$ of compact subsets in the myope topology. Suppose $L \in {}^\star\mathcal{K}$ is near standard. Then its standard part is*

$$\mathrm{st}(L) = \{s \in S : \mathrm{m}(s) \cap L \neq \emptyset\} = \{{}^\circ t : t \in L\}$$

3 Remaining proofs

Proof of Theorem 1 We have already noted that the myope topology is finer than the Lawson topology. To see the converse, let $F \in \mathcal{F}$ and take $G_1, \ldots, G_n \in \mathcal{G}$. Suppose

$$K \in \mathcal{K}^F_{G_1,\ldots,G_n} := \{H \in \mathcal{K} : H \cap F = \emptyset,\, H \cap G_1 \neq \emptyset, \ldots, H \cap G_n \neq \emptyset\}$$

By local compactness of S, there is some $L \in \mathcal{K}$ satisfying $K \subseteq L^{\mathrm{o}} \subseteq L \subseteq F^{\mathrm{c}}$. Then $L_i := G_i{}^{\mathrm{c}} \cap L \in \mathcal{K}$. Moreover, $K \cap L^{\mathrm{oc}} = \emptyset$ and $K \cap L_i{}^{\mathrm{c}} = K \cap G_i \neq \emptyset$. Hence

$$K \in \mathcal{K}^{L^{\mathrm{oc}}}_{L_1{}^{\mathrm{c}},\ldots,L_n{}^{\mathrm{c}}}$$

Suppose now that

$$H \in \mathcal{K}^{L^{\mathrm{oc}}}_{L_1{}^{\mathrm{c}},\ldots,L_n{}^{\mathrm{c}}}$$

Then $H \cap L^{\mathrm{oc}} = \emptyset$, so $H \cap F = \emptyset$. Moreover, $H \cap G_i = H \cap L_i{}^{\mathrm{c}} \neq \emptyset$ since $H \subseteq L$. Hence

$$H \in \mathcal{K}^F_{G_1,\ldots,G_n}$$

completing the proof. ∎

Proof of Proposition 2 Let $\emptyset \neq I \in \widehat{\mathcal{F}}$ be irreducible. We may assume $I \in \mathcal{K}$ since $\hat{S} = \{\Delta\}^-$. By [3, Lemma 2.23], $I = \mathrm{sat}\{s\}$ for some $s \in S$. To see that this point is unique, it is enough to note that $\mathrm{sat}\{s\} = \mathrm{sat}\{t\}$ if, and only if, $\{t\}^- = \{s\}^-$. ∎

Proof of Theorem 4 As already noted Fell's topology is generated by the two families $\widehat{\mathcal{F}}^L$, $L \in \widehat{\mathcal{K}}$, and $\widehat{\mathcal{F}}_G$, $G \in \hat{\mathcal{G}}$. We know that $\hat{S}$ is locally compact (Corollary 3). Hence Fell's topology is compact Hausdorff [1]. We may take $G \neq \emptyset$ in the latter subbase family Then $G = \hat{S} \setminus K$ for some $K \in \mathcal{K}$, so

$$\widehat{\mathcal{F}}_G = \{H \in \widehat{\mathcal{F}} : H \cap \hat{S} \setminus K \neq \emptyset\}$$

Hence

$$\widehat{\mathcal{F}}^G \cap \mathcal{K} = \{H \in \mathcal{K} : H \cap S \setminus K \neq \emptyset\} = \mathcal{K}_{K^{\mathrm{c}}}$$

(where the complementation is w.r.t. S). This takes care of the second family in the generator of Fell's topology, and now we switch attention to the first.

Let $L \in \widehat{\mathcal{K}}$. Then $\widehat{\mathcal{F}}^L$ is a Scott open ideal in $\widehat{\mathcal{F}}$ (i.e., $\widehat{\mathcal{F}}^L$ is a *lower* set: if $H \in \widehat{\mathcal{F}}^L$ and $H' \in \widehat{\mathcal{F}}, H' \subseteq H$, then also $H' \subseteq \widehat{\mathcal{F}}^L$; which is *directed*: if $H_1, H_2 \in \widehat{\mathcal{F}}^L$ then $H_1 \cup H_2 \in \widehat{\mathcal{F}}^L$; and *Scott open*: if $\bigcap_t H_t \in \widehat{\mathcal{F}}^L$, where $\{H_t\}_t \subseteq \widehat{\mathcal{F}}$, then already $\bigcap_i H_{t_i} \in \widehat{\mathcal{F}}^L$ for some finitely many $H_{t_1}, \ldots, H_{t_n}$). That $\widehat{\mathcal{F}}^L$ is an ideal (i.e., a lower directed set) is trivial. The fact that it is Scott open follows by standard compactness arguments.

Now if $\hat{S} \in \widehat{\mathcal{F}}^L$ then $\widehat{\mathcal{F}}^L = \widehat{\mathcal{F}}$, so we may assume $\hat{S} \notin \widehat{\mathcal{F}}^L$. Then $\widehat{\mathcal{F}}^L \subseteq \mathcal{K}$. Thus $\widehat{\mathcal{F}}^L$ is a Scott open ideal in $\mathcal{K}$. By [3, Corollary 2.17] and the Lawson duality [5, 2, 3], there is a unique $G \in \mathcal{G}$ such that

$$\widehat{\mathcal{F}}^L = \{K \in \mathcal{K} : K \subseteq G\}$$

Clearly $F := S \setminus G \in \mathcal{F}$, and

$$\widehat{\mathcal{F}}^L = \mathcal{K}^F$$

This completes the proof. ∎

Proof of Proposition 6 Note first that $\mathrm{m}(s) \subseteq \mathrm{m}_{\mathrm{cc}}(s)$, since $S \setminus K \in \mathcal{G}$ for all $K \in \mathcal{K}$. Next let $r \in \mathrm{m}_{\mathrm{cc}}(s) \setminus \mathrm{m}(s)$. Suppose $r \in {}^\star K$ for some $K \in \mathcal{K}$. Then $r \in \mathrm{m}(t)$ for some $t \in K$ by Robinson's characterization of compactness, cf. [4]. Clearly $t \neq s$. Thus we may choose $K \in \mathcal{K}$ such that $t \in K^{\circ}$ while $s \in K^{c}$. Then, by the former, $r \in {}^\star K^{\circ} \subseteq {}^\star K$ and, by the latter, $r \in {}^\star S \setminus {}^\star K$. This is clearly impossible, so $r \notin {}^\star K$ for all $K \in \mathcal{K}$. Hence $r \in \mathrm{m}(\Delta)$, and we have proved the first inclusion in

$$\mathrm{m}_{\mathrm{cc}}(s) \setminus \mathrm{m}(s) \subseteq \mathrm{m}(\Delta) \subseteq \mathrm{m}_{\mathrm{cc}}(s) \setminus \mathrm{m}(s)$$

The remaining inclusion and the last statement of the proposition are trivial. ∎

Acknowledgement

The content of Theorem 1 grew out in discussions with Adrian Baddeley. I am moreover grateful to Keith Wicks for some helping comments to an earlier version of the manuscript.

References

[1] Fell, J. M. G.: A Hausdorff topology for the closed subsets of a locally compact non-Hausdorff space. *Procedings of the American Mathematical Society* **13**, 1962, 472-476.

[2] Gierz, G., Hofmann, K. H., Keimel, K. Lawson, J. D., Mislove, M. & Scott, D. S.: *A Compendium of Continuous Lattices.* Springer, 1980.

[3] Hofmann, Karl H. & Mislove, Michael W.: Local compactness and continuous lattices. In *Continuous Lattices (Proc. Bremen 1979)* (eds. Banaschewski, B. & Hoffmann, R.-E.), Springer LNM 871, 1981, 209-248.

[4] Hurd, Albert E. & Loeb, Peter A.: *An Introduction to Nonstandard Real Analysis.* Academic Press, 1985.

[5] Lawson, Jimmie D.: The duality of continuous posets. *Houston Journal of Mathematics* **5**, 1979, 357-386.

[6] Matheron, G.: *Random Sets and Integral Geometry.* Wiley, 1975.

[7] Narens, Louis: Topologies of closed subsets. *Transactions of the Americal Mathematical Society* **174**, 1972, 55-76.

[8] Norberg, Tommy: Existence theorems for measures on continuous posets, with applications to random set theory. *Mathematica Scandinavica* **64**, 1989, 15-51.

[9] Norberg, Tommy: On Vervaat's sup vague topology. *Arkiv för matematik* **28**, 1990, 139-144.

[10] Norberg, Tommy & Vervaat, Wim: Capacities on non-Hausdorff spaces. Report No 1989-11, Department of Mathematics, CTH & GU, 1989.

[11] Stroyan, K. D. & Bayod, J. M.: *Foundations of Infinitesimal Stochastic Analysis.* North-Holland, 1986.

[12] Vervaat, Wim: Random upper semicontinuous functions and extremal processes. Report MS-R8801, CWI Amsterdam, 1988.

[13] Wattenberg, Frank: Topologies on the set of closed subsets. *Pacific Journal of Mathematics* **68**, 1977, 537-551.

[14] Wicks, Keith R.: *Fractals and Hyperspaces.* Springer LNM 1492, 1991.

Nonlinear theories of generalized functions

Michael Oberguggenberger
Institut für Mathematik und Geometrie, Universität Innsbruck,
A-6020 Innsbruck, Austria

Abstract

The purpose of this article is to present an overview of a branch of nonlinear analysis that has taken up rapid development during the past decade: the theory of "multiplication of distributions" or better the "nonlinear theory of generalized functions", as it goes far beyond the former topic by now. It has a significant overlap with nonstandard analysis; but within the boundaries of its scope, it takes a more general standpoint. We are interested in situations that involve
(a) nonlinear operations;
(b) differentiation;
(c) singular objects (like measures or discontinuous functions).

1 Introduction

The purpose of this article is to present an overview of a branch of nonlinear analysis that has taken up rapid development during the past decade: the theory of "multiplication of distributions" or better the "nonlinear theory of generalized functions", as it goes far beyond the former topic by now. It has a significant overlap with nonstandard analysis; but within the boundaries of its scope, it takes a more general standpoint. We are interested in situations that involve
(a) nonlinear operations;
(b) differentiation;
(c) singular objects (like measures or discontinuous functions).

The linear theory of distributions we take as a basis for our investigations, building upon the foundations laid down by L.Schwartz [34] using duality methods, but also keeping an eye on the sequential approach due to J.Mikusiński [19; 2] and G.Temple [36]. Distribution theory can successfully handle (b) and (c) simultaneously, while classical nonlinear analysis deals with (a) and (b), but predominantly avoids (c). It was realized in the early fifties (L.Schwartz [35], H.König [13]) that the combination of (a) and (b) with (c) entails that the differentiation operators and the nonlinear operations cannot be simultaneously consistent with their classical counterparts. We shall present a framework in which (a), (b), and (c) can be treated in full generality, with a wealth of applications to nonlinear partial differential equations, and in which the consistency problem can be understood and dealt with. This new way of thinking about generalized functions is chiefly due to E.E.Rosinger [26-31] and J.F.Colombeau [4-6,8], whose approach parallels ideas from nonstandard analysis, most prominently the earlier constructions of D.Laugwitz and C.Schmieden [33,14,15]. Exactly this "convergence of ideas" and a comparison thereof will be our theme.

The core of the circle of ideas is putting sequences back into their proper, important place. One may say that generalized functions are sequences, or families, of smooth functions, viewed as members of larger algebraic entities. To explain this viewpoint, let us look at the familiar example of the Dirac measure $\delta(x)$ in $\mathbb{R}^n$. In linear distribution theory, all its families of regularizations are crystallized in the single object $\delta(x)$, and this conceptual simplicity is at the basis of the success of distribution theory in linear analysis. At the same time, this neglect of the regularization process appears as one of the reasons for the failure of classical distribution theory in handling nonlinear operations. This is reflected in the language of physicists, where the meaning of, say, the square $\delta^2(x)$ is considered "ambiguous". In short, to establish a nonlinear theory of generalized functions it is necessary to keep more information on the regularization process. Thus there will be various objects in a nonlinear theory of generalized functions which correspond to the classical Dirac measure, each representing different nonlinear properties. In the most general situation, *all* families of smooth functions approximating the Dirac measure might be taken into account. However, factorization comes as an important tool in reducing this immense amount of information. As we shall see, the factorization can be done in various ways so as to study analytical properties, for example stability, or to produce desired logical structures, as is done in the ultrapower constructions of nonstandard analysis. Thus we can formulate more precisely now: *Generalized functions are members of factor algebras (or factor spaces) of families of smooth functions.*

In keeping our framework on such a general level we shall be able to better understand the structural constraints as well as options and to produce different nonlinear properties, for example, a variety of consistency results with classical operations. We should like to mention that even this framework can be profitably further extended: A nonlinear theory of generalized functions can be based on Dedekind order completion as well as uniform completion of spaces of smooth functions (joint work in progress with E.E.Rosinger [24]).

The plan of exposition is as follows: In Section 2 we discuss various problems which arise when imbedding the space of distributions into differential algebras. In Section 3, a number of important examples of differential algebras of generalized functions are presented, and basic properties are discussed. Section 4 serves to make a comparison between these algebras, while in Section 5 we sketch a few of the by now numerous applications. Let us emphasize an aspect frequently arising in the applications: there is a general *transfer process* that produces generalized solutions out of classical solutions; in the case of ultrapowers it actually coincides with the transfer principle of nonstandard analysis.

2 Constraints on imbedding the distributions into algebras

Given an open subset Ω of $\mathbb{R}^n$, $\mathcal{D}'(\Omega)$ denotes the space of distributions on Ω, the dual of the space $\mathcal{D}(\Omega)$ of infinitely differentiable, compactly supported functions on Ω; $L^\infty_{loc}(\Omega)$ is the algebra of locally bounded functions and $\mathcal{C}^k(\Omega)$, $0 \leq k \leq \infty$, the algebra of k-times continuously differentiable functions on Ω. We should like to construct an associative, commutative algebra $(\mathcal{A}(\Omega), +, \circ)$ with the following properties:

(i) there is a linear imbedding of $\mathcal{D}'(\Omega)$ into $\mathcal{A}(\Omega)$, and the constant function 1 is the unity in $\mathcal{A}(\Omega)$;

(ii) there are n derivation operators $\partial_1, ..., \partial_n$ on $\mathcal{A}(\Omega)$, that is, linear maps satisfying the Leibnitz rule;

(iii) $\partial_j \mid \mathcal{D}'(\Omega)$ coincides with the usual partial derivative, $j = 1, ..., n$;

(iv) $\circ \mid L^\infty_{loc}(\Omega) \times L^\infty_{loc}(\Omega)$ coincides with the usual product of functions.

As already alluded to in the introduction, this seemingly benign list of differential algebraic properties and consistency requirements cannot be satisfied simultaneously. In fact, if we demand (i) and (ii) in an associative, commutative algebra then (iii) and (iv) mutually exclude each other. To see this, consider the case $\Omega = \mathbb{R}$ and let H be the Heaviside function ($H(x) = 1$ for $x > 0$, $H(x) = 0$ for $x < 0$). By (iv) we have that $H^m = H$ for all $m \in \mathbb{N}$ in $\mathcal{A}(\mathbb{R})$, while the Leibnitz rule together with associativity and commutativity gives

$$\partial H = \partial(H^m) = mH^{m-1}\partial H = mH\partial H$$

so

$$H\partial H = \frac{1}{m}\partial H \qquad \text{for all} \quad m \in \mathbb{N}$$

whence $\partial H = 0$ contradicting (iii), according to which we should have $\partial H = \delta$.

A slightly more complicated example, due to L.Schwartz [35], shows similarly that $\circ \mid \mathcal{C}^k(\Omega) \times \mathcal{C}^k(\Omega)$ cannot coincide with the usual product for finite $k \in \mathbb{N}$, or else (iii) will be contradicted. In conclusion, as soon as we insist on imbedding $\mathcal{D}'(\Omega)$ into an associative, commutative differential algebra, the new operations, when restricted to their classical domains of definition, can never be equal to the corresponding classical operations (failure of *strong consistency*). The result can nevertheless be infinitely close to the classical one, in a sense to be made precise below (possibility of *weak consistency*).

We point out that J.F.Colombeau [4] was the first to show that one can achieve (i), (ii), (iii) and

(v) $\circ \mid \mathcal{C}^\infty(\Omega) \times \mathcal{C}^\infty(\Omega)$ coincides with the usual product.

In the setting of associative, commutative differential algebras, this appears to be the best possible result concerning the differential-algebraic structure. If one is willing to give up associativity and commutativity, then (i) through (iv) can be achieved at the same time. The corresponding construction is due to H. König [11,12] and constitutes the earliest nonlinear theory of generalized functions in the literature. However, we shall remain in the associative, commutative setting here.

We close this section by noting that the difficulties with the product $H\partial H$, or $H\delta$, are similarly apparent in the approach using regularizations. In fact, let H_ε and δ_ε be smooth functions such that $H_\varepsilon \to H$, $\delta_\varepsilon \to \delta$ as $\varepsilon \to 0$, say in the sense of distributions. Assume that $H_\varepsilon(x) \equiv 0$ for $x \leq -\varepsilon$, $H_\varepsilon(x) \equiv 1$ for $x \geq \varepsilon$. If δ_ε has its support in $(-\infty\,, -\varepsilon)$, then $H_\varepsilon\delta_\varepsilon \equiv 0$; if it has its support in (ε, ∞), then $H_\varepsilon\delta_\varepsilon \equiv \delta_\varepsilon$. Thus $\lim_{\varepsilon\to 0} H_\varepsilon\delta_\varepsilon$ takes on the value 0 in the first, the value δ in the second case. This example conspicuously shows how the nonlinear properties depend on the regularization process.

3 Construction of algebras of generalized functions

To begin with, we shall investigate algebras of generalized functions on $\mathbb{R}^n$; the case of generalized functions on open sets $\Omega \subset \mathbb{R}^n$ will be indicated later. As the starting point we take an infinite index set E and consider the power set

$$\mathcal{X}(\mathbb{R}^n) = (\mathcal{C}^\infty(\mathbb{R}^n))^E.$$

We shall refer to the elements $(u_\varepsilon)_{\varepsilon\in E}$ of $\mathcal{X}(\mathbb{R}^n)$ as "sequences" of smooth functions, by a slight abuse of language. In fact, let us actually take E as the interval $(0,1)$ for the time being.

Given any sequence $(\varphi_\varepsilon)_{\varepsilon\in E}$ of compactly supported smooth functions converging to the Dirac measure as $\varepsilon \to 0$, we have an imbedding of the space of distributions into $\mathcal{X}(\mathbb{R}^n)$:

$$\begin{aligned} \iota : \mathcal{D}'(\mathbb{R}^n) &\to \mathcal{X}(\mathbb{R}^n) \\ w &\to (w * \varphi_\varepsilon)_{\varepsilon\in E} \end{aligned} \tag{1}$$

Here the star denotes convolution. For example, one could take φ_ε of the form $\varphi_\varepsilon(x) = \varepsilon^{-n}\varphi(x/\varepsilon)$ with $\varphi \in \mathcal{D}(\mathbb{R}^n), \int \varphi(x)dx = 1$. With the operations defined componentwise, $\mathcal{X}(\mathbb{R}^n)$ is an associative, commutative differential algebra, and

$$\partial_j(w * \varphi_\varepsilon) = (\partial_j w) * \varphi_\varepsilon, \quad j = 1, ..., n$$

whence

$$\partial_j \iota(w) = \iota(\partial_j w)$$

for $w \in \mathcal{D}'(\mathbb{R}^n)$. Thus the algebra $\mathcal{X}(\mathbb{R}^n)$ satisfies properties (i), (ii), and (iii) of Section 2; but certainly (v) or even (iv) do not hold. We also have the standard copy imbedding

$$\begin{array}{rcl} \sigma : \mathcal{C}^\infty(\mathbb{R}^n) & \to & \mathcal{X}(\mathbb{R}^n) \\ f & \to & (f)_{\varepsilon \in E} \end{array}$$

We note that σ is a morphism of differential algebras, so condition (v) would follow from the requirement that $\iota \mid \mathcal{C}^\infty(\mathbb{R}^n) = \sigma$, which fails to obtain here.

Remark 1. The dependence of the imbedding ι on the choice of φ_ε can be made to disappear by changing the index set to $E = \mathcal{D}(\mathbb{R}^n)$ and taking

$$w \to (w * \varphi)_{\varphi \in \mathcal{D}(\mathbb{R}^n)}$$

as the imbedding of $\mathcal{D}'(\mathbb{R}^n)$ into $\mathcal{X}(\mathbb{R}^n)$. This does not change any of the differential–algebraic properties but would make $\mathcal{D}'(\mathbb{R}^n)$ "canonically imbedded".

From the rather crude algebra $\mathcal{X}(\mathbb{R}^n)$ we shall now construct more refined algebras by means of factorization. The advantages of this procedure have been indicated in the Introduction and will also be seen below; for example, it might be possible to identify $\iota(f)$ and $\sigma(f)$ for $f \in \mathcal{C}^\infty(\mathbb{R}^n)$ to produce property (v). Thus we shall take subalgebras $\mathcal{X}_s \subset \mathcal{X}$ and ideals $\mathcal{I} \subset \mathcal{X}_s$ and pass to the factor algebra $\mathcal{A} = \mathcal{X}_s/\mathcal{I}$. As a preliminary attempt, consider

$$\begin{array}{rcl} \mathcal{Y}(\mathbb{R}^n) & = & \{(u_\varepsilon)_{\varepsilon \in E} \in \mathcal{X}(\mathbb{R}^n) : u_\varepsilon \text{ converges in } \mathcal{D}'(\mathbb{R}^n) \text{ as } \varepsilon \to 0\} \\ \mathcal{J}(\mathbb{R}^n) & = & \{(u_\varepsilon)_{\varepsilon \in E} \in \mathcal{X}(\mathbb{R}^n) : u_\varepsilon \to 0 \text{ in } \mathcal{D}'(\mathbb{R}^n) \text{ as } \varepsilon \to 0\}. \end{array}$$

With this choice we can realize $\mathcal{D}'(\mathbb{R}^n)$ as a factor space

$$\mathcal{D}'(\mathbb{R}^n) \cong \mathcal{Y}(\mathbb{R}^n)/\mathcal{J}(\mathbb{R}^n).$$

But $\mathcal{Y}(\mathbb{R}^n)$ is not an algebra, and $\mathcal{J}(\mathbb{R}^n)$ is not an ideal, in whatever subalgebra of $\mathcal{X}(\mathbb{R}^n)$ that contains it. This reflects the fact that it is hopeless to try to define the structure of a reasonable algebra on $\mathcal{D}'(\mathbb{R}^n)$ itself. One might say that $\mathcal{Y}(\mathbb{R}^n)$ is too small (it does not even contain polynomials of distributions), while $\mathcal{J}(\mathbb{R}^n)$ is too large: we have lost too much information by identifying the elements of $\mathcal{J}(\mathbb{R}^n)$ with zero.

Before presenting the most important positive examples let us have a brief look at the case of open subsets of $\mathbb{R}^n$.

Remark 2. If Ω is an open subset of $\mathbb{R}^n$, we can similarly define

$$\mathcal{X}(\Omega) = (\mathcal{C}^\infty(\Omega))^E.$$

The space of compactly supported distributions $\mathcal{E}'(\Omega)$ is imbedded by

$$\begin{array}{rcl} \iota_0 : \mathcal{E}'(\Omega) & \to & \mathcal{X}(\Omega) \\ w & \to & (w * \varphi_\varepsilon)_{\varepsilon\in E} \end{array}$$

where $(\varphi_\varepsilon)_{\varepsilon\in E} \subset \mathcal{D}(\mathbb{R}^n)$ is some fixed delta-sequence as above. An ad hoc cut-off method allows to imbed $\mathcal{D}'(\Omega)$ as well: Take $\chi_\varepsilon \subset \mathcal{D}(\Omega)$ such that $\chi_\varepsilon \equiv 1$ eventually for small ε, on every compact subset of Ω, and set

$$\iota(w) = ((\chi_\varepsilon w) * \varphi_\varepsilon)_{\varepsilon\in E} \tag{2}$$

for $w \in \mathcal{D}'(\Omega)$. We note that there are natural restriction maps

$$\mathcal{X}(\Omega_2) \to \mathcal{X}(\Omega_1)$$

for $\Omega_1 \subset \Omega_2$ open. With these restriction maps, the family $\mathcal{X}(\Omega)$, $\Omega \subset \mathbb{R}^n$ open, becomes a sheaf. However, ι is not a sheaf morphism (of vector spaces), nor is it true that $\partial_j\iota(w) = \iota(\partial_j w)$, nor is $\iota \mid \mathcal{E}'(\Omega)$ equal with ι_0. The remedy for this is *localization*, and this leads us to our first example for the factorization process.

Example 1. Here we identify sequences in $\mathcal{X}(\Omega)$ that vanish eventually on every compact set, putting

$$\begin{array}{rcl} \mathcal{I}_{loc}(\Omega) & = & \{(u_\varepsilon)_{\varepsilon\in E} \in \mathcal{X}(\Omega) : \forall K \subset \Omega \quad \text{compact} \quad \exists\eta > 0 \\ & & \text{such that} \quad u_\varepsilon \mid K \equiv 0 \quad \text{for} \quad 0 < \varepsilon < \eta\} \end{array}$$

and letting

$$\mathcal{L}(\Omega) = \mathcal{X}(\Omega)/\mathcal{I}_{loc}(\Omega).$$

This algebra is in the spirit of the theory of C.Schmieden and D.Laugwitz [33,15], who based their foundation of infinitesimal analysis on the sequential approach and identification by means of the Fréchet filter on E as above (see also the discussion in [23, Section 23] for how far one can get with this approach in multiplication of distributions). This algebra was later reintroduced by Yu.V.Egorov [9,10].

We note that (i) and (ii) obviously hold in $\mathcal{L}(\Omega)$, while property (iii) for the imbedding ι, defined by (2), follows from the fact that

$$\partial_j(\chi_\varepsilon w) * \varphi_\varepsilon = (\chi_\varepsilon \partial_j w) * \varphi_\varepsilon$$

eventually for small ε on every compact set. Also, ι is a sheaf morphism. A simple algebraic argument [23, Example 21.6] yields an alternative imbedding for which (i), (ii), and (v) are true.

Example 2. This is the familiar ultrapower construction of the algebra of internal smooth functions ${}^*\mathcal{C}^\infty(\Omega)$. Let $\mathcal{U}$ be a free ultrafilter on E containing all terminal sets $(0,\eta)$, $\eta > 0$, and set

$$\mathcal{I}_{\mathcal{U}}(\Omega) = \{(u_\varepsilon)_{\varepsilon\in E} \in \mathcal{X}(\Omega) : \{\varepsilon \in E : u_\varepsilon \equiv 0\} \in \mathcal{U}\}$$

Then

$${}^*\mathcal{C}^\infty(\Omega) = \mathcal{X}(\Omega)/\mathcal{I}_{\mathcal{U}}(\Omega).$$

We note here right away that ${}^*\mathcal{C}^\infty(\Omega)$ is not a sheaf (in the standard sense). For example, consider the open cover $\mathcal{W} = \{(-m,m) : m \in \mathbb{N}\}$ of $\mathbb{R}$ and the nonstandard function f defined as the class of $\left(\psi(. - \frac{1}{\varepsilon})\right)_{\varepsilon\in E}$ in ${}^*\mathcal{C}^\infty(\Omega)$, where ψ is smooth and has compact support. Then $f \mid (-m,m) \equiv 0$ for every $m \in \mathbb{N}$, but f is not zero in ${}^*\mathcal{C}^\infty(\mathbb{R})$, violating one of the sheaf properties. Of course, ${}^*\mathcal{C}^\infty(\mathbb{R})$ is a $*$-sheaf: we have to test the sheaf axioms using ${}^*W$; the behavior of f at infinitely large values matters, and indeed $f \mid (-\nu,\nu)$ is not zero for certain $\nu \in {}^*\mathbb{N}\backslash\mathbb{N}$. The lack of the sheaf property causes some trouble concerning (iii): While for $\Omega = \mathbb{R}^n$ the imbedding ι defined by (1) commutes with derivatives, this is not the case for $\Omega \neq \mathbb{R}^n$ where we have to use formula (2). The contributions of $\partial_j\chi_\varepsilon$ remain present as long as we do not factor out what happens off compact sets for small $\varepsilon \in E$. Consequently, we can assert that ${}^*\mathcal{C}^\infty(\Omega)$ has property (iii) only when $\Omega = \mathbb{R}^n$. Again, this difficulty can be removed by localization.

Example 3. Define

$$\begin{aligned}\mathcal{I}_{\mathcal{U},loc}(\Omega) &= \{(u_\varepsilon)_{\varepsilon\in E} \in \mathcal{X}(\Omega) : \forall K \subset \Omega \text{ compact} \\ &\qquad \text{it holds that } \{\varepsilon \in E : u_\varepsilon \mid K \equiv 0\} \in \mathcal{U}\}\end{aligned}$$

and let

$${}^*\mathcal{C}^\infty_{loc}(\Omega) = \mathcal{X}(\Omega)/\mathcal{I}_{\mathcal{U},loc}(\Omega).$$

The family ${}^*\mathcal{C}^\infty_{loc}(\Omega), \Omega \subset \mathbb{R}^n$ open, forms a sheaf in the standard sense, and each of its members is a differential algebra satifying (i), (ii), (iii). In discussion with T.D.Todorov we realized that these algebras can also be introduced in purely nonstandard fashion as factor algebras of ${}^*\mathcal{C}^\infty(\Omega)$. Indeed, let

$$\begin{aligned}\mathcal{K}(\Omega) &= \{u \in {}^*\mathcal{C}^\infty(\Omega) : \forall K \subset \Omega \text{ compact}, u \mid {}^*K \equiv 0\} \\ &= \{u \in {}^*\mathcal{C}^\infty(\Omega) : u(x) = 0 \text{ for all nearstandard } x \in {}^*\Omega\}.\end{aligned}$$

A basic property of iterated factorization entails that

$${}^*\mathcal{C}^\infty_{loc}(\Omega) = {}^*\mathcal{C}^\infty(\Omega)/\mathcal{K}(\Omega).$$

Algebras of this type are discussed in T.D.Todorov's paper [40]. We shall have a glimpse at more elaborate versions (in polysaturated models) in Remark 3, Section 4.

Example 4. This is one of the versions of the construction of J.F.Colombeau [5,6]. Let

$$\begin{aligned} \mathcal{X}_M(\Omega) &= \{(u_\varepsilon)_{\varepsilon\in E} \in \mathcal{X}(\Omega) : \forall K \subset \Omega \text{ compact}, \forall \alpha \in \mathbb{N}_0^n \\ &\quad \exists p > 0,\ \exists \eta > 0 \text{ such that} \\ &\quad \sup_{x\in K} \mid \partial^\alpha u_\varepsilon(x) \mid \leq \varepsilon^{-p},\ 0 < \varepsilon < \eta\}, \\ \mathcal{N}(\Omega) &= \{(u_\varepsilon)_{\varepsilon\in E} \in \mathcal{X}(\Omega) : \forall K \subset \Omega \text{ compact}, \forall \alpha \in \mathbb{N}_0^n \\ &\quad \forall q > 0,\ \exists \eta > 0 \text{ such that} \\ &\quad \sup_{x\in K} \mid \partial^\alpha u_\varepsilon(x) \mid \leq \varepsilon^{q},\ 0 < \varepsilon < \eta\} \end{aligned}$$

and set

$$\mathcal{G}(\Omega) = \mathcal{X}_M(\Omega)/\mathcal{N}(\Omega).$$

Apart from the fact that it seems natural from the standpoint of asymptotic analysis to consider the elements of $\mathcal{N}(\Omega)$ negligible (this has been decidedly done e.g. by V.P.Maslov and V.A.Tsupin [17,18]), $\mathcal{G}(\Omega)$ and its variants currently are the only known algebras enjoying all of the properties (i), (ii), (iii), and (v). To see this, take $\varphi \in \mathcal{S}(\mathbb{R}^n)$, a rapidly decreasing smooth function, with $\int \varphi(x)dx = 1$ and all moments vanishing:

$$\int x^\alpha \varphi(x)dx = 0, \text{ for all } \alpha \in \mathbb{N}_0^n, \mid \alpha \mid \geq 1.$$

The imbedding $\iota_0 : \mathcal{E}'(\Omega) \to \mathcal{G}(\Omega)$,

$$w \to \text{ class of } (w * \varphi_\varepsilon)_{\varepsilon\in E}$$

with $\varphi(x) = \varepsilon^{-n}\varphi(x/\varepsilon)$ has the property that $\iota_0 \mid \mathcal{D}(\Omega) = \sigma$, because $(w*\varphi_\varepsilon - w)_{\varepsilon\in E}$ belongs to $\mathcal{N}(\Omega)$, as follows from Taylor expansion. From there an easy exercise in sheaf theory gives $\iota : \mathcal{D}'(\Omega) \to \mathcal{G}(\Omega)$ with $\iota \mid \mathcal{C}^\infty(\Omega) = \sigma$.

Example 5. This is a nonstandard version of $\mathcal{G}(\Omega)$; it was introduced in [20,23] and by T.D.Todorov [37,40]. Let

$$\begin{aligned} \mathcal{X}_{M,\mathcal{U}}(\Omega) &= \{(u_\varepsilon)_{\varepsilon\in E} \in \mathcal{X}(\Omega) : \forall K \subset \Omega \text{ compact}, \forall \alpha \in \mathbb{N}_0^n \\ &\quad \exists p > 0 \text{ such that } \{\varepsilon \in E : \sup_{x\in K} |\partial^\alpha u_\varepsilon(x)| \leq \varepsilon^{-p}\} \in \mathcal{U}\}, \\ \mathcal{N}_{\mathcal{U}}(\Omega) &= \{(u_\varepsilon)_{\varepsilon\in E} \in \mathcal{X}(\Omega) : \forall K \subset \Omega \text{ compact}, \forall \alpha \in \mathbb{N}_0^n \\ &\quad \forall q > 0 \text{ it holds that } \{\varepsilon \in E : \sup_{x\in K} |\partial^\alpha u_\varepsilon(x)| \leq \varepsilon^{q}\} \in \mathcal{U}\}, \end{aligned}$$

where $\mathcal{U}$ is a free ultrafilter on E as in Examples 2 and 3. Finally, set

$$\mathcal{G}_{\mathcal{U}}(\Omega) = \mathcal{X}_{M,\mathcal{U}}(\Omega)/\mathcal{N}_{\mathcal{U}}(\Omega).$$

We note that the corresponding ring of constants $\mathcal{G}_{\mathcal{U}}(\mathbb{R}^0)$ is a field, and actually coincides with the field ${}^\rho\mathbb{R}$ introduced by A.Robinson [25], where ρ is the fixed infinitesimal number given as the class of $(\varepsilon)_{\varepsilon\in E}$ in ${}^*\mathbb{R}$. See T.D. Todorov [40] for a further discussion of this correspondence. We remark that $\mathcal{G}_{\mathcal{U}}(\Omega)$ can also be introduced as a factor algebra in ${}^*\mathcal{C}^\infty(\Omega)$.

Example 6. This construction is due to E.E.Rosinger [29-31]. Define the "nowhere dense ideal" by

$$\begin{aligned}\mathcal{I}_{nd}(\Omega) &= \{(u_\varepsilon)_{\varepsilon\in E} \in \mathcal{X}(\Omega) : \exists \Gamma \subset \Omega \text{ closed, nowhere dense, and} \\ &\quad \forall x \in \Omega\backslash\Gamma\ \exists \eta > 0\ \exists V \subset \Omega\backslash\Gamma,\ V \text{ neighborhood of } x, \\ &\quad \text{such that } u_\varepsilon \mid V \equiv 0,\ 0 < \varepsilon < \eta\}\end{aligned}$$

and set

$$\mathcal{R}(\Omega) = \mathcal{X}(\Omega)/\mathcal{I}_{nd}(\Omega).$$

It is clear that $\mathcal{R}(\Omega)$ is a differential algebra and that it contains $\mathcal{C}^\infty_{nd}(\Omega)$, the algebra of classes of smooth functions defined off nowhere dense, closed subsets of Ω. To show that $\mathcal{D}'(\Omega)$ is contained in $\mathcal{R}(\Omega)$ requires a completely different line of argument; the imbeddings (1) and (2) will not work, as all delta-sequences $(\varphi_\varepsilon)_{\varepsilon\in E}$ with supports shrinking to $\{0\}$ belong to $\mathcal{I}_{nd}(\mathbb{R}^n)$.

For further details as well as additional examples we refer to E.E.Rosinger [29-31] and [23, Section 21].

We close this section with a discussion of some generalities concerning all examples given. The *ring of constants* has already been mentioned; it can be introduced as the space of generalized functions on $\mathbb{R}^0 = \{0\}$. For example,

$$\begin{aligned}\mathcal{X}_\mathcal{M}(\mathbb{R}^0) &= \{(c_\varepsilon)_{\varepsilon\in E} \in \mathbb{R}^E : \exists p > 0,\ \exists \eta > 0 \text{ such that } \mid c_\varepsilon \mid \leq \varepsilon^{-p},\ 0 < \varepsilon < \eta\} \\ \mathcal{N}(\mathbb{R}^0) &= \{(c_\varepsilon)_{\varepsilon\in E} \in \mathbb{R}^E : \forall q > 0,\ \exists \eta > 0 \text{ such that } \mid c_\varepsilon \mid \leq \varepsilon^{q},\ 0 < \varepsilon < \eta\}\end{aligned}$$

and $\mathcal{G}(\mathbb{R}^0) = \mathcal{X}_M(\mathbb{R}^0)/\mathcal{N}(\mathbb{R}^0)$ is the ring of constants in Colombeau's theory. Similarly, $^*\mathbb{R} = \mathcal{X}(\mathbb{R}^0)/\mathcal{I}_\mathcal{U}(\mathbb{R}^0)$ is the field of hyperreal numbers, and so on. All these rings are partially ordered in the obvious way ($c \geq 0$ iff it is the class of some nonnegative sequence $(c_\varepsilon)_{\varepsilon\in E}$). Further, $\mathbb{R}$ is always contained as its standard copy, and one has infinitely large and infinitely small numbers.

Now let $\mathcal{X}_s(\Omega)$ be a subalgebra of $\mathcal{X}(\Omega)$, $\mathcal{I}(\Omega)$ an ideal therein, and $\mathcal{A}(\Omega) = \mathcal{X}_s(\Omega)/\mathcal{I}(\Omega)$ the factor algebra. Given $\psi \in \mathcal{D}(\Omega)$ and $u \in \mathcal{A}(\Omega)$, where u is the class of $(u_\varepsilon)_{\varepsilon\in E}$, we can define the integral

$$\int u(x)\psi(x)dx = \text{class of } (\int u_\varepsilon(x)\psi(x)dx)_{\varepsilon\in E}$$

as a member of the corresponding ring of constants $\mathcal{A}(\mathbb{R}^0)$. With the aid of this integral, a relation of *infinitesimality* can be introduced in $\mathcal{A}(\Omega)$. Indeed, given $u \in \mathcal{A}(\Omega)$ we shall write

$$u \approx 0$$

provided

$$\forall\psi \in \mathcal{D}(\Omega),\ \forall r \in \mathbb{R},\ r > 0 : \left|\int u(x)\psi(x)dx\right| \leq r.$$

In the special case of $^*\mathcal{C}^\infty(\Omega)$, this is the familiar relation of distributional infinitesimality, while in $\mathcal{G}(\Omega)$ it is the so-called association relation. The above relation can be considered in all the Examples 1–5, but not in $\mathcal{R}(\Omega)$ because there $u \in \mathcal{I}_{nd}(\Omega)$ does not imply $u \approx 0$.

As a preliminary application of this concept, we note that *weak consistency* of the product in $\mathcal{A}(\Omega)$ with e.g. the product of continuous functions can be formulated. Indeed, if f, g are members of $\mathcal{C}^0(\Omega)$ then their product in $\mathcal{A}(\Omega)$ is infinitely close to their classical product, i.e.

$$\iota(f)\iota(g) \;-\; \iota(fg) \approx 0.$$

This is an immediate consequence of the weak convergence of $(f * \varphi_\varepsilon)(g * \varphi_\varepsilon)$ to fg, where φ_ε is the regularizing sequence appearing in the definition of ι.

We have claimed in the Introduction that there will be various objects with different nonlinear properties, each corresponding to the Dirac measure. This can now be made precise. Such an object $\theta \in \mathcal{A}(\mathbb{R})$ is meant to be infinitely close to $\delta(x)$, that is,

$$\theta - \iota(\delta) \approx 0.$$

We illustrate the variety by noting that for every $c \in \mathbb{C}$ there is such a $\theta \in \mathcal{A}(\mathbb{R})$ which in addition satisfies

$$\theta^2 - c\theta \approx 0$$

(see[23, Example 10.6]; complex-valued versions of our algebras $\mathcal{A}(\mathbb{R})$ are needed here).

Finally, we observe that the structure of the algebras $\mathcal{A}(\Omega)$ generally goes much beyond that of differential algebras. For example, one can form nonlinear functions of the elements of $\mathcal{A}(\Omega)$. More precisely, if F is an infinitely differentiable function on $\mathbb{R}$ and $u \in \mathcal{A}(\Omega)$, then $F(u)$ is a well-defined element of $\mathcal{A}(\Omega)$ in the cases $\mathcal{A}(\Omega) = \mathcal{L}(\Omega)$, ${}^*\mathcal{C}^\infty(\Omega)$, ${}^*\mathcal{C}^\infty_{loc}(\Omega)$, $\mathcal{R}(\Omega)$. In the cases $\mathcal{A}(\Omega) = \mathcal{G}(\Omega)$, $\mathcal{G}_{\mathcal{U}}(\Omega)$ this is also true, provided F is polynomially bounded together with all its derivatives, i.e. $F \in \mathcal{O}_M(\mathbb{R})$. This observation is important for applications to partial differential equations with smooth, non-polynomial nonlinearities.

4 Comparison

In this section we are going to compare the algebras introduced in Examples 1–6 with respect to certain elementary properties. Let us start right away with a systematic table. Numbers (i)–(v) refer to the properties defined in Section 2.

	(i) $\supset \mathcal{D}'$	(ii) differential algebra	(iii) $\partial^\alpha \mid \mathcal{D}'$ consistent	(v) $\mathcal{C}^\infty$ subalgebra via ι	ring of constants is field	superposition by smooth F possible	sheaf
$\mathcal{L}$	yes	yes	either-or?		no	yes	yes
${}^*\mathcal{C}^\infty$	yes	yes	either-or??		yes	yes	no
${}^*\mathcal{C}^\infty_{loc}$	yes	yes	either-or?		yes	yes	yes
$\mathcal{G}$	yes	yes	yes	yes	no	only $\mathcal{O}_M$	yes
$\mathcal{G}_u$	yes	yes	yes	yes	yes	only $\mathcal{O}_M$	yes
$\mathcal{R}$	yes	yes	no?	yes	no	yes	yes

Proofs of these properties are either simple or else can be found e.g. in [23, Sections 9,21,23]. The "no" with question mark in the last line means that so far no imbedding of $\mathcal{D}'(\Omega)$ into $\mathcal{R}(\Omega)$ is known which produces property (iii); the "either-or" in lines 1–3 means that imbeddings are known which either produce (iii) or else (v). The one giving (iii) is the imbedding ι discussed in Section 3, formula (1), (2). The algebraic argument of [23, Example 21.6], alluded to in Example 1 above, works also for ${}^*\mathcal{C}^\infty$ and ${}^*\mathcal{C}^\infty_{loc}$ and yields an alternative imbedding with property (v). The double question mark in line 2 refers to the fact that we can assert property (iii) for ${}^*\mathcal{C}^\infty(\Omega)$ only when $\Omega = \mathbb{R}^n$. We shall elaborate the either-or situation in more detail in Remark 3 below.

Aside from the elementary properties listed in the table above, each of the algebras has various distinguishing, advanced properties. For example, ${}^*\mathcal{C}^\infty$ is part of the nonstandard universe, and so the fundamental methods of nonstandard analysis can be applied, like the transfer principle or saturation arguments, as we shall see below in Remark 3.

Next, $\mathcal{G}(\Omega)$ appears to be the only algebra allowing for a remarkable regularity theory. Let us explain this with the notion of elliptic regularity. Take a constant coefficient partial differential operator $P(\partial)$ which is elliptic or, more generally, hypoelliptic. These operators possess the following classical regularity property: If $w \in \mathcal{D}'(\Omega)$ and $P(\partial)w \in \mathcal{C}^\infty(\Omega)$, then $w \in \mathcal{C}^\infty(\Omega)$. Such a result never holds in any of the algebras above: Not even $u \in \mathcal{A}(\Omega)$, $P(\partial)u = 0$ implies $u \in \mathcal{C}^\infty(\Omega)$, because we can take u as an element of the ring of constants minus the real numbers. However, in the setting of Colombeau's theory we can construct a substitute for $\mathcal{C}^\infty(\Omega)$ to formulate regularity: Define $\mathcal{G}^\infty(\Omega)$ as the algebra of those elements $u \in \mathcal{G}(\Omega)$ all whose derivatives have the same local order, i.e. the representatives $(u_\varepsilon)_{\varepsilon\in E}$ of u are required to satisfy that:

$\forall K \subset \Omega$ compact $\exists p > 0$ such that $\forall \alpha \in \mathbb{N}_0^n$ $\exists \eta > 0$ with $\sup_{x\in K} \mid \partial^\alpha u_\varepsilon(x) \mid \leq \varepsilon^{-p}$, $0 < \varepsilon < \eta$.

One can show [23, Section 25] that $\mathcal{D}'(\Omega) \cap \mathcal{G}^\infty(\Omega) = \mathcal{C}^\infty(\Omega)$, and one has the regularity result: If $u \in \mathcal{G}(\Omega)$ and $P(\partial)u \in \mathcal{G}^\infty(\Omega)$, then $u \in \mathcal{G}^\infty(\Omega)$.

A distinguishing feature of $\mathcal{R}(\Omega)$ is the fact that it includes $\mathcal{C}^\infty_{nd}(\Omega)$, as we have noted in Example 6. This is of interest in the theory of shock wave solutions to quasilinear hyperbolic systems. Shock waves are functions smooth off a finite set of smooth curves or surfaces and hence belong to $\mathcal{C}^\infty_{nd}(\Omega)$. In fact, a quasilinear conservation law

$$\partial_t u + \partial_x F(u) = 0 \tag{3}$$

can have shock wave solutions in the sense of equality in $\mathcal{R}(\Omega)$, while in all the other algebras (3) has to be replaced by

$$\partial_t u + \partial_x F(u) \approx 0$$

in order to accommodate shock wave solutions. This difficulty comes from an interplay of the differential-algebraic structure, the Rankine-Hugoniot conditions for the jump of a shock wave, and the distributional convergence properties of the members of the ideal under consideration; details can be found in [23, Sections 19 and 22] and [22, Proposition 3.5].

Finally, a remarkable fact deserves attention: the sheaves $\mathcal{A}(\Omega) = \mathcal{L}(\Omega)$, ${}^*\mathcal{C}^\infty_{loc}(\Omega)$, $\mathcal{R}(\Omega)$ are *flabby*. That is, if $\Omega_1 \subset \Omega_2$ are open sets and $u \in \mathcal{A}(\Omega_1)$, then there exists $v \in \mathcal{A}(\Omega_2)$ such that $v \mid \Omega_1 = u$. Indeed, as the boundary $\partial\Omega_1$ is closed, there is $\gamma \in \mathcal{C}^\infty(\Omega_2)$ such that $\partial\Omega_1 = \{x \in \Omega_2 : \gamma(x) = 0\}$. Take $\chi \in \mathcal{C}^\infty(\mathbb{R})$, $\chi \equiv 0$ on $(-1,1)$, $\chi \equiv 1$ on $(-2,2)^C$. If $(u_\varepsilon)_{\varepsilon\in E}$ is a representative of u, define v as the class of the sequence of functions

$$v_\varepsilon(x) = \chi\left(\frac{1}{\varepsilon}\gamma(x)\right) u_\varepsilon(x).$$

It is immediately verified that $u_\varepsilon \equiv v_\varepsilon$ on every compact subset of Ω_1 for sufficiently small ε, so the sequence $(u_\varepsilon \mid \Omega - v_\varepsilon)_{\varepsilon\in E}$ belongs to any of the defining ideals for the algebras above, hence $v \mid \Omega_1 = u$. The sheaves $\mathcal{G}(\Omega)$ and $\mathcal{G}_{\mathcal{U}}(\Omega)$ are not flabby, because the order of growth ε^{-p} of u_ε can increase beyond bounds near $\partial\Omega_1$, whereas v_ε should have bounded order on and near $\partial\Omega_1$.

Remark 3. The either-or situation with properties (iii) and (v) in the table above can be nicely demonstrated by means of ${}^*\mathcal{C}^\infty(\mathbb{R}^n)$. At the same time, this is a good opportunity to discuss what saturation can do for us. We choose the index set E so large and the ultrafilter appropriately so that our nonstandard universe (built on ${}^*\mathbb{R}$, say) becomes polysaturated (see e.g. T.Lindstrøm [16] for the terminology). The following argument we learned from H.Akiyama [1]. Due to polysaturation, there is a vector space $\mathcal{H}$ of hyperfinite dimension ν such that

$$\mathcal{D}(\mathbb{R}^n) \subset \mathcal{H} \subset {}^*\mathcal{D}(\mathbb{R}^n)$$

(see [16, Proposition III.3.9]). Consider the inner product $(\varphi,\psi) \to \int \varphi(x)\bar{\psi}(x)dx$ inherited from ${}^*\mathcal{D}(\mathbb{R}^n)$ on $\mathcal{H}$. Choose an orthonormal basis $\{\psi_1, ..., \psi_\nu\}$ in $\mathcal{H}$ and define $\Delta \in {}^*\mathcal{D}(\mathbb{R}^{2n})$ by

$$\Delta(x,y) = \sum_{j=1}^{\nu} \psi_j(x)\bar{\psi}_j(y)$$

Define an imbedding κ of $\mathcal{D}'(\mathbb{R}^n)$ into ${}^*\mathcal{C}^\infty(\mathbb{R}^n)$ by assigning to $w \in \mathcal{D}'(\mathbb{R}^n)$ the nonstandard function

$$x \to \langle {}^*w, \Delta(x,.)\rangle$$

where the brackets denote duality in ${}^*\mathcal{D}'(\mathbb{R}^n)$, ${}^*\mathcal{D}(\mathbb{R}^n)$. It is immediate to check that

$$\int \Delta(x,y)\, {}^*\varphi(y)\, dy = {}^*\varphi(x) \tag{4}$$

for all $\varphi \in \mathcal{D}(\mathbb{R}^n)$, $x \in {}^*\mathbb{R}^n$; hence $\kappa \mid \mathcal{D}(\mathbb{R}^n) = \sigma$, and this is almost property (v). A partition of unity argument shows that equation (4) holds also for $\varphi \in \mathcal{C}^\infty(\mathbb{R}^n)$ at all finite $x \in {}^*\mathbb{R}^n$. Therefore, passing to ${}^*\mathcal{C}^\infty_{loc}(\mathbb{R}^n)$, the map κ defines an imbedding $\mathcal{D}'(\mathbb{R}^n) \to {}^*\mathcal{C}^\infty_{loc}(\mathbb{R}^n)$ with $\kappa \mid \mathcal{C}^\infty(\mathbb{R}^n) = \sigma$. Thus we have property (v). However, it appears that nothing can be said about derivatives; we do not know whether $\kappa(\partial_j w) = \partial_j \kappa(w)$ for $w \in \mathcal{D}'(\mathbb{R}^n)$ (note that integration by parts does not work because $\partial_x \Delta(x,y) \neq -\partial_y \Delta(x,y)$, otherwise Δ could not have $*$-compact support).

On the other hand, using a different saturation argument, T.D.Todorov [38,39] has constructed a nonstandard function $\theta \in {}^*\mathcal{D}(\mathbb{R}^n)$ with the property

$$\int \theta(x-y)\, {}^*\varphi(y)\, dy = {}^*\varphi(x)$$

for all $\varphi \in \mathcal{C}^0(\mathbb{R}^n)$ and standard $x \in \mathbb{R}^n$. The map $\lambda : \mathcal{D}'(\mathbb{R}^n) \to {}^*\mathcal{C}^\infty(\mathbb{R}^n)$ which assigns to each $w \in \mathcal{D}'(\mathbb{R}^n)$ the nonstandard function ${}^*w * \theta$ is an imbedding which commutes with derivatives and hence satisfies (iii). This time we cannot infer property (v), because $({}^*\varphi * \theta)(x) = {}^*\varphi(x)$ is only known to hold for standard $x \in \mathbb{R}^n$, so ${}^*\varphi * \theta$ and ${}^*\varphi$ need not be equal as nonstandard functions.

5 Applications

We begin with a general observation. Let $T(\partial)$ be a possibly nonlinear differential operator with smooth coefficients on $\Omega \subset \mathbb{R}^n$. Given two factor algebras $\mathcal{A}_1(\Omega) = \mathcal{X}_1/\mathcal{I}_1$, $\mathcal{A}_2(\Omega) = \mathcal{X}_2/\mathcal{I}_2$ with $\mathcal{X}_1, \mathcal{X}_2 \subset \mathcal{X}(\Omega)$, we can define

$$T(\partial) : \mathcal{A}_1(\Omega) \to \mathcal{A}_2(\Omega)$$

provided $T(\partial)$, acting componentwise, maps $\mathcal{X}_1$ into $\mathcal{X}_2$ and

$$T(\partial)(s + \mathcal{I}_1) - T(\partial)(s) \in \mathcal{I}_2, \forall s \in \mathcal{X}_1. \tag{5}$$

An element $u = s + \mathcal{I}_1$ of $\mathcal{A}_1(\Omega)$ is a solution to the equation $T(\partial)u = 0$ iff

$$T(\partial)(s + \mathcal{I}_1) \subset \mathcal{I}_2.$$

Thus $\mathcal{I}_1$ determines the amount that the representative s can be perturbed and still defines a solution, that is, the size of $\mathcal{I}_1$ corresponds to *stability*. On the other hand, the size of $\mathcal{I}_2$ corresponds to *exactness*, i.e. the degree how much $T(\partial)(s)$ is allowed to differ from zero and yet is considered a solution.

We note that (5) is valid if in particular $T(\partial)$ is polynomially nonlinear and $\mathcal{A}_1(\Omega) = \mathcal{A}_2(\Omega)$ is one of the differential algebras listed in Section 3.

Let us now see how solutions effectively are constructed in a simple, illuminating example: the Cauchy problem for the semilinear wave equation with generalized functions as initial data. Denoting the coordinates of $\mathbb{R}^2$ by (t, x), we consider the problem

$$\begin{aligned}(\partial_t^2 - \partial_x^2)u + F(u) = 0 \text{ on } \mathbb{R}^2 \\ u \mid \{t = 0\} = a, \partial_t u \mid \{t = 0\} = b \text{ on } \mathbb{R}.\end{aligned} \tag{6}$$

Suppose we are given a type of factor algebra $\mathcal{A}(\mathbb{R}^n) = \mathcal{X}_s(\mathbb{R}^n)/\mathcal{I}(\mathbb{R}^n)$, $n = 1, 2$. For the purpose of this discussion, we take the same algebra on both sides of the differential equation, assuming that $T(\partial) = \partial_t^2 - \partial_x^2 + F(.)$ maps $\mathcal{A}(\mathbb{R}^2)$ into itself, so that any solution $u \in \mathcal{A}(\mathbb{R}^2)$ will satisfy the equation in the differential-algebraic sense of $\mathcal{A}(\mathbb{R}^2)$, while the restriction to $\{t = 0\}$ is defined componentwise on representatives. We now put forth a list of three basic hypotheses, which roughly amount to requiring that problem (6) with classical $\mathcal{C}^\infty$-initial data has unique $\mathcal{C}^\infty$-solutions, and possibly admits certain $\mathcal{C}^\infty$-estimates; more precisely:

(H1) if $a_\varepsilon, b_\varepsilon \in \mathcal{C}^\infty(\mathbb{R})$ then there is a unique solution $u_\varepsilon \in \mathcal{C}^\infty(\mathbb{R}^2)$;

(H2) if $(a_\varepsilon)_{\varepsilon\in E}, (b_\varepsilon)_{\varepsilon\in E}$ belong to $\mathcal{X}_s(\mathbb{R})$ then the solution sequence $(u_\varepsilon)_{\varepsilon\in E}$ belongs to $\mathcal{X}_s(\mathbb{R}^2)$;

(H3) if $(u_\varepsilon)_{\varepsilon\in E}$ and $(v_\varepsilon)_{\varepsilon\in E}$ belong to $\mathcal{X}_s(\mathbb{R}^2)$ and

$$\begin{aligned}(T(\partial)u_\varepsilon - T(\partial)v_\varepsilon)_{\varepsilon\in E} \in \mathcal{I}(\mathbb{R}^2), \\ (u_\varepsilon - v_\varepsilon \mid \{t = 0\})_{\varepsilon\in E}, (\partial_t u_\varepsilon - \partial_t v_\varepsilon \mid \{t = 0\})_{\varepsilon\in E} \in \mathcal{I}(\mathbb{R})\end{aligned}$$

then $(u_\varepsilon - v_\varepsilon)_{\varepsilon\in E} \in \mathcal{I}(\mathbb{R}^2)$.

As an immediate consequence of these hypotheses we have the following result:

Proposition: *Let $a, b \in \mathcal{A}(\mathbb{R})$ and suppose that (H1)–(H3) hold. Then problem (6) has a unique solution $u \in \mathcal{A}(\mathbb{R}^2)$.*

Proof: Let $(a_\varepsilon)_{\varepsilon\in E}, (b_\varepsilon)_{\varepsilon\in E}$ be representatives of the initial data $a, b \in \mathcal{A}(\mathbb{R})$. According to (H1) and (H2), the classical solution sequence $(u_\varepsilon)_{\varepsilon\in E}$ can serve as a representative for a solution $u \in \mathcal{A}(\mathbb{R}^2)$. (H3) translates directly into uniqueness of generalized solutions. ∎

All what remains to do is to verify hypotheses (H1)–(H3). Now (H1) is known to hold if F is smooth and e.g. F' is bounded. In the case $\mathcal{X}_s(\mathbb{R}^n) = \mathcal{X}(\mathbb{R}^n)$, (H2) is trivially true, while finite propagation speed for the wave equation and uniqueness of classical solutions gives (H3) for $\mathcal{I}_{loc}$, $\mathcal{I}_{\mathcal{U}}$, $\mathcal{I}_{\mathcal{U},loc}$ of Section 3. Thus we have unique solutions in $\mathcal{A}(\mathbb{R}^2) = \mathcal{L}(\mathbb{R}^2)$, $^*\mathcal{C}^\infty(\mathbb{R}^2)$, $^*\mathcal{C}^\infty_{loc}(\mathbb{R}^2)$. The case of $\mathcal{G}(\mathbb{R}^2)$, $\mathcal{G}_{\mathcal{U}}(\mathbb{R}^2)$ requires some more work: here the estimates entering in $\mathcal{X}_M$ and $\mathcal{N}$ have to be verified; these are classical a priori bounds for $\mathcal{C}^\infty$-solutions. Thus we also have unique solutions in $\mathcal{G}(\mathbb{R}^2)$ and $\mathcal{G}_{\mathcal{U}}(\mathbb{R}^2)$.

Let us pause here for a moment and reflect upon what we have achieved: We have produced generalized solutions, almost without effort, by building up from classical $\mathcal{C}^\infty$-solutions, going to sequences, and then factoring. As pointed out in the Introduction, this can be viewed as a *transfer process* (and in the case of $^*\mathcal{C}^\infty(\mathbb{R}^2)$ it actually is nothing but *the* transfer principle at work).

Now, as $\mathcal{D}'(\mathbb{R}) \subset \mathcal{A}(\mathbb{R})$ we have in particular unique solutions to the nonlinear problem (6) when the initial data are arbitrary distributions. This, of course, goes much beyond what could be obtained or even formulated in the classical, distributional setting. However, in certain cases classical solutions do exist. For example, if a, b belong to $\mathcal{C}^0(\mathbb{R})$ then there is a unique solution $v \in \mathcal{C}^0(\mathbb{R}^2)$. What is the relation of v to the generalized solution $u \in \mathcal{A}(\mathbb{R}^2)$ with initial data $\iota(a), \iota(b)$, where ι is the imbedding of $\mathcal{D}'$ into $\mathcal{A}$? Well, we generally can assert that u is infinitely close to v, that is, $u \approx \iota(v)$. This is just a translation of the fact that v depends continuously on the initial data. However, it is in general impossible to have u equal to $\iota(v)$ in $\mathcal{A}(\mathbb{R}^2)$; this is caused by reasons similar to the negative results of Section 2. We do have $u = \iota(v)$ in $\mathcal{G}(\mathbb{R}^2)$ and $\mathcal{G}_{\mathcal{U}}(\mathbb{R}^2)$ provided a, b belong to $\mathcal{C}^\infty(\mathbb{R})$, thanks to the property $\iota \mid \mathcal{C}^\infty = \sigma$. Finally, it is possible to achieve strong consistency $u = \iota(v)$ for merely continuous a, b if one is willing to discard the differential algebraic structure and work in vector spaces instead (to be sure, different ones on both sides of the equation, and thus exploiting more of the power of the general factorization framework). Such a construction is worked out in [21] and [23, Section 24], it involves factoring $\mathcal{X}(\mathbb{R}^2)$ by the space of zero-sequences in $\mathcal{C}^0(\mathbb{R} : L^1_{\mathrm{loc}}(\mathbb{R}))$.

We mention that so-called delta waves can be accommodated as well: These are solutions with initial data measures or distributions with discrete support. Here various cases are known where the classical, approximate solutions with regularized initial data converge weakly to a limit $v \in \mathcal{D}'(\mathbb{R}^2)$, and then the generalized solution $u \in \mathcal{A}(\mathbb{R}^2)$ is again infinitely close to v.

We now list a number of further applications that have been obtained during the past few years, or are in the process of being worked out. References will be omitted, but can be found in [23, Sections 16–20, and Sections 22,26,27], together with further details. We also ask the reader to look at the monographs of H.A.Biagioni [3], J-F.Colombeau [5,6,8] and E.E.Rosinger [28–31] as well as the survey article of J.F.Colombeau [7].

(1) The results above on the Cauchy problem with singular data are not confined to the wave equation; they hold more generally for semilinear hyperbolic systems and equations in one space dimension, and have also been extended to several space dimensions.

(2) It may happen that instead of the initial data some coefficients are singular: the same methods allow to handle hyperbolic equations with discontinuous coefficients, as e.g. arising in transmission problems in layered media.

(3) Quasilinear hyperbolic equations and systems. In the case of a conservation law

$$\partial_t u + \partial_x F(u) = 0$$

smooth approximations can be constructed from the parabolic perturbation

$$\partial_t u + \partial_x F(u) = \mu \partial_x^2 u$$

with $\mu \approx 0$; these approximate solutions then satisfy $\partial_t u + \partial_x F(u) \approx 0$, and thus one can study shock waves (as indicated in Section 4), singular data, and asymptotic behavior in terms of μ. Our algebras provide also a setting for the study of shock waves in nonconservative systems, where the usual weak solution concept fails. There is the possibility to employ both equality and the infinitesimality relation in order to fix the jump conditions. A typical system that has been studied is

$$\begin{aligned} \partial_t u + u\partial_x u &= \partial_x v \\ \partial_t v + u\partial_x v &\approx \partial_x u. \end{aligned}$$

(4) A number of special equations have been investigated: transport equations, like the Carleman system, the Korteweg - de Vries equation, Schrödinger equations, the p-system from gas dynamics, and so on.

(5) Regularized derivatives. Given a generalized function u on $\mathbb{R}^n$ with some representative $(u_\varepsilon)_{\varepsilon \in E}$, the *regularized derivatives* $\tilde{\partial}_j u$, $j = 1, ..., n$ can be defined by means of an additional convolution with a delta-sequence:

$$\tilde{\partial}_j u_\varepsilon = \partial_j (u_\varepsilon * \varphi_\varepsilon)$$

for $\varepsilon \in E$. The Cauchy problem for an evolution equation in $\mathbb{R}^{n+1}$ can be rewritten by replacing the space derivatives with regularized derivatives while keeping the usual time derivative:

$$\begin{aligned} \partial_t u &= \sum_{|\alpha| \leq m} a_\alpha \, \tilde{\partial}_x^\alpha u \\ u \mid \{t = 0\} &= u_0. \end{aligned}$$

This has the effect that for each fixed ε the original partial differential equation is turned into an ordinary integro-differential equation, which is readily solved. This way arbitrary evolution equations, irrespective of their type (be it hyperbolic, parabolic,...) have unique generalized solutions, backward and forward in time. This approach also offers the opportunity to construct unconditionally stable difference schemes (cf. E.E.Rosinger [32]).

(6) The theory of pseudodifferential operators can be transplanted into the setting of $\mathcal{G}(\Omega)$ and used to prove regularity results involving $\mathcal{G}^\infty(\Omega)$.

(7) The Cauchy problem for analytic partial differential equations (Cauchy-Kowalewski type), which generally can only be solved locally in the classical setting, admits global solutions in the algebra $\mathcal{R}(\Omega)$. This relies on the fact that classical solutions can be constructed off a nowhere dense, closed set which carries the singularities.

(8) Quantum field theory. Each renormalization prescription determines an object in an algebra of generalized functions; the otherwise symbolic computations involved thus acquire a differential-algebraic meaning.

(9) Generalized stochastic processes can be accommodated by viewing their paths as generalized functions; this yields a reduction to smooth paths for fixed $\varepsilon \in E$.

This list of applications may suffice. We hope to have shown that going into the structure of factor algebras is a rewarding venture. On the one hand, it delivers a variety of generalized functions and supplies a wealth of new objects which can be used for precise modelling. On the other hand, it clarifies the issue of nonlinear operations with them and unifies various approaches, from weak solutions to nonstandard analysis. Finally, it provides a general framework for dealing with nonlinear differential equations and, by taking a fresh viewpoint, raises a range of new questions.

6 References

[1] H. Akiyama: *Applications of nonstandard analysis to diffusion on manifolds,* this volume.

[2] P. Antosik, J. Mikusiński, R. Sikorski: *Theory of Distributions. The Sequential Approach,* Elsevier, Amsterdam 1973.

[3] H.A. Biagioni: *A Nonlinear Theory of Generalized Functions,* Lecture Notes Math. 1421, Springer, Berlin 1990.

[4] J.F. Colombeau: *Une multiplication générale des distributions,* C.R. Acad. Sci. Paris, Sér.I, 296 (1983), 357–360.

[5] J.F. Colombeau: *New Generalized Functions and Multiplication of Distributions,* North Holland, Amsterdam 1984.

[6] J.F. Colombeau: *Elementary Introduction to New Generalized Functions,* North Holland, Amsterdam 1985.

[7] J.F. Colombeau: *Multiplication of distributions,* Bull. Am. Math. Soc. 23 (1990), 251–268.

[8] J.F. Colombeau: *Multiplication of Distributions. A tool in mathematics, numerical engineering and theoretical physics,* Lecture Notes Math. 1532, Springer, Berlin 1992.

[9] Yu. V. Egorov: *On a new theory of generalized functions,* (Russian), Vestnik Moskov. Univ. Ser.I, 1989, No.4, 96–99.

[10] Yu. V. Egorov: *A contribution to the theory of generalized functions,* Russian Math. Surveys 45:5 (1990), 1–49. Translated from: Uspekhi Mat. Nauk 45:5 (1990), 3–40.

[11] H. König: *Neue Begründung der Theorie der "Distributionen" von L. Schwartz,* Math. Nachr. 9 (1953), 129–148.

[12] H. König: *Multiplikation von Distributionen. I,* Math. Ann. 128 (1955), 420–452.

[13] H. König: *Multiplikation und Variablentransformation in der Theorie der Distributionen,* Arch. Math. 6 (1955), 391–396.

[14] D. Laugwitz: *Eine Einführung der δ-Funktionen,* Sitzungsb. Bayer. Akad. Wiss., Math.-nat. Kl. 1959, 41–59.

[15] D. Laugwitz: *Anwendung unendlich kleiner Zahlen. I. Zur Theorie der Distributionen,* J. reine angew. Math. 207 (1961), 53–60.

[16] T. Lindstrøm: *An invitation to nonstandard analysis,* In: N. Cutland (Ed.), Nonstandard Analysis and its Applications,. Cambridge Univ. Press, Cambridge 1988, 1–105.

[17] V. P. Maslov, V. A. Tsupin: *δ-shaped Sobolev generalized solutions of quasilinear equations,* Russion Math. Surveys 34:1 (1979), 231–232. Translated from: Uspekhi Mat. Nauk 34:1 (1979), 235–236.

[18] V. P. Maslov, V. A. Tsupin: *Necessary conditions for the existence of infinitesimally narrow solitons in gas dynamics,* Sov. Phys. Dokl. 24 (5) (1979), 354–356. Translated from: Dokl. Akad. Nauk SSSR 246 (1979), 298–300.

[19] J. Mikusiński: *Sur la méthode de généralisation de M. Laurent Schwartz sur la convergence faible,* Fund. Math. 35 (1948), 235–239.

[20] M. Oberguggenberger: *Products of distributions: nonstandard methods,* Zeitschrift Anal. Anw. 7 (1988), 347–365. Corrections to this article: Zeitschr. Anal. Anw. 10 (1991), 263–264.

[21] M. Oberguggenberger: *Semilinear wave equations with rough initial data,* In: P. Antosik, A. Kamiński (Eds.), Generalized Functions and Convergence,. World Scientific Publ., Singapore 1990, 181–203.

[22] M. Oberguggenberger: *Case study of a nonlinear, nonconservative, non-strictly hyperbolic system,* Nonlinear Anal. 19 (1992), 53–79.

[23] M. Oberguggenberger: *Multiplication of Distributions and Applications to Partial Differential Equations,* Pitman Research Notes Math. 259, Longman, Harlow 1992.

[24] M. Oberguggenberger, E. E. Rosinger: *Solution of Continuous Nonlinear PDEs through Order Completion,* Preprint 1992/1993.

[25] A. Robinson: *Function theory on some nonarchimedean fields,* Am. Math. Monthly 80 (6), Part II: Papers in the Foundations of Mathematics (1973), 87–109.

[26] E. E. Rosinger: *Scufundarea distribulitor $\mathcal{D}'_n$ în algebre pseudotopologice,* Stud. cerc. mat. 18 (1966), 687–729.

[27] E. E. Rosinger: *Spatii pseudotopologice. Scufundarea distribulitor $\mathcal{D}'_n$ în algebre,* Stud. cerc. mat. 20 (1968), 553–582.

[28] E. E. Rosinger: *Distributions and Nonlinear Partial Differential Equations,* Lecture Notes Math. 684, Springer, Berlin 1978.

[29] E. E. Rosinger: *Nonlinear Partial Differential Equations. Sequential and Weak Solutions,* North Holland, Amsterdam 1980.

[30] E. E. Rosinger: *Generalized Solutions of Nonlinear Partial Differential Equations,* North Holland, Amsterdam 1987.

[31] E. E. Rosinger: *Non-Linear Partial Differential Equations. An Algebraic View of Generalized Solutions,* North Holland, Amsterdam 1990.

[32] E. E. Rosinger: *Nonlinear Equivalence, Reduction of PDEs to ODEs and Fast Convergent Numerical Methods,* Pitman Research Notes Math. 77, Pitman, Boston 1982.

[33] C. Schmieden, D. Laugwitz: *Eine Erweiterung der Infinitesimalrechnung,* Math. Zeitschr. 69 (1958], 1–39.

[34] L. Schwartz: *Théorie des distributions,* Nouvelle ed., Hermann, Paris 1966.

[35] L. Schwartz: *Sur l'impossibilité de la multiplication des distributions,* C. R. Acad. Sci. Paris 239 (1954), 847–848.

[36] G. Temple: *Theories and applications of generalized functions,* J. London Math. Soc. 28 (1953), 134–148.

[37] T. D. Todorov: *Colombeau's generalized functions and nonstandard analysis,* In: B. Stanković, E. Pap, S. Pilipović, V. S. Vladimirov Eds.), Generalized Functions, Convergence Structures, and Their Applications. Plenum Press, New York 1988, 327–339.

[38] T. D. Todorov: *A nonstandard delta function,* Proc. Am. Math. Soc. 110 (1990), 1143–1144.

[39] T. D. Todorov: *Pointwise kernels of Schwartz distributions,* Proc. Am. Math. Soc. 114 (1992), 817–819.

[40] T. D. Todorov: *Nonstandard asymptotic analysis and nonlinear theory of generalized functions,* Preprint 1993

A NONSTANDARD APPROACH TO THE PETTIS INTEGRAL

Horst Osswald
Mathematisches Institut, Universität München
Theresienstr. 39, D-80333 München

Abstract

We shall prove lifting theorems for Pettis measurable stochastic processes and Pettis integrable functions on Loeb spaces with values in nonseparable locally convex spaces. We apply the results to give a nonstandard proof of

[1] Vitali's convergence theorem for uniformly Pettis integrable functions with values in weakly complete spaces.

We will show for functions with weakly compact range:

[2] the existence of conditional expectations in adabted Loeb spaces,

[3] Keisler's Fubini theorem,

[4] the existence of solutions to special cases of nondeterministic vector valued Peano-Caratheodory differential equations.

1 Introduction

It is well known that each Borel measurable function f on a finite Loeb space $(\Omega, \mathrm{L}(\mathcal{A}), \hat{\mu})$ with values in a Banach space has a lifting (i. e., an internal representation) if and only if the range of f is separable valued $\hat{\mu}$-almost everywhere. Therefore Loeb-theory seems to be restricted only to functions with values in separable spaces.

By means of linear continuous functionals we introduce a notion of "lifting" in order to study functions with values in nonseparable spaces. Our procedure provides a nonstandard approach to Pettis-measurable functions and stochastic processes and to the Pettis integral.

It seems to be natural and is useful later to extend the notion of Pettis integrability (Pettis [15]) in the following way: Fix a topological vector space $\mathbb{B}$ (over $\mathbb{R}$ or $\mathbb{C}$) and a separable Banach space $\mathbb{D}$ such that there exists a **separating** set $\mathbb{B}'$ (i. e., for each $a \in \mathbb{B}$, $a \neq 0$, there exists a $\varphi \in \mathbb{B}'$ with $\varphi(a) \neq 0$) of continuous linear operators from $\mathbb{B}$ into $\mathbb{D}$. A function f from a measure space $(\Lambda, \mathcal{B}, \nu)$ into $\mathbb{B}$ is now called **Pettis integrable**, if $\varphi \circ f$ is Bochner integrable for all $\varphi \in \mathbb{B}'$ and if for every set $B \in \mathcal{B}$ there exists a vector $a_B \in \mathbb{B}$ such that for all $\varphi \in \mathbb{B}'$

$$\varphi(a_B) = \mathcal{B} - \int_B \varphi \circ f d\nu \text{ (in the sense of the Bochner integral).}$$

Then a_B is called the **Pettis integral** of f over B, denoted by $\mathcal{P} - \int_B f d\nu$.

Among other things we will show that the Pettis integral of a Pettis integrable function f on a Loeb space can be written —up to an infinitesimal error— as the internal integral of a *finitely valued function F, where F is an appropriate lifting of f. Important is that F does not depend on $\varphi \in \mathbb{B}'$. This result is a nonstandard version of a result due to R. F. Geitz [5], which (the nonstandard version) we obtain here in a comparatively simple way.

In the case when $\mathbb{B}$ is a Banach space, $\mathbb{D} = \mathbb{C}$ and $\mathbb{B}'$ is the topological dual of $\mathbb{B}$ the result shows in an intuitive manner the difference between Pettis- and Bochner integrability: (The result for the Bochner integral can be found in [12].)

A function $f : \Omega \to \mathbb{B}$ is Bochner (Pettis) integrable if and only if there exists an internal measurable *simple function

$F : \Omega \to {}^*\mathbb{B}$ such that (a), (b) and (c) hold:

(a) $\|F(\omega) - f(\omega)\| \approx 0$ for $\hat{\mu}$-almost all $\omega \in \Omega$
($\|\varphi \circ F(\omega) - \varphi \circ f(\omega)\| \approx 0$ for all $\varphi \in \mathbb{B}'$ and $\hat{\mu}$-almost all $\omega \in \Omega$).
Notice that $\hat{\mu}$-a.a. depends on φ.

(b) $\|\cdot\| \circ F$ is S-integrable.
($\varphi \circ F$ is S-integrable for all $\varphi \in \mathbb{B}'$).

(c) $\int_A F d\mu$ is nearstandard in the norm (weak) topology for all $A \in \mathcal{A}$.

Whenever F is such a function, then

$$\int_\Omega F d\mu \approx \mathcal{B} - \int_\Omega f d\hat{\mu} \text{ in the norm topology}$$
$$\left(\int_\Omega F d\mu \approx \mathcal{P} - \int_\Omega f d\hat{\mu} \text{ in the weak topology}\right).$$

Our lifting results can be used as a tool to prove vector valued versions of well known results (due to Anderson [2], Keisler [9] , Loeb [10]) in scalar valued Loeb theory.

Acknowledgement

I would like to thank Peter Loeb and Jerry Uhl for many stimulating comments.

2 Preliminaries

We will work with a polysaturated enlargement of a standard model $\mathcal{M}$. We assume that $\mathbb{B}$, $\mathbb{C}$ and $\mathbb{D}$ are elements of $\mathcal{M}$, so every subset of $\mathbb{B}$, $\mathbb{C}$ and $\mathbb{D}$ and every function on these sets with values in these sets and so on belong to $\mathcal{M}$, so $\mathbb{B}' \in \mathcal{M}$.

For undefined notions and for the basic of nonstandard analysis consult the book of S. Albeverio, J. E. Fenstad, R Høegh-Krohn and T. Lindstrøm [1] or the book of A.E. Hurd, P.A. Loeb [8]. If there is no risk of confusion, we omit the left upper index *. For example, we will always write $\|\cdot\|$, φ, with $\varphi \in \mathbb{B}'$, instead of ${}^*\|\cdot\|$, ${}^*\varphi$.

In order to fix the terminology, I will give a short and simple introduction to scalar valued Loeb spaces. The idea comes from P. Loeb's functional approach to nonstandard measure theory [11].

Let $(\Omega, \mathcal{A}, \mu)$ be an internal measure space such that $\mu : \mathcal{A} \longrightarrow {}^*[0, \infty[$ is limited. An internal or external subset $N \subset \Omega$ is called a μ-**nullset**, if for a ll standard $\epsilon > 0$ there exists $A \in \mathcal{A}$ with $N \subset A$ and $\mu(A) < \epsilon$. Let $\mathcal{N}$ denote the set of all μ-nullsets. Using the saturation assumption, it is easy to check that the union of a sequence in $\mathcal{A} \cap \mathcal{N}$ of length at most card$(\mathcal{M})$ is a subset of an element of $\mathcal{A} \cap \mathcal{N}$. Moreover, note that $\mathcal{N}$ is closed under countable unions.

Let $B \subset \Omega$ be external or internal and let $A \in \mathcal{A}$. Then A is called a μ-**approximation** of B, if the symmetric difference $A \bigtriangleup B$ of A and B is a μ-nullset. It is easy to see that the set

$$L(\mathcal{A}) := \{B \subset \Omega | B \text{ has a } \mu\text{-approximation }\}$$

is a σ-algebra and that $\hat{\mu}$ is well defined and σ-additive, where $\hat{\mu}(B) := {}^0\mu(A))$ for all $B \in L(\mathcal{A})$ and all μ-approximations A of B. $(\Omega, L(\mathcal{A}), \hat{\mu})$ is called the **Loeb space** over $(\Omega, \mathcal{A}, \mu)$.

Whenever $\mathcal{A} = {}^*\mathcal{P}(\Omega) = \{A \subset \Omega | A \text{ is internal }\}$, then we will write $L(\Omega)$ instead of $L(\mathcal{A})$.

3 Weak nonstandard hulls of $\mathbb{B}$

The notion "nonstandard hull" for topological vector spaces was developed by W. A. J. Luxemburg [13] and extensively studied by C. W. Henson and L. C. Moore. See for example [7]. We use weak nonstandard hulls in order to show that not only functions with values in $\mathbb{B}$, but also functions with values in the algebraic dual $\mathbb{B}'^{\sim}$ of $\mathbb{B}'$ have ${}^*\mathbb{B}$-valued liftings.

Fix a linear space $\mathbb{B}_0$ over $\mathbb{C}$ between $\mathbb{B}$ and the linear space $s^*\mathbb{B}$, where

$$s^*\mathbb{B} := \{a \in {}^*\mathbb{B} | \varphi(a) \text{ is nearstandard in the norm topology on } \mathbb{D} \text{ for all } \varphi \in \mathbb{B}'\}.$$

Define for all $a \in {}^*\mathbb{B}$

$$\bar{a} := \{b \in {}^*\mathbb{B} | \varphi(a) \approx \varphi(b) \text{ in the norm topology on } \mathbb{D} \text{ for all } \varphi \in \mathbb{B}'\}.$$

An element $a \in {}^*\mathbb{B}$ is called **n̂earstandard** and $\bar{a}$ is called the **ŝtandard part** of a, if $\bar{a} \cap \ BanB_0 \neq \emptyset$. Define

$$\hat{\mathbb{B}} := \{\bar{a} | a \text{ is n̂earstandard }\}.$$

Notice that $\hat{\mathbb{B}}$ depends on $\mathbb{B}_0$. $\hat{\mathbb{B}}$ is a linear space over $\mathbb{C}$ with the natural definitions of $+$ and $\cdot$ and $\langle\hat{\mathbb{B}},\mathbb{B}'\rangle$ is a dual pair with respect to the bilinear mapping $(\bar{a},\varphi)\longmapsto {}^0\varphi(a)$. Define for all $\hat{\text{n}}$earstandard $a\in {}^*\mathbb{B}$

$$\hat{\varphi}:\bar{a}\longmapsto {}^0\varphi(a)$$

$\hat{\mathbb{B}}$ is called a **weak nonstandard hull** of $\mathbb{B}$. Often we identify $a\in\mathbb{B}$ with $\bar{a}\in\hat{\mathbb{B}}$, because $a\longmapsto\bar{a}$ is an embedding of $\mathbb{B}$ into $\hat{\mathbb{B}}$. We will write $a\approx_w\bar{b}$ or $a\approx_w b$, if $\bar{a}=\bar{b}$. For every subset $L\subset {}^*\mathbb{B}$ we define

$$\hat{L}:=\{\bar{a}\in\hat{\mathbb{B}}|\bar{a}\cap L\neq\emptyset\}.$$

Examples 3.1 *(α) Let $\mathbb{D}:=\mathbb{C}$ and let $\mathbb{B}'$ be the topological dual of $\mathbb{B}$. Notice that for all $\alpha\in s^*\mathbb{B}$ the function $\Phi_\alpha:\mathbb{B}'\longrightarrow\mathbb{C}$, defined by $\Phi_\alpha(\varphi)={}^0\varphi(\alpha)$, belongs to the algebraic dual $\mathbb{B}'^{\sim}$ of $\mathbb{B}'$ and that for each $\Phi\in\mathbb{B}'^{\sim}$ there exists an $\alpha\in s^*\mathbb{B}$ such that $\Phi=\Phi_\alpha$. (See Henson and Moore [7].)*

(β) If $\mathbb{B}_0=\mathbb{B}$ then $\hat{\mathbb{B}}=\mathbb{B}$. Thus $\mathbb{B}$ itself is a special example of a weak nonstandard hull of $\mathbb{B}$. Further examples for nonstandard characterizations of spaces between $\mathbb{B}$ and $\mathbb{B}'^{\sim}$ can be found in the paper of S. Heinrich [6].

(γ) Suppose that L is an internal set of $\hat{\text{n}}$earstandard points. Then $\hat{L}$ is a weakly compact subset of $\hat{\mathbb{B}}$. Because $\hat{L}$ is embeddable into $\mathbb{B}'^{\sim}$, where $\mathbb{B}'^{\sim}$ is here the set of all linear operators on $\mathbb{B}'$ with values in $\mathbb{D}$, one may copy the proof of W. A. J. Luxemburg's result [13] that the standard part of an internal set of nearstandard points in a ***regular*** *space X (i. e. X is a Hausdorff space, in which every point has a neighbourhood base of closed sets) is compact.*

For the remainder of this paper we fix a nonstandard hull $\hat{\mathbb{B}}$ of $\mathbb{B}$.

4 Pettis measurable stochastic processes

In the following fix a Loeb probability space $(\Lambda,L(\mathcal{A}),\hat{\mu})$ and let S be a Hausdorff-space with countable base $(O_n)_{n\in\mathbb{N}}$ and $\mathbb{D}$ a separable Banach space with countable base $(U_n)_{n\in\mathbb{N}}$. We denote the set of functions f from S into $\mathbb{D}$ with $f[X]\subset Y$ by $W(X,Y)$. Let $W_{(n,m)}:=W(O_n,\bar{U}_m)$, where $\bar{U}_m$ is the closure of U_m. We identify functions $f:A\times B\longrightarrow C$ with f', where $f':A\longrightarrow\{g|g:B\longrightarrow C\}$ with $f'(a)(b)=f(a,b)$. An internal function $F:\Omega\longrightarrow X$ is called **$\mathcal{A}$-measurable**, if F has only *finitely many values and if for all $a\in X$

$$\{\omega\mid F(\omega)=a\}\in\mathcal{A}.$$

The next result, which is an extension of Keisler's "uniform lifting theorem" for functions with values in locally compact spaces S (se e [9]) to a larger class of spaces, is due to D. Ross [16], where he used capacity arguments. Because his result is important in the following and because we want to have statement (+), we give the proof, which uses arguments similar to the original proof of the Loeb- Anderson lifting theorem (see [9]) for random variables on a Loeb-space. The proof here is different from Ross' and Keisler's proof.

Lemma 4.1 *Let $f : \Omega \times S \longrightarrow \mathbb{D}$. Then the following statements (1) and (2) are equivalent:*

(1) (a) $f(\cdot, s) : \Omega \longrightarrow \mathbb{D}$ is measurable for all $s \in S$ and

(b) $f(\omega, \cdot) : S \longrightarrow \mathbb{D}$ is continuous for $\hat{\mu}$-almost all $\omega \in \Omega$.

(2) There exists an internal function $F : \Omega \times {}^*S \longrightarrow {}^*\mathbb{D}$ such that

(a') $F : \Omega \longrightarrow \{G : {}^*S \longrightarrow {}^*\mathbb{D} | G$ is internal $\}$ is $\mathcal{A}$-measurable and

(b') for $\hat{\mu}$-almost all $\omega \in \Lambda$

$$F(\omega, s) \approx f(\omega, {}^0 s) \text{ for all nearstandard } s \in {}^*S.$$

More precisely, we will prove for the implication "(1) $\Longrightarrow$ (2) (b')" that for all $n, m \in \mathbb{N}$ and $\hat{\mu}$-almost all $\omega \in \Omega$

$$(+) \text{ if } f(\omega) \in W_{(n,m)}, \text{ then } F(\omega) \in {}^*W_{(n,m)}.$$

Proof: Because the implication "(2) $\Longrightarrow$ (1)" is well known, we only prove the implication "(1) $\Longrightarrow$ (2)". Let D b e a countable dense subset of S. Since

$$f^{-1}[W_{(n,m)}] = \bigcap_{p \in D \cap O_n} \{f(\cdot, p) \in \bar{U}_m\} \text{ and } \{f(\cdot, p) \in \bar{U}_m\} \in L(\mathcal{A}),$$

$f^{-1}[W_{(n,m)}] \in L(\mathcal{A})$. It follows that there exists a μ-approximation $A'_{(n,m)}$ of $f^{-1}[W_{(n,m)}]$. Let $(\mathbb{N} \times \mathbb{N}, \vdash)$ be a well ordering of order type $(\mathbb{N}, <)$. As in the Loeb-Anderson lifting theorem we modify A'_α to A_α such that A_α remains a μ-approximation of $f^{-1}[W_\alpha]$ and for each finite sequence $(\alpha_1, \ldots, \alpha_k)$ in $\mathbb{N} \times \mathbb{N}$

$$(++) \; A_{\alpha_1} \cap \ldots \cap A_{\alpha_k} \neq \emptyset \Longrightarrow {}^*W_{\alpha_1} \cap \ldots \cap {}^*W_{\alpha_k} \neq \emptyset.$$

To this end define by recursion

$$A_\alpha := A'_\alpha \setminus \bigcup \{A_{\alpha_1} \cap \ldots \cap A_{\alpha_k} | \alpha_1 \vdash \ldots \vdash \alpha_k \vdash \alpha \text{ and } W_{\alpha_1} \cap \ldots \cap W_{\alpha_k} \cap W_\alpha = \emptyset\}.$$

Using (++) and a saturation argument we obtain an $\mathcal{A}$-measurable function

$$F : \Lambda \longrightarrow \{G : {}^*S \longrightarrow {}^*\mathbb{D} | G \text{ internal } \}$$

such that for all $\alpha \in \mathbb{N} \times \mathbb{N}$ $F[A_\alpha] \subset {}^*W_\alpha$. It follows that for all $\omega \notin \bigcup_{\alpha \in \mathbb{N} \times \mathbb{N}} A_\alpha \Delta f^{-1}[W_\alpha]$ such that $f(\omega, \cdot)$ is continuous and all $\beta \in \mathbb{N} \times \mathbb{N}$

$$f(\omega) \in W_\beta \Longleftrightarrow \omega \in f^{-1}[W_\beta] \Longleftrightarrow \omega \in A_\beta \Longrightarrow F(\omega) \in {}^*W_\beta,$$

thus $f(\omega, {}^0 s) \approx F(\omega, s)$ for all nearstandard $s \in {}^*S$. ∎

Lemma 4.2 *Let $L \subset {}^*\mathbb{B}$ be internal. For each mapping $h : S \longrightarrow \hat{L}$ there exists an internal mapping $H : {}^*S \longrightarrow L$ such that for all $\varphi \in \mathbb{B}'$ and all $(n, m) \in \mathbb{N} \times \mathbb{N}$*

$$\hat{\varphi} \circ h \in W_{(n,m)} \Longrightarrow \varphi \circ H \in {}^*W(O_n, G) \text{ for all open } G \supset \bar{U}_m$$

Proof: Set

$$X := \{(\varphi, n, m) | \hat{\varphi} \circ h \in W_{(n,m)}\}.$$

Because for each finite sequence $(\varphi_1, n_1, m_1), \ldots, (\varphi_k, n_k, m_k)$ in X and open $G_1 \supset \bar{U}_{m_1}, \ldots, G_k \supset \bar{U}_{m_k}$

$$O_{n_1} \cap \ldots \cap O_{n_k} \neq \emptyset \Longrightarrow \varphi_1^{-1}[{}^*G_1] \cap \ldots \cap \varphi_k^{-1}[{}^*G_k] \cap L \neq \emptyset,$$

by saturation, there exists an internal function $H : {}^*S \longrightarrow L$ such that for all $(\varphi, n, m) \in X$ and all open $G \supset \bar{U}_m$:

$$H[{}^*O_n] \subset \varphi^{-1}[{}^*G], \text{ i.e., } \varphi \circ H \in {}^*W(O_n, G). \blacksquare$$

Before we give the main result in this section, we define $f : \Omega \times S \longrightarrow \hat{\mathbb{B}}$ to be **Pettis measurable**, if for all $\varphi \in \mathbb{B}'$

(1) for $\hat{\mu}$-almost all $\omega \in \Omega$, where $\hat{\mu}$-a. a. depends on φ (!), $\hat{\varphi} \circ f(\omega, \cdot) : S \longrightarrow \mathbb{D}$ is continuous and

(2) $\hat{\varphi} \circ f(\cdot, s) : \Omega \longrightarrow \mathbb{D}$ is measurable for all $s \in S$.

An $\mathcal{A}$-measurable function $F : \Omega \longrightarrow \{G : {}^*S \longrightarrow L | G \text{ internal }\}$ (then, by definition, F has *finitely many values) is called a **weak lifting** of $f : \Omega \times S \longrightarrow \hat{L}$, if for all $\varphi \in \mathbb{B}'$ and $\hat{\mu}$-almost all $\omega \in \Omega$ (again $\hat{\mu}$-a. a. depends an φ) and all nearstandard $a \in {}^*S$

$$\varphi \circ F(\omega, a) \approx \hat{\varphi} \circ f(\omega, {}^0a).$$

The next result is an extension of the Keisler Ross lifting theorem for continuous stochastic processes to processes with nonseparable range.

Theorem 4.3 *Let $L \subset {}^*\mathbb{B}$ be internal. Then $f : \Omega \times S \longrightarrow \hat{L}$ is Pettis measurable, if and only if f has a weak lifting $F : \Omega \times {}^*S \longrightarrow L$.*

Proof: Because the "if"-part of the theorem is well known, we only prove the "only if"-part. By Lemma 4.1, for each $\varphi \in \mathbb{B}'$ there exists an $\mathcal{A}$-measurable function $G_\varphi : \Omega \times {}^*S \longrightarrow {}^*\mathbb{D}$ such that for all $\alpha \in \mathbb{N} \times \mathbb{N}$ and $\hat{\mu}$-a. a. $\omega \in \Omega$ ($\hat{\mu}$-a. a. depends on φ):

$$\hat{\varphi} \circ f(\omega) \in W_\alpha \Longrightarrow \varphi \circ F(\omega) \in {}^*W_\alpha.$$

For all $n \in \mathbb{N}$ and $\varphi \in \mathbb{B}'$ there exists $X_{n,\varphi} \in \mathcal{A}$ such that $\mu(\Omega) - 1/n < \mu(X_{n,\varphi})$ and

$$X_{n,\varphi} \subset \{\omega \in \Omega \mid \forall \alpha \in \mathbb{N} \times \mathbb{N}(\hat{\varphi} \circ f(\omega) \in W_\alpha \\ \Longrightarrow G_\varphi(\omega) \in {}^*W_\alpha \text{ and } \hat{\varphi} \circ f(\omega) \text{ is continuous })\}.$$

Notice that for all $\omega \in X_{n,\varphi}$ $\hat{\varphi} \circ f(\omega, {}^0s) \approx G_\varphi(\omega, s)$ for all nearstandard $s \in {}^*S$. Our aim is to find an $\mathcal{A}$-measurable function $F : \Omega \longrightarrow \{G : {}^*S \longrightarrow L | G \text{ internal}\}$ such that for all $\varphi \in \mathbb{B}'$ and all $n, m, k \in \mathbb{N}$ and all open $G \supset \bar{U}_m$

$$(+)\ F[G_\varphi^{-1}[{}^*W_{(n,m)}] \cap X_{k,\varphi}] \subset \{H : {}^*S \longrightarrow L | H \text{ internal and } \varphi \circ H \in {}^*W(O_n, G)\}.$$

By saturation, it suffices to show that for all finite $\Phi \subset \mathbb{B}'$ and $E \subset \mathbb{N}$

$$\begin{aligned}&\bigcap\{G_\varphi^{-1}[{}^*W_{(n,m)}] \cap X_{k,\varphi} | n, m, k \in E \text{ and } \varphi \in \Phi\} \neq \emptyset \Longrightarrow \\ &\bigcap\{\{H : {}^*S \longrightarrow L | H \text{ internal and } \varphi \circ H \in {}^*W(O_n, G)\}| \\ &\quad n, m \in E, \varphi \in \Phi \text{ and } G \supset \bar{U}_m \text{ is open }\} \neq \emptyset.\end{aligned}$$

Let $\omega \in \Omega$ be an element of the left hand side. Then $G_\varphi(\omega) \in {}^*W_{(n,m)}$ and $\omega \in X_{k,\varphi}$ for all $\varphi \in \Phi$ and all $n, m, k \in E$. It follows that $\hat{\varphi} \circ f(\omega) \in W_{(n,m)}$ for all $\varphi \in \Phi$ and all $n, m, k \in E$. By Lemma 4.2, there exists $H : {}^*S \longrightarrow L$ such that for all $\varphi \in \Phi$ and all $n, m \in E$ and all open $G \supset \bar{U}_m$

$$\varphi \circ H \in {}^*W(O_n, G).$$

This proves that the right hand side is nonempty and therefore a function F with the property $(+)$ exists. It is easy to check that $\hat{\varphi} \circ f(\omega, {}^0s) \approx \varphi \circ F(\omega, s)$ for all $\omega \in \bigcup X_{n,\varphi}$ and all nearstandard $s \in {}^*S$. ∎

5 Bochner integrable functions

Recall that $f : \Omega \longrightarrow \mathbb{D}$ is called **Bochner integrable**, if there exists a sequence $(f_n)_{n \in \mathbb{N}}$ of $L(\mathcal{A})$-measurable simple functions $f_n : \Omega \longrightarrow \mathbb{D}$ converging to f in measure and such that

$$\lim \int_\Omega \|f_n - f_m\| d\hat{\mu} = 0.$$

Then the **Bochner integral** of f over $B \in L(\mathcal{A})$ is defined by

$$\mathcal{B} - \int_B f d\hat{\mu} = \lim \int_B f_n d\hat{\mu}.$$

The internal counterparts to the Bochner integrable functions are the internal S-Bochner integrable functions on Ω with values in ${}^*\mathbb{D}$. An internal $\mathcal{A}$-measurable function $F : \Omega \longrightarrow {}^*\mathbb{D}$ (recall that F is then *finitely valued) is called S-**Bochner-integrable**, if $\|F\|$ is S-**integrable**, i. e., for all $k \in {}^*\mathbb{N} \setminus \mathbb{N}$

$$\int_{A_k} \|F\| d\mu \approx 0, \text{ where } A_k := \{\|F\| \geq k\}.$$

Recall that $\|F\|$ is S-integrable, if and only if

$$\forall A \in \mathcal{N} \cap \mathcal{A}(\int_A \|F\| d\mu \approx 0) \text{ and } \int_\Omega \|F\| d\mu \text{ is limited.}$$

In [12] there is a proof of the next result:

Proposition 5.1 *A function $f : \Omega \longrightarrow \mathbb{D}$ is Bochner integrable, if and only if f has an S-Bochner integrable* **lifting** $F : \Omega \longrightarrow {}^*\mathbb{D}$ *(i. e., $\{f \not\approx F\}$ is a μ-nullset), in which case*

$$\mathcal{B} - \int_B f d\hat{\mu} \approx \int_A F d\mu \text{ for all } B \in L(\mathcal{A}) \text{ and all } \mu - \text{approximations } A \text{ of } B.$$

It is easy to see that we get as in the scalar valued case:

If $F : \Omega \longrightarrow {}^*\mathbb{D}$ is $\mathcal{A}$-measurable and a lifting of a Bochner integrable function f, then there exists $H \in {}^*\mathbb{N} \setminus \mathbb{N}$ such that

$$F|_H : \omega \longmapsto \begin{cases} F(\omega) & , \text{if } \|F(\omega)\| \leq H \\ 0 & , \text{otherwise} \end{cases} \quad \text{for all } \omega \in \Omega$$

is S-Bochner integrable. Of course, $F|_H$ remains a lifting of f.

6 Dunford and Pettis integrable functions

Now we will prove lifting theorems for Dunford and Pettis integrable functions.

A function $f : \Omega \longrightarrow \hat{\mathbb{B}}$ is called **Dunford integrable**, if $\hat{\varphi} \circ f$ is Bochner integrable for all $\varphi \in \mathbb{B}'$. An $\mathcal{A}$-measurable function $F : \Omega \longrightarrow {}^*\mathbb{B}$ is called S-**Dunford integrable**, if $\varphi \circ F$ is S-Bochner integrable for all $\varphi \in \mathbb{B}'$.

The proof of the next result — it is a lifting theorem for Dunford integrable functions — is a simplification, due to P. Loeb, of a proof for the same proposition in an earlier version of this paper.

Proposition 6.1 *Let L be an internal subset of ${}^*\mathbb{B}$ and $f : \Omega \longrightarrow \hat{L}$. Then f is Dunford integrable, if and only if f has an S-Dunford integrable weak lifting $F : \Omega \longrightarrow L$.*

Proof: The "if" part follows from 5.1. Now we prove the "only if" part. By Theorem 4.3, f has a weak lifting $G : \Omega \longrightarrow L$. For every $\varphi \in \mathbb{B}'$ there exists an $H_\varphi \in {}^*\mathbb{N} \setminus \mathbb{N}$ such that $(\varphi \circ G)|_{H_\varphi}$ is S-Bochner integrable. Hence, for every $\varphi \in \mathbb{B}'$ there exists a μ-nullset $N_\varphi \in \mathcal{A}$ such that

$$1_{\Omega \setminus N_\varphi} \cdot \varphi \circ G = (\varphi \circ G)|_{H_\varphi}.$$

By saturation, there exists an $N \in \mathcal{A} \cap \mathcal{N}$ with $\bigcup\{N_\varphi | \varphi \in \mathbb{B}'\} \subset N$. Since $\varphi \circ (1_{\Omega \setminus N} \cdot G) = 1_{\Omega \setminus N} \cdot \varphi \circ G$, the function $F := 1_{\Omega \setminus N} \cdot G$ s atisfies the assertion. ∎

The next result is a lifting theorem for Pettis integrable functions. A Dunford integrable function $f : \Omega \longrightarrow \hat{\mathbb{B}}$ is called **Pettis integrable**, if for all $B \in L(\mathcal{A})$ there exists a vector $a_B \in \hat{\mathbb{B}}$ such that for all $\varphi \in \mathbb{B}'$

$$\hat{\varphi}(a_B) = \mathcal{B} - \int_B \hat{\varphi} \circ f d\hat{\mu}.$$

The vector a_B is called the **Pettis integral** of f over B, written

$$a_B = \mathcal{P} - \int_B f d\hat{\mu}.$$

An S-Dunford integrable function $F : \Omega \longrightarrow {}^*\mathbb{B}$ is called S-**Pettis integrable**, if $\int_A F d\mu$ is n̂earstandard for all $A \in \mathcal{A}$.

Theorem 6.2 *Let $L \subset {}^*\mathbb{B}$ be internal and $f : \Omega \longrightarrow \hat{L}$. Then f is Pettis integrable, if and only if there exists an S-Pettis integrable weak lifting $F : \Omega \longrightarrow L$ of f, in which case*

$$\mathcal{P} - \int_B f d\hat{\mu} \approx_w \int_A F d\mu \text{ for all } B \in L(\mathcal{A}) \text{ and all } \mu - \text{approximations } A \text{ of } B.$$

Proof: Let f be Pettis integrable. Then f is Dunford integrable, hence, by 6.1, f has an S-Dunford integrable weak lifting $F : \Omega \longrightarrow L$. We obtain for all $\varphi \in \mathbb{B}'$, for all $B \in L(\mathcal{A})$ and all μ-approximations A of B

$$\hat{\varphi}(\mathcal{P} - \int_B f d\hat{\mu}) = \mathcal{B} - \int_B \hat{\varphi} \circ f d\hat{\mu} \approx \int_A \varphi \circ F d\mu = \varphi(\int_A F d\mu).$$

Hence $\int_A F d\mu$ is n̂earstandard and $\mathcal{P} - \int_A f d\hat{\mu} \approx_w \int_A F d\mu$.

Now assume that F is an S-Pettis integrable weak lifting of f. By 6.1 f is Dunford integrable. Let $B \in L(\mathcal{A})$. Then there exists a μ-approximation A of B. Define

$$a_B := \overline{\int_A F d\mu} \in \hat{\mathbb{B}}.$$

We obtain for all $\varphi \in \mathbb{B}'$

$$\hat{\varphi}(a_B) \approx \varphi(\int_A F d\mu) = \int_A \varphi \circ F d\mu \approx \int_B \hat{\varphi} \circ f d\hat{\mu}.$$

Hence, f is Pettis integrable. ■

Remark: Theorem 6.2 reminds of a result in the standard theory of Pettis integrability, due to R. F. Geitz [5]. He proved that a bounded function $f : \Omega \longrightarrow \mathbb{B}$, where $\mathbb{B}$ is a Banach space and Ω is a perfect finite measure space, is Pettis integrable, if and only if there exists a uniformly bounded sequence $(f_n)_{n \in \mathbb{N}}$ of measurable simple functions f_n such that for all $\varphi \in \mathbb{B}'$

$$\lim \varphi \circ f_n = \varphi \circ f \ \ \hat{\mu}\text{-a. e.} \qquad (\hat{\mu} - \text{a. e. depends on } \varphi)$$

In Theorem 6.2, F can be understood as an idealization of Geitz's sequence, which exists for all Pettis integrable functions on Loeb spaces. Notice that Loeb spaces belong to the class of perfect spaces.

7 Examples

In this section, we will give 6 examples to show, how the preceding results can be used.

(α) The set of values of the indefinite Bochner integral of a Bochner integrable function $f : \Omega \longrightarrow \mathbb{D}$ (i. e., $\{\mathcal{B} - \int_B f d\hat{\mu} | B \in L(\mathcal{A})\}$) is norm compact.

(β) Suppose that $f : \Omega \longrightarrow \hat{\mathbb{B}}$ is Pettis integrable. Then the set of values of the indefinite Pettis integral of f is weakly compact.

Both results are well known. See [4]. Here they follow immediately from Proposition 5.1 and Theorem 6.2 respectively, and from W. A. J. Luxemburgs result, quoted in 3.1(γ).

(γ) **Convergence theorem for uniformly Pettis integrable functions**

In this example we combine the results in the preceding sections with nonstandard techniques, applicable to uniformly integrable functions. See for example N. Cutland [3].

A sequence $(f_n)_{n\in\mathbb{N}}$ of Pettis integrable functions $f_n : \Omega \longrightarrow \hat{\mathbb{B}}$ is called **uniformly Pettis integrable**, if for all $\varphi \in \mathbb{B}'$

$$\lim_{k\to\infty} \int_{\{\|\hat{\varphi}\circ f_n\| \geq k\}} \|\hat{\varphi} \circ f_n\| d\hat{\mu} = 0 \text{ uniformly in } n \in \mathbb{N}.$$

A sequence $(F_n)_{n\in\mathbb{N}}$ of S-Pettis integrable functions $F_n : \Omega \longrightarrow {}^*\mathbb{B}$ is called **uniformly S-Pettis integrable**, if for all $\varphi \in \mathbb{B}'$

$$\lim_{k\to\infty} {}^0\!\!\int_{\{\|\varphi\circ F_n\| \geq k\}} \|\varphi \circ F_n\| d\hat{\mu} = 0 \text{ uniformly in } n \in \mathbb{N}.$$

Lemma 7.1 *A sequence $(F_n)_{n\in\mathbb{N}}$ of S-Pettis integrable functions $F_n : \Omega \longrightarrow {}^*\mathbb{B}$ is uniformly S-Pettis integrable, if and only if $(F_n)_{n\in\mathbb{N}}$ can be extended to an internal sequence $(F_n)_{n\in{}^*\mathbb{N}}$ such that F_n is S-Dunford integrable for all $n \in {}^*\mathbb{N}$*

Proof: First we will prove the "only if" part. By the hypothesis, there exists a standard function $g : \mathbb{B}' \times \mathbb{N} \longrightarrow \mathbb{N}$ such that for all $n \in \mathbb{N}$ and all $\varphi \in \mathbb{B}'$

$$\forall k \geq g(\varphi, n) \forall l \in \mathbb{N} \int_{\{\|\varphi\circ F_l\| \geq k\}} \|\varphi \circ F_l\| d\mu < n^{-1}.$$

By saturation, $(F_n)_{n\in\mathbb{N}}$ can be extended to an internal sequence $(F_n)_{n\in{}^*\mathbb{N}}$ such that all F_n are $\mathcal{A}$-measurable and for all $n \in \mathbb{N}$ and all $\varphi \in \mathbb{B}'$

$$\forall k \geq {}^*g(\varphi, n), k \in {}^*\mathbb{N}, \forall l \in {}^*\mathbb{N} \int_{\{\|\varphi\circ F_l\| \geq k\}} \|\varphi \circ F_l\| d\mu < n^{-1}.$$

Hence, if $k \in {}^*\mathbb{N} \setminus \mathbb{N}$, then $k \geq {}^*g(\varphi, n)$ for all $n \in \mathbb{N}$ and $\varphi \in \mathbb{B}'$, thus all $F_l, l \in {}^*\mathbb{N}$, are S-Dunford integrable.

To prove the "if part", assume that the assertion is wrong. Then there exist $\varphi \in \mathbb{B}'$, a standard $\epsilon > 0$ and arbitrary large $k \in \mathbb{N}$ such that

$$\exists n \in {}^*\mathbb{N} \int_{\{\|\varphi \circ F_n\| \geq k\}} \|\varphi \circ F_n\| d\mu \geq \epsilon.$$

By saturation there exists $K \in {}^*\mathbb{N} \setminus \mathbb{N}$ such that

$$\exists n \in {}^*\mathbb{N} \int_{\{\|\varphi \circ F_n\| \geq K\}} \|\varphi \circ F_n\| d\mu \geq \epsilon,$$

which contradicts the hypothesis that $\varphi \circ F_n$ is S-Bochner integrable for all $n \in {}^*\mathbb{N}$. ∎

Let $(f_n)_{n \in \mathbb{N}}$ be a sequence of Pettis measurable functions $f_n : \Omega \longrightarrow \hat{\mathbb{B}}$. We say that (f_n) converges to $f : \Omega \longrightarrow \hat{\mathbb{B}}$ **in measure**, if $(\hat{\varphi} \circ f_n)$ converges to $\varphi \circ f$ in measure. Note that f is then Pettis measurable.

Using Theorem 6.2 and Lemma 7.1 we obtain immediately

Proposition 7.2 *Suppose that $\hat{\mathbb{B}}$ is weakly complete. If (f_n) is a sequence of uniformly Pettis integrable functions $f_n : \Omega \longrightarrow \hat{\mathbb{B}}$, converging to $f : \Omega \longrightarrow \hat{\mathbb{B}}$ in measure, then f is Pettis integrable and for all $B \in L(\mathcal{A})$*

$$\mathcal{P} - \int_B f d\hat{\mu} = \lim \mathcal{P} - \int_B f_n d\hat{\mu} \text{ in the weak topology.}$$

In the remainder of this paper we make the following assumptions:

L is an internal *convex subset of *$\mathbb{B}$, containing only n̂earstandard points and the origin 0. Let $(\Omega, \mathcal{A}, \mu)$ be an internal probability space, where Ω is a *finite set. Because then $\mathcal{A}$ is also *finite, we obtain for all $\omega \in \Omega$

$$[\omega] := \bigcap \{A \in \mathcal{A} | \omega \in A\} \in \mathcal{A}.$$

$[\omega]$ is called the **atom** of ω. Then the measure μ results from an $\mathcal{A}$-measurable weighted counting measure c_μ, given by

$$c_\mu(\omega) := \mu([\omega]) \cdot |[\omega]|^{-1} \text{for all } \omega \in \Omega,$$

where $|\cdot|$ denotes the internal *finite cardinality of $\cdot$. It is easy to see —write A as the disjoint union of all atoms of A— that

$$\mu(A) = \sum_{\omega \in A} c_\mu(\omega) \text{ for all } A \in \mathcal{A}.$$

Let us identify μ and c_μ. Then

$$\mu(A) = \sum_{\omega \in A} \mu(\omega) \text{ and } \int_A F d\mu = \sum_{\omega \in A} F(\omega) \cdot \mu(\omega) \text{ for all } A \in \mathcal{A},$$

where F is an $\mathcal{A}$-measurable function on Ω with value in $^*\mathbb{B}$ or $^*\mathbb{D}$.

(δ) Keisler's Fubini Theorem

Assume that $\mathcal{A} = {}^*\mathcal{P}(\Omega)$ (so $L(\mathcal{A}) = L(\Omega)$), and suppose that $(\Lambda, L(\Lambda), \hat{\nu})$ is a second Loeb space over an internal probability space $(\Lambda, {}^*\mathcal{P}(\Lambda), \nu)$ with Λ *finite. Define

$$\mu \cdot \nu := \sum_{(\omega,\lambda) \in A} \mu(\omega) \cdot \nu(\lambda) \text{ for all internal subsets } A \subset \Omega \times \Lambda.$$

R. A. Anderson [2] has shown that the Loeb product $(\Omega \times \Lambda, L(\Omega \times \Lambda), \widehat{\mu \times \lambda})$ of Ω and Λ is an extension of the completion $(\Omega \times \Lambda, L(\Omega \times \Lambda), \hat{\mu} \otimes \hat{\nu})$ of the product of Ω and Λ in the standard sense. Because this extension may be strict —examples were given by D. Norman and D. H. Hoover— standard Fubini Theorem cannot be applied to the Loeb product. So, H. J. Keisler [9] proved a Fubini Theorem for the Loeb product. The next Proposition is a vector valued version of Keisler's result.

Proposition 7.3 *Suppose that $f : \Omega \times \Lambda \longrightarrow \hat{L}$ is Pettis integrable w. r. to $\widehat{\mu \cdot \nu}$. Then there exists a Pettis integrable function $h : \Omega \longrightarrow \hat{L}$ such that for all $\varphi \in \mathbb{B}'$*

$$\hat{\varphi} \circ h(\cdot) = \mathcal{B} - \int_\Lambda \hat{\varphi} \circ f(\cdot, \lambda) d\hat{\nu}(\lambda) \;\; \hat{\mu}\text{-a. e.} \qquad (\hat{\mu}\text{-a. e. depends on } \varphi)$$

and

$$\mathcal{P} - \int_\Omega h d\hat{\mu} = \mathcal{P} - \int_{\Omega \times \Lambda} f d\widehat{\mu \cdot \nu}.$$

Proof: By Theorem 6.2, there exists an S-Pettis integrable weak lifting $F : \Omega \times \Lambda \longrightarrow L$ of f. Since for all $\omega \in \Omega$

$$H(\omega) := \sum_{\lambda \in \Lambda} F(\omega, \lambda) \cdot \nu(\lambda)$$

is a *convex combination of elements $F(\omega, \lambda) \in L, H(\omega) \in L$. Hence, $h(\omega) := \overline{H(\omega)} \in \hat{L}$ is well defined. Fix $\varphi \in \mathbb{B}'$. Using Keisler's Fubini Theorem, it is easy to see that for $\hat{\mu}$-almost all $\omega \in \Omega, \varphi \circ F(\omega, \cdot)$ is an S-Bochner integrable lifting of $\hat{\varphi} \circ f(\omega, \cdot)$. Hence, by Proposition 5.1, for $\hat{\mu}$-almost all $\omega \in \Omega$

$$\hat{\varphi} \circ h(\omega) = {}^0\varphi \circ H(\omega) = \mathcal{B} - \int_\Lambda \hat{\varphi} \circ f(\omega, \lambda) d\hat{\nu}(\lambda).$$

It remains to prove that h is Pettis integrable: Fix $\varphi \in \mathbb{B}'$. Since $\varphi \circ F$ is S-Bochner integrable

$$\sum_{\omega \in A} \|\varphi \circ H(\omega)\| \cdot \mu(\omega) \leq \sum_{(\omega,\lambda) \in A \times \Lambda} \|\varphi \circ F(\omega, \lambda)\| \cdot \mu(\omega) \cdot \nu(\lambda) \begin{cases} \approx 0, \text{ if A is a } \mu\text{-nullset} \\ \text{is limited in any case.} \end{cases}$$

Hence, $\varphi \circ H$ is S-Bochner integrable and of course a lifting of $\hat{\varphi} \circ h$. Since

$$\int_A H d\mu = \sum_{\omega \in A} H(\omega) \cdot \mu(\omega) \in L, \int_A H d\mu \text{ is } \hat{\text{n}}\text{earstandard.}$$

By Theorem 6.2, h is Pettis integrable and we obtain

$$\mathcal{P} - \int_{\Omega \times \Lambda} f d\widehat{\mu \times \nu} \approx_w \sum_{(\omega,\lambda) \in \Omega \times \Lambda} F(\omega, \lambda) \cdot \mu(\omega) \cdot \nu(\lambda) \approx_w \mathcal{P} - \int_\Omega h d\hat{\mu}. \blacksquare$$

(ϵ) Existence of conditional expectation

Stochastic filtrations, which result from internal ones, play an important role in the nonstandard approach to stochastic analysis. Se e for example [1] or [9]. We assume now that $\mu(A) \neq 0$ for all $A \in \mathcal{A}$.

Proposition 7.4 *Let $f : \Omega \longrightarrow \hat{L}$ be Pettis integrable and let $\mathcal{B}$ be an internal subalgebra of $\mathcal{A}$. Then there exists a Pettis integrable function $g : \Omega \longrightarrow \hat{L}$ w. r. to $(L(\mathcal{B}), \hat{\mu})$ such that*

$$\mathcal{P} - \int_B f d\hat{\mu} = \mathcal{P} - \int_B g d\hat{\mu} \text{ for all } B \in L(\mathcal{B}).$$

Proof: By Theorem 6.2, f has an S-Pettis integrable weak lifting $F : \Omega \longrightarrow L$. Recall the definition of the atom $[\omega]$ of ω and define

$$G(\omega) := \int_{[\omega]} F d\mu \cdot \mu([\omega])^{-1} \text{ for all } \omega \in \Omega.$$

Since L is *convex, $G(\omega) \in L$ for all $\omega \in \Omega$. It is easy to see that for all $A \in \mathcal{B}$

$$\sum_{\omega \in A} G(\omega) \cdot \mu(\omega) = \sum_{\omega \in A} F(\omega) \cdot \mu(\omega).$$

Moreover, G is $\mathcal{B}$-measurable. It follows that G is S-Dunford integrable and a weak lifting of $g : \Omega \longrightarrow \hat{L}$, where g is defined by $g(\omega) := \overline{G(\omega)}$. Hence, by Proposition 6.1, g is Dunford integrable. Since $\sum_{\omega \in A} G(\omega) \cdot \mu(\omega)$ is n̂earstandard for all $A \in \mathcal{B}$, by Theorem 6.2, g is Pettis integrable w. r. to $(L(\mathcal{B}), \hat{\mu})$ and we obtain for all $B \in L(\mathcal{B})$ and all μ-approximations $A \in \mathcal{B}$ of B

$$\mathcal{P} - \int_B f d\hat{\mu} \approx_w \sum_{\omega \in A} F(\omega) \cdot \mu(\omega) = \sum_{\omega \in A} G(\omega) \cdot \mu(\omega) \approx_w \mathcal{P} - \int_B g d\hat{\mu}. \blacksquare$$

In the standard theory of Pettis integrability, W. R. Pestman [14] proved the existence of conditional expectation for Pettis integrable functions with values in metrizable convex compact sets and extended this result to totally summable functions.

A proof of the existence of conditional expectations with values in measure convex sets can be found in the lecture notes of G. Winkler [18].

(ζ) Nondeterministic Peano Caratheodory integral equations

In this application, we use the well known close relationship between Loeb and Lebesgue measure. Let H be an infinite natural number. Define

$$\Delta t := H^{-1}, T := \{0, \Delta t, 2 \cdot \Delta t, \ldots, 1\},$$

$$\nu(A) := |A| \cdot \Delta t \text{ for all internal } A \subset T.$$

$(T, L(T), \hat{\nu})$ denotes the Loeb space over $(T, {}^*\mathcal{P}(T), \nu)$.

The next result is due to H. J. Keisler [9]. Here λ denotes Lebesgue-measure and Leb([0, 1]) the Lebesgue-measurable subsets of [0, 1]).

Theorem 7.5 *A function $g : \Omega \times [0,1] \longrightarrow \mathbb{D}$ is $L(\Omega)\bar{\otimes}Leb([0,1])-$measurable, if and only if $\tilde{g} : (\omega, t) \longmapsto g(\omega, {}^0t)$ is $L(\Omega \times T)-$measurable.*

A function $f : \Omega \times [0,1] \longrightarrow \hat{\mathbb{B}}$ is called a **Pettis integrable process**, if for all $\varphi \in \mathbb{B}'$ and $\hat{\mu}-$almost all $\omega \in \Omega$ ($\hat{\mu}-$a. a. depends again on φ)

$$\hat{\varphi} \circ f(\omega, \cdot) \text{ is Bochner integrable}$$

and if there exists a function $x : \Omega \times [0,1] \longrightarrow \hat{\mathbb{B}}$ such that for all $\varphi \in \mathbb{B}'$

$$\hat{\varphi} \circ x(\omega, \cdot) = \int_0^{(\cdot)} \hat{\varphi} \circ f(\omega, s) d\lambda(s) \text{ for } \hat{\mu} - \text{a lmost all } \omega \in \Omega.$$

The mapping $x = x(\circ, \cdot)$ is then called the **Pettis integral process** of f, written

$$x = \mathcal{P} - \int_0^{(\cdot)} f(\circ, s) d\lambda(s).$$

The Pettis integral process of f is only unique w. r. to weak equivalence. Two functions $x, y : \Omega \times [0,1] \longrightarrow \hat{\mathbb{B}}$ are called **weakly equivalent**, if for all $\varphi \in \mathbb{B}'$

$$\hat{\varphi} \circ x(\omega, \cdot) = \hat{\varphi} \circ y(\omega, \cdot) \text{ for } \hat{\mu} - \text{almost all } \omega \in \Omega.$$

For the next proposition, fix $\psi \in \mathbb{B}'$ and assume that there exists a closed subset S of $\mathbb{D}$ such that

$$\psi[L] \subset {}^*S, \text{ so } \hat{\psi}[\hat{L}] \subset S.$$

Because we are able to write the Pettis integral as a sum, we may apply a method, developed by H. J. Keisler, and used by many authors in order to solve differential equations —see for example [9]—, to handle integral equations of the form below. The method is working as follows:

First represent the differential equation by a *finite difference equation. In this step, it is crucial to characterize standard measurable and integrable functions by objects which behave like finite entities. Here we use the preceding lifting theorems.

The second step may be the simple part of that method: Solve the difference equation.

In the third step show that this solution is —up to an infinitesimal error— a standard solution of the equation, we have started from.

By means of the preceding lifting results this method may be applied to the following

Proposition 7.6 *Suppose that*

$$f : \Omega \times [0,1] \times S \longrightarrow \hat{L} \text{ and}$$
$$g : \Omega \times [0,1] \longrightarrow \hat{L}$$

and that

(a) f is Pettis measurable w. r. to $(L(\Omega)\bar{\otimes}Leb[0,1], \hat{\mu} \otimes \lambda)$ and

(b) g is Dunford integrable w. r. to $(L(\Omega)\bar{\otimes}Leb[0,1], \hat{\mu} \otimes \lambda)$ and

(c) for all $\varphi \in \mathrm{IB}'$ and $\hat{\mu}$-almost all $\omega \in \Omega$ ($\hat{\mu}$-a. a. depends on φ)

$$\sup_{s \in S} \|\hat{\varphi} \circ f(\omega, \cdot, s)\| \leq \|\hat{\varphi} \circ g(\omega, \cdot)\|.$$

The assertion is: There exists a Pettis measurable function (w. r. to $(L(\Omega), \hat{\mu})$) $x : \Omega \times [0,1] \longrightarrow \hat{L}$ such that

$$x(\omega, t) = \mathcal{P} - \int_0^t f(\omega, s, \hat{\psi} \circ x(\omega, s)) d\lambda(s),$$

i. e., x is the Pettis integral process of $f(\circ, \cdot, \hat{\psi} \circ x(\circ, \cdot))$.

References

[1] S. Albeverio, J. E. Fenstad, R. Høegh Krohn, T. L. Lindstrøm, *Nonstandard methods in stochastic analysis and mathematical physics;* Academic Press (1986)

[2] R. A. Anderson, *A nonstandard representation of Brownian motion and Ito-integration;* Israel J. Math. 15 (1976)

[3] N. J. Cutland, *Nonstandard measure theory and its application;* Bull. London Math. Soc. 15 (1983)

[4] J. Diestel, J. Uhl, *Progress in vector measures 1977 - 83;* Lecture notes in Math. 1033 (1983)

[5] R. F. Geitz, *Pettis integration;* Proc. Amer. Math. Soc. 82 (1983)

[6] S. Heinrich, *Ultrapowers of locally convex spaces;* Mathematische Nachrichten 118 (1984)

[7] C. W. Henson, L. C. Moore, *The nonstandard theory of topological vector spaces;* Trans. Amer. Math. Soc. 172 (1972)

[8] A. E. Hurd, P. A. Loeb, *An introduction to nonstandard real analysis;* Academic Press (1985)

[9] H. J. Keisler, *An infinitesimal approach to stochastic analysis;* Memoirs Amer. Math Soc. (1984)

[10] P. A. Loeb, *Conversion from standard to nonstandard measure spaces and applications to probability theory;* Trans. Amer. Mat. Soc. 211 (1975)

[11] P. A. Loeb, *A functional approach to nonstandard measure theory;* Contemporary Math. 26 (1984)

[12] P. A. Loeb, H. Osswald, *Nonstandard integration theory on topological vector spaces;* to appear

[13] W. A. J. Luxemburg, *A general theory of monads;* Applications of model theory to algebra, analysis and probability, Holt Rinehart and Winston, New York (1969)

[14] W. R. Pestman, *Convergence of martingales with values in locally convex Suslin spaces;* Math. Institut, Rijsuniversiteit Groningen, preprint (1981)

[15] B. J. Pettis, *On integration on vector spaces;* Trans. Amer. Math. Soc. 44 (1936)

[16] D. A. Ross, *Lifting theorems in nonstandard measure theory;* Proc. Amer. Math. Soc. 109 (1990)

[17] M. Talagrand, *Pettis integrability;* Memoirs Amer. Math. Soc (1984)

[18] G. Winkler, *Choquet order and simplices;* Springer Lecture notes in Math. 1145 (1986) .

A COUNTEREXAMPLE TO THE SPECTRAL MAPPING THEOREM REVISITED FROM A NONSTANDARD POINT OF VIEW

H. Ploss
Zentrum für Fernstudien der J. K. Universität Linz
Belruptstr. 10 A – 6900 Bregenz

Abstract

We show that a classical counterexample to the spectral mapping theorem for the exponential function is not very surprising if we consider the example in a suitable nonstandard setting.

Keywords: Spectral mapping theorem, C_0–semigroups

1 The Counterexample

We recall briefly the example of *Zabcyk* and refer to [5], [1] and [3]. [3] is also a general reference for one-parameter semigroups of operators.

For every $n \in \mathbb{N}$ consider the n-dimensional Hilbert space $\mathbb{C}^n$ and operators $A_n \in L(\mathbb{C}^n)$ defined by the matrices

$$A_n = \begin{pmatrix} 0 & 1 & \cdots & & 0 \\ 0 & 0 & 1 & \cdots & 0 \\ \vdots & & & \ddots & \\ 0 & & \cdots & & 1 \\ 0 & & \cdots & & 0 \end{pmatrix}$$

Then $\sigma(A_n) = \{0\}$, $\|A_n\| = 1$, ${A_n}^n = 0$.

Consider now the Hilbert space $E := \bigoplus_{n\in\mathbb{N}} \mathbb{C}^n$ and the operator $A := \bigoplus_{n\in\mathbb{N}} (A_n + 2\pi i n)$ with maximal domain in E.

Analogously we define $T = \big(T(t)\big)_{t\geq 0}$ by

$$T(t) := \bigoplus_{n\in\mathbb{N}} \left(e^{2\pi i n t} \exp(t\, A_n)\right).$$

Since $\|\exp(t\,A_n)\| \leq e^t$ for every $n \in \mathbb{N}$, $t \geq 0$ and since $t \longmapsto T(t)\,x$ is continuous on each component $\mathbb{C}^n$ it follows that T is a strongly continuous semigroup of operators. Its generator is the operator A as defined above.

The spectrum of A and the resolvent of A are given by

$$\begin{aligned} \sigma(A) &= \{2\pi i n : n \in \mathbb{N}\} \quad \text{and} \\ \mathcal{R}(\lambda, A) &= \bigoplus_{n\in\mathbb{N}} \mathcal{R}(\lambda, A_n + 2\pi i n),\ \lambda \in \varrho(A) = \mathbb{C} \setminus \sigma(A) \end{aligned}$$

respectively.

Therefore for the spectral bound $s(A) := \sup\{\text{Re}\lambda : \lambda \in \sigma(A)\}$ of the generator A we have $s(A) = 0$. On the other hand one can compute that the growth bound

$$\omega(T) := \inf\{w \in \mathbb{R} : \|T(t)\| \leq M_w\,e^{wt},\ \text{for all } t \geq 0 \text{ and suitable } M_w\}$$

of the semigroup T is in the example given by $\omega(T) = 1$.

From this it follows (e.g.[3]) that the *spectral mapping theorem* (SMT)

$$\sigma\big(T(t)\big) \setminus \{0\} = \exp\big(t.\sigma(A)\big) = \{e^{\lambda t} : \lambda \in \sigma(A)\}$$

cannot be true in the example of *Zabcyk*. We notice that in [1] the spectra of both the generator and the semigroup are computed explicitly by means of

Gearhart's Theorem: Let $T = \big(T(t)\big)_{t\geq 0}$ be a semigroup with generator A in a Hilbert space. Then $e^\lambda \in \varrho\big(T(1)\big)$ iff

i) $\lambda + 2\pi i n \in \varrho(A)$ for all $n \in \mathbb{Z}$, and

ii) $\sup\big\{\|\mathcal{R}(\lambda + 2\pi i n, A)\| : n \in \mathbb{Z}\big\} < \infty$

In particular: $\{e^\lambda\colon |\lambda| \leq 1\} = \sigma\big(T(1)\big)$ is shown in [1].

2 The example from a nonstandard point of view

As a general reference for our nonstandard terminology we refer to [L]. For $n \in {}^*\mathbb{N}$ denotes $l_2(n)$ the space of all internal sequences $x_1, \ldots, x_n$ of hypercomplex numbers given the metric

$$d(x,y) = \|x - y\|_2 = \Big(\sum_{i=1}^{n} |x_i - y_i|^2\Big)^{\frac{1}{2}}$$

We fix $\omega \in {}^*\mathbb{N}_\infty$ and define

$$X := \bigoplus_{n=1}^{\omega} l_2(n) \qquad \Big(\equiv l_2(\Omega) \text{ with } \Omega = \tfrac{\omega(\omega+1)}{2} \Big)$$

For every n, $1 \leq n \leq \omega$, consider the internal linear (over ${}^{s}tar\mathbb{C}$) operators $A_n : l_2(n) \longrightarrow l_2(n)$ defined by matrices as in *part* 1. and define

$$A := \bigoplus_{n=1}^{\omega} (A_n + 2\pi i n).$$

(The internal operator $A : X \longrightarrow X$ can be regarded as an $\Omega \times \Omega$-matrix.)

For $\lambda \in {}^{s}tar\mathbb{C} \setminus \{2\pi i n : 1 \leq n \leq \omega\}$ there exists

$$\begin{aligned} \mathcal{R}(\lambda, A) &:= (\lambda - A)^{-1} \qquad \text{given by} \\ \mathcal{R}(\lambda, A) &= \bigoplus_{n=1}^{\omega} \mathcal{R}(\lambda, A_n + 2\pi i n) \qquad \text{with} \end{aligned}$$

$$\mathcal{R}(\lambda, A_n + 2\pi i n) = \mathcal{R}(\lambda - 2\pi i n, A_n) = \sum_{k=0}^{n-1} (\lambda - 2\pi i n)^{-k-1} {A_n}^k, \; 1 \leq n \leq \omega.$$

By means of an easy estimation we see that for λ with $\operatorname{st}(\operatorname{Re}\lambda) > 1$, $\mathcal{R}(\lambda, A)$ is bounded.

But we claim that for $n \in {}^{*}\mathbb{N}_{\infty}$, $n \leq \omega$ and λ satisfying $|\lambda - 2\pi i n| \leq 1$ ($\lambda \neq 2\pi i n$) $\mathcal{R}(\lambda, A)$ is unbounded.

To verify this consider $e_n = (0, \ldots, 1)^T \in l_2(n)$ or e_n embedded in X respectively. Then with $\mu := \lambda - 2\pi i n$ we have

$$\|\mathcal{R}(\lambda, A)\, e_n\| = \|\mathcal{R}(\mu, A_n)\, e_n\| = \Big(\sum_{k=1}^{n} |\mu|^{-2k} \Big)^{\frac{1}{2}}.$$

Therefore it is natural to say that $\lambda = 2\pi i n + \mu$, for $n \in {}^{*}\mathbb{N}_{\infty}$, $n \leq \omega$, $|\mu| \leq 1$ is a *spectral value of* A and consequently the *spectral bound of* A is equal to 1.

For $t \in {}^{*}\mathbb{R}_{+}$ let

$$T_n(t) := \exp(2\pi i n\, t) \sum_{k=0}^{n-1} \frac{{A_n}^k t^k}{k!}, \qquad 1 \leq n \leq \omega.$$

$$\text{Then} \quad \|T_n(t)\| \leq \sum_{k=0}^{n-1} \frac{\|A_n\|^k t^k}{k!} \leq \sum_{k=0}^{\omega} \frac{t^k}{k!} \approx \mathrm{e}^t.$$

Consider $T(t) := \bigoplus_{n=1}^{\omega} T_n(t)$.

Because of $\|T(t)\| = \max\{\|T_n(t)\| : 1 \leq n \leq \omega\}$, $T(t) : X \longrightarrow X$ is for finite t a bounded operator. Clearly $\big(T(t)\big)_{t \in {}^{s}tar\mathbb{R}_{+}}$ inherits the semigroup property

$$\begin{aligned} T(0) &= \mathrm{Id}_X, \\ T(t+s) &= T(t)\, T(s). \end{aligned}$$

For simplicity we choose $t = 1$. Then by direct computation $\big[$ or using the formula $\left(e^{\lambda t} - T(t)\right) f = (\lambda - A) \int_0^t e^{\lambda(t-s)} T(s) f\, ds$, f in the underlying Banach space, and a transfer argument $\big]$ we conclude that with $\lambda = 2\pi i n + \mu$

$$\|T(1)\, \mathcal{R}(\lambda, A)\, e_n - e^{\mu}\, \mathcal{R}(\lambda, A)\, e_n\| \quad \text{is finite, } 1 \leq n \leq \omega.$$

Now let $f_n := \dfrac{\mathcal{R}(\lambda, A)\, e_n}{\|\mathcal{R}(\lambda, A)\, e_n\|}$. Then $\|f_n\| = 1$ and for $n \in {}^*\mathbb{N}_\infty$, $n \leq \omega$, $|\mu| \leq 1$

$$\|T(1)\, f_n - e^{\mu}\, f_n\| \approx 0.$$

Therefore given $\varepsilon \in \mathbb{R}_+$, there exists $N \in \mathbb{N}$ such that

$$(*) \qquad \|T(1)\, f_n - e^{\mu}\, f_n\| \leq \varepsilon\,, \text{ for } N \leq n \leq \omega,\ |\mu| \leq 1.$$

In order to get a standard setting let S be the set of all $x_1, x_2, \ldots, x_m, 0, \ldots, 0$ where both m and all x_i's are finite and take the *nonstandard hull*

$$\left(\operatorname{pns}(X, S)/_{\approx}\,,\ \widehat{d}\,\right)$$

which is isomorphic to l_2 (see e.g. [L, III,3]).

Because of $T(t) : \operatorname{pns}(X, S) \longrightarrow \operatorname{pns}(X, S)$, for finite $t \in {}^*\mathbb{R}_+$, we can define $\widehat{T}(t)\colon l_2 \longrightarrow l_2$ in the usual way. Moreover it is easy to see that

$$T(t)\, x \approx x, \text{ for } x \in \operatorname{pns}(X, S) \text{ and } t \text{ positive infinitesimal.}$$

From this it follows that $\left(\widehat{T}(t)\right)_{t \geq 0}$ is a strongly continuous semigroup on l_2, see [4].

Clearly this semigroup coincides with the semigroup in the example of *Zabcyk*. Using $(*)$ we conclude that for $|\mu| \leq 1$

$$e^{\mu} \in a\sigma\left(\widehat{T}(1)\right) \qquad \text{(approximate point spectrum).}$$

That means we have verified the result in [1], but we did not need *Gearhart's* Theorem.

3 Final Remark

We have shown that from a nonstandard point of view the example of *Zabcyk* is not *really* a counterexample to the SMT. We have only to approximate the standard setting from above. In this way we find the points in the *spectrum of the generator* in the extended complex plane, which – via the exponential function – contribute to the spectrum of the semigroup. We are working to transfer these arguments into a more general situation.

References

[1] Hejtmanek, J., Kaper, H.: 1986, "Counterexample to the Spectral Mapping Theorem for the Exponential Function", *Proc. Amer. Math. Soc.* Vol. no. **96**, pp. 563–568.

[2] Lindstrøm, T.: 1988, "An Invitation to Nonstandard Analysis", in Cutland, N. (ed): "Nonstandard Analysis and its Applications", *London Math. Soc. Stud. Texts* Vol. no. **10**, pp. 1–105

[3] Nagel, R.: 1985, "One–Parameter Semigroups of Positive Operators", *Lect. Notes Math.* Vol. no. **1184**, Springer, Berlin–Heidelberg– New York

[4] Wolff, M.: 1984, "Spectral Theory of Group Representations and their Nonstandard Hulls", *Israel. J. Math.* Vol. no. **48**, pp. 205–224.

[5] Zabcyk, J.: 1975, "Note on C_o-Semigroups", *Bull. Acad. Polon. Sci.* Vol. no. **23**, pp. 895–898.

Nonstandard polynomials in several variables

Hermann Render
Department of Mathematics, University of Duisburg, Lotharstr. 65
D-47057 Duisburg, Germany

Abstract

C. Impens has investigated nonstandard polynomials in one variable. In this note we show that many of his results carry over to polynomials in several variables. The concept of complex and real microcontinuity of nonstandard polynomials is intimately related to polynomial approximation in function theory.

1 Introduction

In his book "Non-standard Analysis" A. Robinson has given a very nice discussion of *-holomorphic functions of one complex variable. Of fundamental importance is the concept of *S-continuity* or *complex microcontinuity*: If U is an open subset of $\mathbb{C}^n$ then an internal function $f: {}^*U \to {}^*\mathbb{C}$ is called *S-continuous* or *complex microcontinuous* at $w \in U$ if $z \approx w$ implies $f(z) \approx f(w)$. Let fin ${}^*\mathbb{C}$ be the set of all finite numbers of ${}^*\mathbb{C}$. Then the set

$$S_f := \{w \in U : f(w) \in \text{fin}\,{}^*\mathbb{C} \text{ and } f \text{ is complex microcontinuous at } w\} \quad (1)$$

is called the *set of standard complex microcontinuity*. It is well-known that the standard part $st\, f: S_f \to \mathbb{C}$ of the internal function f, defined by $st\, f(w) := st\,(f(w))$, is a continuous function; in the last equality st is the standard part map of the Hausdorff space $\mathbb{C}$. We denote the set of all holomorphic functions on an open subset U of $\mathbb{C}^n$ by $H(U)$, see [7] for the precise definition. An element $f \in {}^*H(U)$ is called a *-holomorphic function. In analogy to results due to A. Robinson and C. Impens in the one-dimensional case we prove that a function $f \in {}^*H(U)$ is complex microcontinuous at $w \in U$ satisfying $f(w) \in \text{fin}\,{}^*\mathbb{C}$ iff $f(z) \in \text{fin}\,{}^*\mathbb{C}$ for all $z \approx w$. If $Q: {}^*\mathbb{C}^n \to {}^*\mathbb{C}$ is a nonstandard polynomial (for definition see section 2) then $Q(z) = \sum_{\nu \in E} a_\nu (z-w)^\nu$ is complex microcontinuous at $w \in \mathbb{C}^n$ iff $|a_\nu|^{1/|\nu|} \in \text{fin}\,{}^*\mathbb{C}$ for all $\nu \in E$. At a first glance one may be tempted to consider the set of standard complex microcontinuity as the nonstandard counterpart of the *domain of convergence* C_P of a power series $P(z)$, i.e. the interior of the set of all $z \in \mathbb{C}^n$ such that the power series $P(z)$ converges. The following simple example shows that this is not the case: Let $p(z) = z(z-3)$ and $Q(z) = p(z)^N$ for some infinitely large $N \in {}^*\mathbb{N}$. Using Theorem 2.3 $(ii) \Rightarrow (i)$ it is easy to see that S_Q consists of all $z \in \mathbb{C}^n$ with $|st\, p(z)| < 1$. Then 0 and 3 are contained in S_Q which is also a subset of $B_1(0) \cup B_1(3)$, where $B_r(w) := \{z \in \mathbb{C} : |z - w| < r\}$. In particular S_Q is not connected.

The main difference between nonstandard polynomials and power series can be expressed in the following way: complex microcontinuity at a point represents like a power series local holomorphy but complex microcontinuity on the set S_Q represents a special kind of polynomial approximation of the standard part $st\, Q$.

The paper is divided in four sections. In the second section we generalize results about complex microcontinuity to the case of several variables. In the next section we consider hyperfinite truncations of power series (in one variable): the domain of convergence is the intersection over all hyperfinite truncations of their standard sets of complex microcontinuity. Moreover we give a standard description of the union of all S_Q where Q runs over all truncations of a power series. The last section is devoted to the question of the structure of S_Q for a nonstandard polynomial in one variable. We give an easier proof of a result of C. Impens that every simply connected open set $U \subset \mathbb{C}$ is the set of standard complex microcontinuity for an appropriate nonstandard polynomial. Using the theorem of Mergelyan we construct a plenty of certain nonstandard polynomials which are microcontinuous on a subset ${}^*K$ where K is a rather general compact subset of $\mathbb{C}$.

Finally we mention some terminology: $\mathbb{R}_+$ denotes the set of all positive real numbers. By $[a, b]$ we mean the set of all real numbers x with $a \leq x \leq b$. The set of all infinitely large hypernatural numbers is denoted by ${}^*\mathbb{N}_\infty$. We say that $z \in {}^*\mathbb{C}^n$ is *infinitesimal near* to $w \in \mathbb{C}^n$, briefly $z \approx x$, if $|z - w| < \epsilon$ for all $\epsilon \in \mathbb{R}_+$. The set $ns\,{}^*U$ of all *nearstandard points* is the union of all *monads* $m(w) := \{z \in {}^*U : z \approx w\}$ with $w \in U$.

2 Complex microcontinuity of nonstandard polynomials

Let $\mathbb{N}$ be the natural numbers including zero and let $n \in \mathbb{N}$ be a fixed natural number greater equal than 1. A *multi-index* $\nu = (\nu_1, \ldots, \nu_n)$ is just an element in $\mathbb{N}^n$ and as usual we define $|\nu| := \nu_1 + \ldots + \nu_n$. It is convenient to embed $\mathbb{N}$ into $\mathbb{N}^n$ by the formula $k \mapsto (k, \ldots, k)$. We can introduce a partial order on $\mathbb{N}^n$ if we define $\mu \leq \nu$ to mean that $\mu_i \leq \nu_i$ for $i = 1, \ldots, n$. An element $\nu \in {}^*\mathbb{N}^n$ is called *infinitely large* if ${}^*\mu\,{}^*\!\leq \nu$ for all $\mu \in \mathbb{N}^n$. The set of all infinitely large multi-indices is equal to ${}^*\mathbb{N}_\infty \times \ldots {}^*\mathbb{N}_\infty$, or more briefly, to ${}^*\mathbb{N}_\infty^n$. Let $z = (z_1, \ldots, z_n) \in \mathbb{C}^n, w = (w_1, \ldots, w_n) \in \mathbb{C}^n$ and $\nu = (\nu_1, \ldots, \nu_n)$ a multi-index. If $a_\nu \in \mathbb{C}$ and $a_\nu \neq 0$ then $a_\nu(z - w)^\nu := a_\nu(z_1 - w_1)^{\nu_1} \cdot \ldots \cdot (z_n - w_n)^{\nu_n}$ is called a *monomial*. A *polynomial* (in n variables) is a function $Q: \mathbb{C}^n \to \mathbb{C}$ of the form $Q(z) = \sum_{\nu \in E} a_\nu z^\nu$ where E is a finite subset of $\mathbb{N}^\nu$ and $a_\nu \in \mathbb{C}$. The set of all polynomials in n variables is denoted by $P(\mathbb{C}^n)$. Every element in ${}^*P(\mathbb{C}^n)$ is called a *nonstandard polynomial.* It follows by the transfer principle that every nonstandard polynomial can be written in the form $Q(z) = \sum_{\nu \in F} b_\nu (z-w)^\nu$, where $w \in {}^*\mathbb{C}^n$ is given and F is a suitable hyperfinite subset of ${}^*\mathbb{N}^n$ and $b_\nu \in {}^*\mathbb{C}$.

Let $w = (w_1, \ldots, w_n) \in \mathbb{C}^n$ and $(r_1, \ldots, r_n) \in \mathbb{R}_+^n$. A *polydisc* $P_r(w)$ is a set of the form $B_{r_1}(w_1) \times \ldots \times B_{r_n}(w_n)$ where $B_{r_i}(w_i)$ are the open balls $\{a \in \mathbb{C} : |a - w_i| < r_i\}$ in the complex plane for $i = 1, \ldots n$. The topological closure of $P_r(w)$ is called the *closed polydisc* and is denoted by $\overline{P_r(w)}$. Note that $\overline{P_r(w)} = \{z \in \mathbb{C}^n : |z_i - w_i| \leq r_i \text{ for } i = 1, \ldots, n\}$. For $r \in \mathbb{R}_+$ we write $P_r(w)$ instead of $P_{(r,\ldots,r)}(w)$.

2.1 Definition. Let $Q(z) = \sum_{\nu \in E} a_\nu (z-w)^\nu$ be a nonstandard polynomial. We define $C_Q := C_Q(w) = \sup\{st|a_\nu|^{1/|\nu|} : \nu \in E \text{ with } |\nu| \in {}^*\mathbb{N}_\infty\}$ and we call $H_Q := 1/C_Q$ the Hadamard number (at the point w). In the case of $C_Q = 0$ we define $H_Q = \infty$.

In the last definition it is assumed that st is the standard part map with respect to the one-point compactification of the Hausdorff space $\mathbb{C}$. Hence, if some number $|a_\nu|^{1/\nu}$ (with $|\nu| \in {}^*\mathbb{N}_\infty$) is not contained in fin ${}^*\mathbb{C}$ then H_Q is equal to 0. The next result shows that nonstandard polynomials behave in a certain way like power series.

2.2 Theorem: *Let $Q(z) = \sum_{\nu \in E} a_\nu (z-w)^\nu$ be a nonstandard polynomial. If $a_{{}^*\mu} \in$ fin ${}^*\mathbb{C}$ for every $\mu \in E \cap \mathbb{N}^n$ then for every $0 < r < H_Q$ and for all $z \in {}^*P_r(w)$ the number $Q(z)$ is finite and*

$$st\, Q(z) = \sum_{\nu \in E \cap \mathbb{N}^n} st\, a_\nu \cdot st\,(z-w)^\nu. \tag{2}$$

Proof: For $H_Q = 0$ the statement is trivial. If $H_Q > 0$ then $C_Q < \infty$. Let $r < H_Q$ and choose $\theta \in \mathbb{R}$ with $C_Q < \theta < 1/r$. Let ${}^*\mathbb{N}_k := \{l \in {}^*\mathbb{N} : l \geq k\}$ for $k \in \mathbb{N}$. Then $\cap_{k=1}^\infty ({}^*\mathbb{N}_k \times {}^*\mathbb{N} \times \ldots \times {}^*\mathbb{N}) \cap E$ is contained in the internal set $\{\nu \in E : |a_\nu|^{1/|\nu|} \leq \theta\}$. By a saturation argument there exists $k \in \mathbb{N}$ such that $|a_\nu|^{1/|\nu|} \leq \theta$ for all $\nu = (\nu_1, \ldots, \nu_n) \in E$ with $k \leq \nu_1$. Using this argument for every component we obtain a finite set $F \subset E \cap \mathbb{N}^n$ such that $|a_\nu|^{1/|\nu|} \leq \theta$ for all $\nu \in E \backslash {}^*F$. We can assume that $E \backslash F$ is contained in the set $\{\nu \in E : k \leq \nu_i \text{ for all } i = 1, \ldots, n\}$ for a given large number $k \in \mathbb{N}$. Moreover there exists $\sigma \in {}^*\mathbb{N}^n$ with $\nu \leq \sigma$ for all $\nu \in E$. We are looking at the inequality

$$|Q(z) - \sum_{\nu \in F} a_\nu (z-w)^\nu| \leq \sum_{\nu \in E \backslash F} |a_\nu||(z-w)^\nu| \tag{3}$$

for $z = (z_1, \ldots, z_n) \in {}^*\mathbb{C}^n$ satisfying $|z_i - w_i| \leq r$ for $i = 1, \ldots, n$ and we obtain as an upper bound

$$\sum_{\nu\in E\setminus F} |a_\nu| r^{|\nu|} \leq \sum_{k\leq\nu\leq\sigma} (\theta\cdot r)^{|\nu|} \leq (\theta\cdot r)^k \sum_{\nu\leq\sigma} (\theta\cdot r)^{|\nu|}. \tag{4}$$

Since $\sum_{\nu\leq\sigma}(\theta\cdot r)^{|\nu|} \leq \prod_{i=1}^n(\sum_{\nu_i=0}^{\sigma_i}(\theta\cdot r)^{\nu_i}) = \prod_{i=1}^n \frac{(\theta r)^{\sigma_i+1}-1}{\theta r-1} \in \text{fin}\,^*\mathbb{C}$ it follows that $Q(z)$ is finite. Eq. 2 follows now from the equations 3 and 4 since we can choose $k\in\mathbb{N}$ arbitrary large. ∎

Let U be an open subset of $\mathbb{C}^n$ and $f\in H(U)$. Let $\gamma_i\colon[a,b]\to\mathbb{C}$ be continuous paths for $i=1,\ldots,n$ and set $\gamma:=(\gamma_1,\ldots,\gamma_n)\colon[a,b]^n\to\mathbb{C}^n$. We define the iterated integral over γ by the formula

$$\int_\gamma f(\xi)d\xi := \int_a^b\cdots\int_a^b f(\gamma((t_1,\ldots,t_n)))\cdot\gamma_1'(t_1)\cdot\ldots\cdot\gamma_n'(t_n)dt_1\ldots dt_n. \tag{5}$$

If $r\in\mathbb{R}_+^n$ and $w=(w_1,\ldots,w_n)\in\mathbb{C}^n$ we define $c_{r_i,w_i}(t):=w_i+r_ie^{it}$ and $c_{r,w}:=(c_{r_1,w_1},\ldots,c_{r_n,w_n})\colon[0,2\pi]^n\to\mathbb{C}^n$. Moreover we denote the function $z=(z_1,\ldots,z_n)\mapsto(z_1-w_1)\cdot\ldots\cdot(z_n-w_n)$ by $(z-w)^1$; hence we consider 1 as the multi-index $(1,\ldots,1)$. Using the Cauchy integral formula in the one-dimensional case it is not very hard to prove the well-known Cauchy integral formula for $f\in H(U)$: If $\overline{P_r(w)}$ is a subset of the open set $U\subset\mathbb{C}^n$ then for all $z\in P_r(w)$

$$f(z)=\frac{1}{(2\pi i)^n}\int_{c_{r,w}}\frac{f(\xi)}{(\xi-z)^1}d\xi \tag{6}$$

C. Impens has proved the equivalence of the following statements for nonstandard polynomials in one variable:

2.3 Theorem: *Let $Q(z)=\sum_{\nu\in E}a_\nu(z-w)^\nu$ be a nonstandard polynomial and $w\in\mathbb{C}^n$. Then the following statements are equivalent:*

(i) There exists $r\in\mathbb{R}_+$ such that $Q(z)\in\text{fin}\,^*\mathbb{C}$ and Q is complex microcontinuous for all $z\in P_r(w)$.

(ii) Q is complex microcontinuous at w and $Q(w)\in\text{fin}\,^*\mathbb{C}$.

(iii) $Q(z)\in\text{fin}\,^*\mathbb{C}$ for all $z\approx w$.

(iv) The nonstandard monomial $a_\nu(z-w)^\nu$ is complex microcontinuous at w for every $\nu\in E$.

(v) $|a_\nu|^{1/|\nu|}\in\text{fin}\,^*\mathbb{C}$ for all $\nu\in E$.

(vi) $\tilde{Q}(z):=\sum_{\nu\in E}|a_\nu|(z-w)^\nu$ is complex microcontinuous at w.

Proof: The equivalence of (i), (ii) and (iii) is valid for every element $f\in{}^*H(U)$: we have only to show $(iii)\Rightarrow(i)$. The underflow principle applied to the set $\{k\in{}^*\mathbb{N}:$

$(\forall u \in {}^*\overline{P_{\frac{1}{k}}(w)})\ (|f(u)| \leq k)\}$ yields $k \in \mathbb{N}$ such that $|f(u)| \leq k$ for all $u \in {}^*\overline{P_{\frac{1}{k}}(w)}$. Now let $r = (1/2k, \ldots, 1/2k)$ and $u \in {}^*P_r(w)$ and $v \in P_r(w)$ with $u \approx v$. Then the Cauchy integral formula shows that for $\rho := 2r$

$$|f(u) - f({}^*v)| \leq \frac{1}{(2\pi)^n} \Big| \int_{c_{\rho,w}} f(\xi) \cdot \frac{(\xi - {}^*v)^1 - (\xi - u)^1}{(\xi - u)^1 (\xi - {}^*v)^1} d\xi \Big|. \tag{7}$$

Since $|\xi_i - u_i| \geq 1/2k$ and $|\xi_i - {}^*v_i| \geq 1/2k$ and $|f(\xi)| \leq k$ on ${}^*\overline{P_\rho(w)}$ we obtain

$$|f(u) - f({}^*v)| \leq \frac{1}{(2\pi)^n} \cdot \frac{1}{(2k)^{2n}} \cdot k \cdot \int_{c_{\rho,w}} |(\xi - {}^*v)^1 - (\xi - u)^1| d\xi. \tag{8}$$

By Lemma 2.4 we know that $|(\xi - {}^*v)^1 - (\xi - u)^1| \leq \epsilon$ for all $\epsilon \in \mathbb{R}_+$. Hence $|f(u) - f({}^*v)| \approx 0$, i.e., that f is complex microcontinuous at $v \in P_r(w)$.

Now let us prove $(i) \Rightarrow (iv)$. Let $c_{r,w}$ be as above. By the transfer principle the meaning of $c_{r,w}$ for $r \in {}^*\mathbb{R}_+$ is clear. Then we can define a map $< \cdot, \cdot >: {}^*P(\mathbb{C}^n) \times {}^*P(\mathbb{C}^n) \to {}^*\mathbb{C}$ by defining:

$$< f, g > := \int_0^{2\pi} \cdots \int_0^{2\pi} f(c_{r,w}(t)) \cdot \overline{g(c_{r,w}(t))} dt, \tag{9}$$

where $t = (t_1, \ldots, t_n) \in {}^*[0, 2\pi]^n$. Moreover we have that

$$< (z - w)^\nu, (z - w)^\mu > = \begin{cases} 0, & \text{for } \nu \neq \mu \\ r^{2|\nu|}(2\pi)^n, & \text{for } \nu = \mu. \end{cases} \tag{10}$$

Define $h_\nu(z) := (z-w)^\nu a_\nu / (r^{|\nu|}(2\pi)^n |a_\nu|)$ for ν with $a_\nu \neq 0$, and $h_\nu(z) = 0$ otherwise. Then Eq. 10 means that the monomials h_ν are orthogonal and we obtain for $\nu \neq (0, \ldots, 0)$

$$|a_\nu| r^{|\nu|} = < Q(z) - a_0, h_\nu(z) > . \tag{11}$$

Obviously we have $|h_\nu(c_{r,w}(t))| \leq 1$, and for $r \approx 0$ it follows by complex microcontinuity at w that $Q(c_{r,w}(t)) - a_0 \approx 0$. Eq. 11 yields $|a_\nu| r^{|\nu|} \approx 0$ for all $r \approx 0$. Since $|a_\nu (z - w)^\nu| \leq |a_\nu| r^{|\nu|}$ with $r := \max\{|z_1 - w_1|, \ldots, |z_n - w_n|\}$ we obtain (iv). The implication $(iv) \Rightarrow (v)$ is very easy: if $M := |a_\nu|^{1/|\nu|}$ is infinitely large then $r := 1/M$ is infinitesimal near to zero. But $|a_\nu| \cdot r^{|\nu|} = 1$, a contradiction to (iv). For $(v) \Rightarrow (vi)$ let $\tilde{Q}(z)$ as in (vi). Since $I := \{|a_\nu|^{1/|\nu|} : \nu \in E\}$ is an internal set with $I \subset \text{fin}\,{}^*\mathbb{C}$ there exists $k \in \mathbb{N}$ such that I is contained in the interval ${}^*[0, k]$. Hence $C_Q < \infty$ and $H_Q = 1/C_Q > 0$. By Theorem 2.2 $\tilde{Q}(z)$ is finite for $z \approx w$. Hence $(iii) \Rightarrow (ii)$ shows that $\tilde{Q}(z)$ is complex microcontinuous at w. The last implication $(vi) \Rightarrow (iii)$ is easy: Let $z \approx w$ and $r := \max\{|z_i - w_i| : i = 1, \ldots, n\}$. Then $w + (r, \ldots, r) \approx w$ and $|Q(z)| \leq \sum_{\nu \in E} |a_\nu| (r, \ldots, r)^\nu = \tilde{Q}(w + (r, \ldots, r)) \approx \tilde{Q}(w)$. ∎

2.4 Lemma: *Let $\xi \in \text{fin}\,{}^*\mathbb{C}^n$ and $h_\xi: {}^*\mathbb{C}^n \to {}^*\mathbb{C}$ be defined by $h_\xi(v) := (\xi_1 - v_1) \cdot \ldots \cdot (\xi_n - v_n)$. Then h_ξ is complex microcontinuous at each $v \in \mathbb{C}^n$.*

Proof: Let $u \approx v \in \mathbb{C}^n$. Then $(\xi_i - u_i) \approx (\xi_i - v_i)$. Since $(\xi_i - u_i) \in \text{fin}\,{}^*\mathbb{C}$ we obtain $h_\xi(u) = \prod_{i=1}^n (\xi_i - u_i) \approx \prod_{i=1}^n (\xi_i - v_i) = h_\xi(v)$. ∎

2.5 Theorem: *Let* $Q(z) = \sum_{\nu \in E} a_\nu (z-w)^\nu$ *be a nonstandard polynomial with* $a_{*\mu} \in \operatorname{fin} {}^*\mathbb{C}$ *for all* $\mu \in E \cap \mathbb{N}^n$. *Then*

$$H_Q(w) = \sup\{r \in \mathbb{R}_+ : Q \text{ is finite on } {}^*P_r(w)\}. \tag{12}$$

Proof: Let $R_Q(w)$ be the number on the right hand side of Eq. 12. Let $0 < r < H_Q(w)$. By Theorem 2.2 $Q(z)$ is finite for all $z \in {}^*P_r(w)$. Hence $H_Q(w) \leq R_Q(w)$. For the other inequality let $0 < r < R_Q(w)$. It is easy to see that $Q(z)$ is finite on ${}^*\overline{P_r(w)}$. Similar to the proof of Theorem 2.3 $(i) \Rightarrow (iv)$ it follows that $|a_\nu| r^{|\nu|} =< Q(z), h_\nu(z) >$ is less than or equal to some $c \in \mathbb{R}_+$. Hence $|a_\nu|^{1/|\nu|} \cdot r \leq c^{1/|\nu|} \approx 1$ for $\nu \in E$ with $|\nu| \in {}^*\mathbb{N}_\infty$. Therefore $st\, |a_\nu|^{1/|\nu|} \leq 1/r$ and $r \leq 1/C_Q(w) = H_Q(w)$. This implies $R_Q(w) \leq H_Q(w)$. The proof is complete. ∎

Let $Q(z) = \sum_{k \in E} a_k (z-w)^k$ be a nonstandard polynomial in one variable with $w \in \mathbb{C}$. By Theorem 2.2 $H_Q(w)$ is always lower equal to the convergence radius of the induced power series $\sum_{k \in E \cap \mathbb{N}} st\, a_k (z-w)^k$. The example $Q(z) = z^N$ with $N \in {}^*\mathbb{N}_\infty$ shows that this inequality may be strict.

3 Truncations of power series

3.1 Definition. Let $P(z) = \sum_{\nu \in \mathbb{N}^n} a_\nu z^\nu$ be a (formal) power series. A *hyperfinite truncation of* $P(z)$ is a nonstandard polynomial $Q(z)$ of the form $\sum_{\nu \in E} {}^*a_\nu z^\nu$ where E is a hyperfinite set with $\mathbb{N}^n \subset E \subset {}^*\mathbb{N}^n$. A *truncation of* $P(z)$ is a hyperfinite truncation with the property that E is of the form $\{\nu \in {}^*\mathbb{N}^n : \nu \leq \sigma\}$ or $\{\nu \in {}^*\mathbb{N}^n : \nu \leq \sigma, \nu \neq \sigma\}$ with $\sigma \in {}^*\mathbb{N}^n_\infty$.

3.2 Proposition: *The domain of convergence of a power series* $P(z)$ *is contained in* S_Q *for every hyperfinite truncation* $Q(z)$ *of* $P(z)$.

Proof: Let $w \in \mathbb{C}^n$ be in the domain of convergence C_P. Then there exists an open polydisc $P_r(0)$ ($r \in \mathbb{R}_+$) with $w \in P_r(0) \subset C_P$. Let $Q(z) = \sum_{\nu \in E} {}^*a_\nu z^\nu$ be a hyperfinite truncation. Choose $\sigma \in {}^*\mathbb{N}^n_\infty$ such that $E \subset \{\nu \in {}^*\mathbb{N}^n : \nu \leq \sigma\}$. We know that $Q_\rho(z) := \sum_{\nu \leq \rho} a_\nu z^\nu$ converges absolutely and uniformly on $P_r(0)$ to $P(z)$. Hence ${}^*Q_\sigma(z) \approx {}^*Q_\sigma(w)$ for $z \approx w$. It follows that $|Q(z) - Q(w)| \leq |Q(z) - {}^*Q_\sigma(z)| + |{}^*Q_\sigma(z) - {}^*Q_\sigma(w)| + |{}^*Q_\sigma(w) - Q(w)|$. Moreover $|Q(z) - {}^*Q_\sigma(z)| \leq 2\sum_{\nu \leq \sigma, \nu \notin \mathbb{N}^n} |{}^*a_\nu||z^\nu| \approx 0$ and a similar formula is valid for w. Hence $|Q(z) - Q(w)| \approx 0$. ∎

3.3 Theorem: *Let* $P(z) = \sum_\nu a_\nu z^\nu$ *be a power series. Then the following statements are equivalent:*

(i) The domain of convergence contains 0.

(ii) Every hyperfinite truncation is complex microcontinuous at 0.

(iii) There exists a hyperfinite truncation which is complex microcontinuous at 0.

Proof: Proposition 3.2 shows $(i) \Rightarrow (ii)$ and $(ii) \Rightarrow (iii)$ is trivial. Now let $Q(z) = \sum_{\nu \in E} {}^*a_\nu z^\nu$ be a hyperfinite truncation. By Theorem 2.3 $(ii) \Rightarrow (v)$ we know that $|{}^*a_\nu|^{1/|\nu|} \in \text{fin}\,{}^*\mathbb{C}$ for all $\nu \in E$. By a saturation argument (cf. the proof of 2.3 $(v) \Rightarrow (vi)$) there exists $\theta \in \mathbb{N}$ with $E \subset \{\nu : |a_\nu|^{1/|\nu|} \leq \theta\}$. Since $\mathbb{N}^n \subset E$ it follows that $|a_\mu|^{1/|\mu|} \leq \theta$ for all $\mu \in \mathbb{N}^n$. Now a routine arument shows that the power series converges for all $z \in \mathbb{C}^n$ with $|z_i| < 1/\theta$ for $i = 1, \dots, n$: $|\sum_\mu a_\mu z^\mu| \leq \sum_\mu (|a_\mu|^{1/|\mu|})^{|\mu|} |z^\mu| \leq \sum_\mu (\theta \cdot r)^{|\mu|}$ with $r := \max\{|z_1|, \dots, |z_n|\}$. ∎

3.4 Proposition: *Let $P(z) = \sum_{k=0}^\infty a_k z^k$ be a power series in one variable. If there exists $w \in \mathbb{C}$ such that $Q(w) \in \text{fin}\,{}^*\mathbb{C}$ for all truncations $Q(z)$ of $P(z)$ then $P(z)$ converges for all $z \in \mathbb{C}$ with $|z| < |w|$.*

Proof: Let $N \in {}^*\mathbb{N}_\infty$ and $Q_N(z) = \sum_{k=0}^N {}^*a_k z^k$. Then $a_N w^N = Q_N(w) - Q_{N-1}(w) \in \text{fin}\,{}^*\mathbb{C}$. Trivially we have $a_k w^k \in \text{fin}\,{}^*\mathbb{C}$ for $k \in \mathbb{N}$. Hence the following statement is true: $(\exists M \in {}^*\mathbb{N})(\forall N \in {}^*\mathbb{N})(|a_N w^N| \leq M)$. By transfer there exists $M \in \mathbb{N}$ with $|a_k w^k| \leq M$ for all $k \in \mathbb{N}$. Hence $|P(z)| \leq \sum_{k=0}^\infty |a_k||z^k| < \infty$. ∎

3.5 Corollary: *Let $P(z) = \sum_{k=0}^\infty a_k z^k$ be a power series in one variable. Then the domain of convergence coincides with the intersection of all S_{Q_N} with $N \in {}^*\mathbb{N}_\infty$ and $Q_N = \sum_{k=0}^N {}^*a_k z^k$.*

C. Impens has given in 3 an example of a truncated nonstandard polynomial P_N such that S_{P_N} is strictly larger than the domain of convergence. We use his example to show that S_{P_N} may not even be connected.

3.6 Example. Let $c \in \mathbb{C}$ and $Q_i := z^{2^i-1}(c+z)^{2^i-1}$ for $i = 0, 1, 2, \dots$ and let $w = z(c+z)$. Define $P_N(z) = \sum_{i=0}^N Q_i(z) = \sum_{i=0}^N w^{2^i-1}$ with $N \in {}^*\mathbb{N}_\infty$. Then $|P_N(w)| \geq |w|^{2^N-1} - \sum_{k=0}^{2^{(N-1)}-1} |w|^k = |w|^{2^N-1} - \frac{|w|^{2^{(N-1)}}-1}{|w|-1}$. Hence we have for $|w| > 1$ the estimation $R_N(w) := \frac{|P_N(w)|(|w|-1)}{|w|^{2^{(N-1)}}} \geq \frac{|w|^{2^{N-1}}}{|w|}(|w|-1) - 1 + \frac{1}{|w|^{2^{(N-1)}}}$. It follows that $R_N(w) \notin \text{fin}\,{}^*\mathbb{C}$ for all $w \in \mathbb{C}$ with $|w| > 1$ and the same holds for $|P_N(w)|$. Hence S_{P_N} is equal to the set $\{z \in \mathbb{C} : |z(c+z)| < 1\}$ and P_N is a truncation of the power series $\sum_i Q_i(z)$, cf. 3. For $c = -3$ it is easy to see that S_{P_N} is contained in $B_1(0)$ and $B_1(3)$ and that 0 and $3 \in S_{P_N}$.

3.7 Definition. Let X be a topological space and $(f_k)_{k \in \mathbb{N}}$ be a sequence of continuous complex-valued functions. Then the *set of local convergence* $C((f_k)_k)$ is the set of all $x \in X$ such that there exists a neighborhood U of x with the property that $(f_k)_k$ converges uniformly on U to some function g defined on U, or equivalently, that $(f_k)_k$ is a Cauchy sequence with respect to uniform convergence on U.

3.8 Theorem: *Let U be an open subset of $\mathbb{C}^n$ and $(f_k)_k$ be a sequence in $H(U)$. Then*

$$\bigcup_{N \in {}^*\mathbb{N}_\infty} S_{{}^*f_N} = \bigcup_{(g_j)_j \text{ subsequence of } (f_k)_k} C((g_j)_j). \tag{13}$$

Proof: For the inclusion part let $N \in {}^*\mathbb{N}_\infty$ and define $g(w) := st\,{}^*f_N({}^*w)$ for $w \in S_{{}^*f_N}$. Since g is continuous it follows that ${}^*f_N(z) \approx {}^*g(z)$ for all $z \approx w \in S_{{}^*f_N}$. As $S_{{}^*f_N}$ is open there exists a sequence of compact sets K_r such that K_r is contained in the interior of K_{r+1} and $\cup_{r=1}^{\infty} K_r = S_{{}^*f_N}$. Hence the following statement is true for any natural number n_r:

$$(\exists N \in {}^*\mathbb{N})(N > n_r)(\forall z \in {}^*K_{n_r})(|{}^*f_N(z) - {}^*g(z)| \leq 1/n_r). \tag{14}$$

By Eq. 14 and the transfer principle we can construct inductively a subsequence $(f_{n_r})_r$ such that $|f_{n_{r+1}}(x) - g(x)| < 1/n_r$ for all $x \in K_{n_r}$. For $x \in S_{{}^*f_N}$ there exists $l \in \mathbb{N}$ and $n_r > l$ such that x is in the interior of $(K_l) \subset K_{n_r}$. It follows that $(f_{n_r})_r$ converges uniformly to g on the neighborhood K_{n_r} of x. Conversely, let $w \in C((f_{n_r})_r)$ for some subsequence $(f_{n_r})_r$ converging uniformly to some function g on a neighorhood U of w. Then ${}^*f_{n_R}$ is complex microcontinuous at w for $R \in {}^*\mathbb{N}_\infty$ and $w \in S_{{}^*f_{n_R}}$. ∎

4 Polynomial approximation.

In this section we consider only functions in one complex variable. An internal function $f\colon {}^*\mathbb{C} \to {}^*\mathbb{C}$ is called *real microcontinuous* at $w \in \mathbb{R}$ if $x \approx w$ and $x \in {}^*\mathbb{R}$ implies $Q(x) \approx Q(w)$. If M is a subset of ${}^*\mathbb{C}$ we call Q to be *M-continuous* if $z \approx w$ with $z, w \in M$ implies $Q(z) \approx Q(w)$.

4.1 Proposition: *Let $Q(z)$ be a nonstandard polynomial in one variable. Then S_Q is simply connected.*

Proof: We show that $\mathbb{C} \setminus S_Q$ does not possess a bounded arc component B: let $\gamma\colon [0,1] \to S_Q$ be a closed path surrounding B and $Sp\,\gamma := \{\gamma(t) : t \in [0,1]\}$. For every $z_0 \in B$ and $z \approx z_0$ we have $|Q(z)| \leq \max_{w \in {}^*Sp\gamma} |Q(w)| = |Q(w_0)| \in \text{fin}\,{}^*\mathbb{C}$ for a suitable $w_0 \in S_Q$. Hence $z_0 \in S_Q$ and S_Q is simply connected (but not necessarily connected). ∎

4.2 Theorem : *Let $U \subset \mathbb{C}$ be an open simply connected set and $f\colon {}^*U \to {}^*\mathbb{C}$ be *-holomorphic. Then there exists a nonstandard polynomial $Q(z)$ such that $f(z) \approx Q(z)$ for all $z \in ns\,{}^*U$.*

Proof: Let $P := P(\mathbb{C})$ be the set of all polynomials in one variable. The Runge approximation theorem [[9], p. 290] shows that the following statement is true:

$$(\forall f \in H(U))(\forall K \in k)(\forall n \in \mathbb{N})(\exists Q \in P)(\forall z \in K)(|f(z) - Q(z)| < \frac{1}{n}). \tag{15}$$

Here k denotes the set of all compact subsets of U. Since U is locally compact we have $ns\,{}^*U = \cup_{K \in k}{}^*K =: cpt\,{}^*U$. Choose $K \in {}^*k$ with $ns\,{}^*U = cpt\,{}^*U \subset K$ and an infinitely large $N \in {}^*\mathbb{N}$. The Transfer principle applied to the above equation completes the proof. ■

4.3 Theorem: *Let $U \subset \mathbb{C}$ be an open simply connected set and $f: U \to \mathbb{C}$ holomorphic. Then there exists a nonstandard polynomial $Q(z)$ with $Q(z) \approx {}^*f(z)$ for all $z \in ns\,{}^*U$ and $S_Q = U$.*

Proof: Let U_n be a sequence of open, simply connected and relatively compact sets such that $U = \cup_{n=1}^{\infty} U_n$ and $K_n := \overline{U}_n \subset U_{n+1}$. For every finite subset $E \subset \mathbb{C} \setminus U$ define $f_{E,n}: K_n \cup E \to \mathbb{C}$ by $f_{E,n}(z) = f(z)$ for $z \in K_n$ and $f_{E,n}(z) = n$ for $z \in E$. Choose $r \in \mathbb{R}_+$ such that $\overline{B_r(z)} \cap K_{n+1} = \emptyset$ for all $z \in E$. Then $U_{E,n} := U_{n+1} \cup \cup_{z \in E} B_r(z)$ is an open simply connected set and $f_{E,n}$ extends in an obvious manner to a holomorphic function on $U_{E,n}$. By Runge s approximation theorem the following statement is true: for all $n \in \mathbb{N}$ and for every finite subset $E \subset \mathbb{C} \setminus U$ there exists $Q \in P(\mathbb{C})$ such that $|Q(z) - f_{E,n}(z)| < \frac{1}{n}$ for all $z \in K_n \cup E$. Choose $N \in {}^*\mathbb{N}_\infty$ and E hyperfinite with $\mathbb{C} \setminus U \subset E \subset {}^*(\mathbb{C} \setminus U)$. Then there exists $Q \in {}^*P(\mathbb{C})$ with ${}^*f_{E,N}(z) \approx Q(z)$ for all $z \in {}^*K_N \cup E$. For every $z \in \mathbb{C} \setminus U \subset E$ we have $Q(z) \approx N$, hence $S_Q \subset U$. The other inclusion is clear since $ns\,{}^*U = \cup_{n=1}^{\infty} {}^*K_n \subset {}^*K_N$ and $Q(z) \approx {}^*f_{E,N}(z) = {}^*f(z)$ for all $z \in {}^*K_N$. ■

Now let us consider polynomial approximation on compact sets. Observe that the nonstandard polynomial Q in the following theorem is necessarily ${}^*K$-continuous since f is continuous.

4.4 Theorem: *Let $K \subset \mathbb{C}$ be compact and $\mathbb{C} \setminus K$ connected. For every continuous function $f: K \to \mathbb{C}$ which is holomorphic in the interior there exists a nonstandard polynomial $Q(z)$ such that $Q(z) \approx {}^*f(z)$ for all $z \in {}^*K$ and $Q({}^*z_0) \notin \text{fin}\,{}^*\mathbb{C}$ for all $z_0 \in \mathbb{C} \setminus K$. In particular, if K is nowhere dense then S_Q is void.*

Proof: Let $D := \mathbb{C} \setminus K$. For every finite subset $E \subset D$ we define a continuous function $f_E: K \cup E \to \mathbb{C}$ by $f_E(z) = f(z)$ for all $z \in K$ and $f_E(z) = |E|$ for $z \in E$ where $|E|$ denotes the cardinality of E. Now $K_E := K \cup E$ is compact with the same interior as K and its complement is still connected. By the Theorem of Mergelyan [[9], p. 423] the following statement is true: for every f_E, for every $n \in \mathbb{N}$ there exists $Q \in P(\mathbb{C})$ such that $|Q(z) - f_E(z)| < \frac{1}{n}$ for all $z \in K_E$. Choose E hyperfinite with $\mathbb{C} \setminus K \subset E \subset {}^*(\mathbb{C} \setminus K)$ and $N \in {}^*\mathbb{N}_\infty$ and use the transfer principle. ■

It is also possible to construct a nonstandard polynomial $Q(z)$ with $Q(z) \approx {}^*f(z)$ for all $z \in {}^*K$ such that (i) $Q(z_0) \in \text{fin}\,{}^*\mathbb{C}$ for all $z_0 \in \mathbb{C} \setminus K$ and (ii) Q is nowhere complex microcontinuous on $\mathbb{C} \setminus K$. For example, divide $\mathbb{C} \setminus K$ into two disjoint dense set D_1, D_2 and define $f_E(z) = i$ for $z \in E \cap D_i$ with $i = 1, 2$. Moreover one can achieve that f_E restricted to any line segment $I \subset \mathbb{C} \setminus K$ is not ${}^*I$-continuous, choose $D_1 := D \cap (\mathbb{R} \setminus \mathbb{Q} \times \mathbb{Q})$ and $D_2 = D \setminus D_1$. Here $\mathbb{Q}$ denotes the set of all rational numbers.

4.5 Example. Let f be a holomorphic function on the open unit disk B_1 such that (i) f can be extended continuously to the closed disk D and (ii) the partial sums

$s_n := \sum_{k=0}^{n} a_k$ are not bounded where $f(z) = \sum_{k=0}^{\infty} a_n z^n$ for all $z \in B_1$, cf. [[6], p. 30]. By Theorem 4.4 there exists $Q(z) = \sum_{k=0}^{N} b_k z^k$ with $Q(z) \approx {}^*f(z)$ for all $z \in {}^*D$. Now Theorem 2.2 shows that

$$st\,(\sum_{k=0}^{N} b_k z^k) = \sum_{k=0}^{\infty} st\, b_k st\, z^k \text{ for all } z \in ns\,{}^*B_1. \tag{16}$$

Hence we have $st\, b_k = a_k$ for every $k \in \mathbb{N}$. Although the left side is meaningful for every $z \in {}^*D$ the right side does not exist for $z = 1$.

We remark that the validity of Eq. 2 only for all $z \in P_r(w)$ (instead of ${}^*P_r(w)$) is not equivalent to complex microcontinuity as our next example shows.

4.6 Example. Let $E := \{z_1, \ldots, z_N\}$ be a hyperfinite set with $P_r(w) \subset E$ and let $P(z) := \prod_{k=1}^{N}(z - z_i)$. Choose $y \approx w$ with $y \notin E$ and choose $c \in {}^*\mathbb{R}$ such that $|c(y - w)^N P(y)|$ is infinitely large. Then $Q(z) := c(z - w)^N P(z)$ is not complex microcontinuous since $Q(y)$ is infinitely large and $Q(w) = 0$. On the other side we have $Q(z_i) = 0$ for all $z_i \in E$.

Let U be an open subset of $\mathbb{C}$ and choose a sequence of open relatively compact subsets U_n with $K_n := \overline{U_n} \subset U_{n+1}$ and $U = \cup_{n=1}^{\infty} K_n$. Endow the set $C(U)$ of all continuous complex-valued functions with the locally convex topology induced by the norms $||f_n|| := \sup_{z \in K_n} |f(z)|$ $(n \in \mathbb{N})$. Choose an internal subspace E of ${}^*C(U)$ such that $P(\mathbb{C}) \subset E \subset {}^*P(\mathbb{C})$. Call $Q \in E$ *finite* if $||Q_n|| \in \text{fin}\,{}^*\mathbb{R}$ for all $n \in \mathbb{N}$. The set of all finite points of E is denoted by $\text{fin}\, E$. It is obvious that $\text{fin}\, E = \{Q \in E : (\forall z \in ns\,{}^*U)\ (f(z) \in \text{fin}\,{}^*\mathbb{C})\}$. Now the equivalence of (iii) and (i) in Theorem 2.3 shows that

$$\text{fin}\, E = \{Q \in E : Q \text{ is microcontinuous at every } z \in U \text{ and } Q(z) \in \text{fin}\,{}^*\mathbb{C}\}.$$

Hence the standard part st is a (linear) map from E into $H(U)$. If U is simply connected then it is also surjective: For $f \in H(U)$ apply the overflow principle to the set $\{n \in {}^*\mathbb{N} : (\exists Q \in E)(\forall z \in {}^*K_n)(|{}^*f(z) - Q(z)| < \frac{1}{n})$. In this case the nonstandard hull $\widehat{E}$ of E (for definition see [1]) coincides with $H(U)$. In the general case $\widehat{E}$ is isomorphic to the algebra of all holomorphic functions on U which are limits of polynomials with respect to compact convergence in U.

References

[1] C.W. Henson, L.C. Moore, *The nonstandard theory of topological vector spaces.* Trans. Amer. Math. Soc. **172** (1972) 405-435.

[2] C. Impens, *Local microcontinuity of nonstandard polynomials.* Israel J. Math. **59** (1987) 81-97.

[3] - , *Propagation of microcontinuity for nonstandard polynomials.* Journal D'Analyse Mathématique **53** (1989) 187-200.

[4] - , *Real functions as traces of infinite polynomials.* Math. Ann. **284** (1989) 63-73.

[5] - , *Standard and nonstandard polynomial approximation.* To appear in J. Math. An. Appl.

[6] E. Landau, D. Gaier, *Darstellung und Begründung einiger neuerer Ergebnisse der Funktionentheorie.* Springer Berlin 1986.

[7] L. Nachbin, *Holomorphic functions, domain of Holomorphy and local porperties.* North-Holland, Amsterdam 1972, Second Edition.

[8] A. Robinson, *Non-Standard Analysis.* North-Holland, Amsterdam 1966.

[9] W. Rudin, *Real and complex analysis.* McGraw-Hill, New Dehli 1974.

An Existence Result for a Class of Partial Differential Equations with Smooth Coefficients

Todor Todorov
Mathematics Department, California Polytechnic State University, San Luis Obispo, California 93407, USA

Abstract

We prove the existence of a solution for a class of linear partial differential equations with smooth coefficients in a space of generalized functions larger than the space of Schwartz distributions. As an example, we show that H. Lewy's equation has a solution in this space whenever its right hand side is a classical smooth function or a Schwartz distribution.
AMS subject classification: 03H05, 30G10, 35A05, 35A08, 46F10.
Keywords: differential operator, fundamental solution, generalized solution, partial differential equation, Schwartz distribution, nonstand analysis, internal function, transfer principle, saturation principle.

0 Introduction

The main purpose of this paper is to show that the equations of the type:

$$P(x, \partial)U = F \tag{0.1}$$

have solutions U in $\widehat{\mathcal{E}}$ for any choice of the right hand side F also in $\widehat{\mathcal{E}}$. Here $\widehat{\mathcal{E}}$ is a class of generalized distributions, larger than the class of Schwartz distributions $\mathcal{D}'$, defined in Section 2 in the framework of nonstandard analysis (Tom Lindstrøm [8]). The result is proved for a class of linear partial differential operators with C^∞-coefficents:

$$P(x, \partial) = \sum_{|\alpha| \leq m} c_\alpha(x)\partial^\alpha, \tag{0.2}$$

called in this paper "regular" (Section 1). As a consequence, we show that H. Lewy's [7] equation in $\mathbb{R}^3$:

$$\frac{\partial U}{\partial x_1} + i\frac{\partial U}{\partial x_2} - 2i(x_1 + ix_2)\frac{\partial U}{\partial x_3} = F \tag{0.3}$$

has a solution U in $\widehat{\mathcal{E}}$ for any choice of F in $\widehat{\mathcal{E}}$, in particular, whenever F is a classical smooth function or a Schwartz distribution. In addition, we show that the class of generalized distributions $\widehat{\mathcal{E}}$ preserves the solvability of the linear partial differential equations with constant coefficients

$$P(\partial)U = F \tag{0.4}$$

in the sense that the equation (0.4) has a solution U in $\widehat{\mathcal{E}}$ for any choice of F in $\widehat{\mathcal{E}}$.

We denote the usual classes of the C^∞-functions and C^∞-functions with compact supports defined on $\mathbb{R}^d$ (d is a natural number) by: $\mathcal{E} = \mathcal{E}(\mathbb{R}^d) = C^\infty(\mathbb{R}^d)$ and $\mathcal{D} = \mathcal{D}(\mathbb{R}^d) = C_0^\infty(\mathbb{R}^d)$, respectively, and by $\mathcal{E}' = \mathcal{E}'(\mathbb{R}^d)$ and $\mathcal{D}' = \mathcal{D}'(\mathbb{R}^d)$ we denote the corresponding classes of Schwartz distributions on $\mathbb{R}^d$ (L. Schwartz [11]). For integration over $\mathbb{R}^d$ we use the Lebesgue integral. As usual, $\mathbb{N}, \mathbb{R}$ and $\mathbb{C}$ will be the systems of the natural, real and complex numbers, respectively, and we use also the notation $\mathbb{N}_0 = \{0\} \cup \mathbb{N}$. For the partial derivatives we write $\partial^\alpha, \alpha \in \mathbb{N}_0^d$. If $\alpha = (\alpha_1, \alpha_2, \ldots, \alpha_d)$ for some $\alpha \in \mathbb{N}_0^d$, we write $|\alpha| = \alpha_1 + \alpha_2 + \ldots + \alpha_d$. If f is a function or Schwartz distribution, we denote $\check{f}(x) = f(-x)$. Finally, we denote by $\widehat{\mathcal{E}} \equiv \widehat{\mathcal{E}}(\mathbb{R}^d)$ the class of "generalized distributions on $\mathbb{R}^n$", defined in Section 2.

Recall that any linear partial differential equation with constant coefficients (0.4) has a solution U in $\mathcal{D}'$ for any choice of F also in $\mathcal{D}'$ (L. Ehrenpreis [5] - B. Malgrance [9]). A general existence result for the linear partial differential equations with C^∞-coefficients was first conjectured, then proved to be false in the settings of distributions (H. Lewy [7]) and hyperfunctions (P. Schapira [10]). In particular, H. Lewy's equation (0.3) is famous for not having solutions in $\mathcal{D}'$ for a large choice of F even in $\mathcal{D}$. This explains why we are looking for solutions in classes of generalized functions larger than the class of Schwartz distributions $\mathcal{D}'$.

The result presented here is related to F. Treves's works [14], [15], where the local existence and uniqueness of the Cauchy problem for the equation (0.1) is proved for operators $P(x, \partial)$ with analytic coefficients, where the right hand side, the Cauchy data and the solution are, in general, analytic functionals. For comparison we mention the following:

a) Our solutions are global, whereas the analytic functionals present, in general, local solutions.

b) The class of generalized $\widehat{\mathcal{E}}$ contains the class of Schwartz distributions $\mathcal{D}'$ while the class of analytic functionals contains neither $\mathcal{D}'$, nor the class of the classical smooth functions $\mathcal{E}$. Hence, in addition to the restriction on the coefficients, F. Treves's result does not include the case when the right hand side of the equation F is in $\mathcal{E}(\Omega) - \mathcal{D}(\Omega)$ (or in $\mathcal{D}'(\Omega) - \mathcal{E}'(\Omega)$).

c) The generalized distributions are localizable (e.g. we can define the concept of "support" of a given generalized distribution), whereas analytic functionals are not.

d) The classes of operators $P(x, \partial)$ in these two different approaches are not comparable (neither of them is larger than the other), although, their intersection is comparatively large: it contains, for example, all elliptic $P(x, \partial)$ with analytic coefficients.

Our results are related also to J.F. Colombeau's work [2], where a general existence result for the linear PDEs with smooth coefficients:

$$P^h(x,\partial)U(x) \sim F(x) \tag{0.5}$$

has been established in the class of the "new generalized functions" $\mathcal{G}(\Omega)$ (J. F. Columbeau [1]). Here $P^h(x,\partial)$ is a regularization of the original operator $P(x,\partial)$ (depending on a function h) and $\sim$ is an equivalence relation in $\mathcal{G}(\Omega)$ called "association". Later this result was improved in (J.F. Colombeau, A. Heibig, M. Obergruggenberger [3]), where the uniqueness of the Cauchy problem for the equation (0.5) was proved in the class of generalized functions $\mathcal{G}_g(\Omega)$ ("g" stands for "global") and the association $\sim$ in (0.5) was replaced by the strict equality in $\mathcal{G}_\}(\Omega)$. In our approach the solutions satisfy the equations in the usual sense, i.e. in the sense of the equality in $\widehat{\mathcal{E}}$ and the operator in the equation is an extension of the original operator $P(x,\partial)$ for $\mathcal{E}$ onto $\widehat{\mathcal{E}}$.

1 Regular differential operators with C^∞–coefficients

We specify and study a class of differential operators, here called "regular". The main result of this section is presented in Lemma (1.7).

(1.1) Definition: Let $P(x,\partial)$ be a linear partial differential operator with C^∞-coefficients (see (0.2) at the beginning of the Introduction). We say that $P(x,\partial)$ is "regular" if its transposed operator $P^t(x,\partial)$, defined by

$$P^t(x,\partial)\varphi(x) = \sum_{|\alpha|\leq m} (-1)^{|\alpha|}\partial^\alpha[c_\alpha(x)\varphi(x)],$$

is injective on $\mathcal{D}$ (i.e. if $\varphi \in \mathcal{D}$ and $P^t(x,\partial)\varphi = 0$ implies $\varphi = 0$)

(1.2) Examples

1. The linear partial differential operators with constant coefficients are regular. To show this assume that $P(\partial)\varphi = 0$ for some $\varphi \in \mathcal{D}$ and take the Fourier transform $\mathcal{F}$. We obtain $P(-iz)\mathcal{F}(\varphi)(z) = 0$ which implies $\mathcal{F}(\varphi) = 0$ and hence $\varphi = 0$. Now the result follows since $P^t(\partial)$ is also a differential operator with constant coefficients.

2. The elliptic linear partial differential operators with analytic (entire) coefficients are regular. This follows immediately from (L. Hörmander [6], Theorem (7.5.6), p. 178) and the fact that the transposed operators are also elliptic with analytic coefficients.

3. The second-order elliptic linear partial differential operators with C^∞-coefficients are regular.

4. H. Lewy's operator:

$$L(x,\partial)=\frac{\partial}{\partial x_1}+\frac{\partial}{\partial x_2}-2i(x_1+ix_2)\frac{\partial}{\partial x_3}$$

in $\mathbb{R}^3$ (H. Lewy [7]) is regular. This follows from the following result (also due to H. Lewy): Let us imbed $\mathbb{R}^3$ in $\mathbb{C}^2$ by the mapping

$$(x_1,x_2,x_3)\to(x_1+ix_2,x_3+i(x_1^2+x_2^2)).$$

Then $\mathbb{R}^3$ becomes the boundary of the domain $D=\{\mathrm{Im} z_2>|z_1|^2\}$, and any solution of $Lu=0$ extends holomorphically into D. Hence we get $u=0$, by the unique continuation for holomorphic functions. Now the result follows since $L^t(x,\partial)=-L(x,\partial)$.

5. By contrast with the above, the operator $P(x,\partial)=\psi\partial^\alpha$ is not regular for $\psi\in\mathcal{D},\psi\neq 0$, since $P^t(x,\partial)\varphi=0$ for any $\varphi\in\mathcal{D}$ such that $\mathrm{supp}\psi\bigcap\mathrm{supp}\varphi=\emptyset$.

6. The operator $P(x,\partial)=x_1\partial/\partial x_2-x_2\partial/\partial x_1$ in $\mathbb{R}^2$ is not regular, since $P(x,\partial)\varphi(x_1^2+x_2^2)=0$ for any $\varphi\in\mathcal{D}(\mathbb{R})$ and $P^t(x,\partial)=-P(x,\partial)$.

(1.3) Notation: Let $P(x,\partial)$ be a fixed linear partial differential operator with C^∞-coefficients. For any $\varphi\in\mathcal{D}$ we define the functional $P_\varphi:\mathcal{D}\to\mathbb{C}$ by

$$P_\varphi(u)=\int_{\mathbb{R}^d}[P(x,\partial)u](x)\varphi(x)dx,\qquad u\in\mathcal{D}.\tag{1.4}$$

We write $\ker P_\varphi$ for the kernel of P_φ.

Notice that the integrating by parts we get

$$P_\varphi(u)=\int_{\mathbb{R}^d}u(x)[P^t(x,\partial)\varphi](x)dx$$

for all $u\in\mathcal{D}$. Hence $P_\varphi=P^t(x,\partial)\varphi$ where $P^t(x,\partial)\varphi$ is considered as an element of $\mathcal{D}'$ (in the sense of Schwartz theory of distributions). It follows, in particular, that $P_\varphi\in\mathcal{D}'$ for any $\varphi\in\mathcal{D}$.

(1.5) Lemma: *Let $P(x,\partial)$ be a linear partial differential operator with C^∞-coefficients. Then $P(x,\partial)$ is regular iff the mapping $\varphi\to P_\varphi$ from $\mathcal{D}$ into $\mathcal{D}'$ is injective.*

Proof: The result follows immediately from $P_\varphi=P^t(x,\partial)\varphi$. ∎

In what follows, we shall work in a nonstandard model of the set of the complex numbers $\mathbb{C}$ with degree of saturation larger than *card* $\mathbb{R}$. In particular, any polysaturated model of $\mathbb{C}$ will suffice (Tom Lindstrøm [8], p. 49). If X is a set of complex numbers or a set of (standard) functions, then ${}^*X$ will be its nonstandard extension and if $f:X\to Y$ is a (standard) mapping, then ${}^*f:{}^*X\to{}^*Y$ will be its nonstandard extension. We shall use the same notation, $*$, for the convolution operator $*: \mathcal{D}' \times \mathcal{D} \to \mathcal{E}$ and its nonstandard extensions $*: {}^*\mathcal{D}' \times {}^*\mathcal{D} \to {}^*\mathcal{E}$. For integration over ${}^*\mathbb{R}^d$ we use the $*$-Lebesgue integral (Tom Lindstrøm [8], Ch. I). We shall systematically apply the saturation and transfer principles in the form presented in (Tom Lindstrøm [8], Chapter III and Chapter IV) where the reader can find more references to the subject.

In particular for the nonstandard extension ${}^*P_\varphi : {}^*\mathcal{D} \to / =^* \mathbb{C}$ of the functional P_φ defined in (1.3) we have

$$ {}^*P_\varphi(u) = \int_{{}^*\mathbb{R}^d} [{}^*P(x,\partial)u]\, {}^*\varphi(x)dx \qquad u \in {}^*\mathcal{D} \tag{1.6} $$

where ${}^*P(x,\partial)$ is the nonstandard extension of the operator $P(x,\partial)$ and ${}^*\varphi$ is the nonstandard extension of φ. Finally, we should mention that $P_\varphi(u) = {}^*P_\varphi({}^*u)$ for all u in $\mathcal{D}$, where ${}^*u$ is the nonstandard extension of u.

(1.7) Lemma: *Let $P(x,\partial)$ be a regular linear partial differential operator with C^∞-coefficients and f be a function in ${}^*\mathcal{E}$. Then for any choice of $k \in \mathbb{N}$ and $\varphi_i \in \mathcal{D}, i = 1,2,\dots,k$, the system of equations:*

$$ {}^*P_{\varphi_i}(u) = \int_{{}^*\mathbb{R}^d} f(x)\, {}^*\varphi_i(x)dx, \qquad i = 1,2,\dots,k, \tag{1.8} $$

has a solution u in ${}^*\mathcal{D}$.

Proof: By induction, consider first the case $k = 1$ of one equation:

$$ {}^*P_\varphi(u) = \int_{{}^*\mathbb{R}^d} f(x)\, {}^*\varphi(x)dx \tag{1.9} $$

If $\varphi = 0$, then any u in ${}^*\mathcal{D}$ is a solution of (1.9). If $\varphi \neq 0$, then the set $\Phi = \mathcal{D} - \ker\, P_\varphi$ is non-empty, by Lemma (1.5), and hence the function

$$ u = \frac{{}^*u_1}{P_\varphi(u_1)} \int_{{}^*\mathbb{R}^d} f(x)\, {}^*\varphi(x)dx, $$

satisfies (1.9) for any choice of $u_1 \in \Phi$. Assume now, that the statement is true for $k-1$. If $\varphi_1, \dots, \varphi_k$ are linearly dependent in $\mathcal{D}$, then (1.8) is equivalent to a system of $k-1$ equations and, by assumption, has a solution. If $\varphi_1, \dots, \varphi_k$ are linearly independent, then the functionals $P_{\varphi_1} \dots, P_{\varphi_k}$ are linearly independent in $\mathcal{D}'$, by Lemma (1.5), and hence the sets

$$ \Phi_i = \left(\bigcap_{j=1}^{k} \ker P_{\varphi_j} \right) - \ker\; P_{\varphi_i}, \qquad i = 1,2,\dots,k $$

are non-empty (N. Dunford and J.T. Schwartz [4], V. 3., Lemma 10, p. 421). It is easy to verify now that the function

$$u = \sum_{i=1}^{k} \frac{{}^*u_i}{P_{\varphi_i}(u_i)} \int\limits_{{}^*\mathbb{R}^d} f(x)\, {}^*\varphi_i(x)dx,$$

is a solution of (1.8) for any choice of $u_i \in \Phi_i$. The proof is complete. ∎

2 Generalized Distributions

In this section we define a class $\widehat{\mathcal{E}}$ of the "generalized distributions" as a factor space (a nonstandard hull) of the class of nonstandard internal smooth functions ${}^*\mathcal{E}$ (Tom Lindstrøm [8]). We show that $\widehat{\mathcal{E}}$ contains an isomorphic copy of the class of Schwartz distributions $\mathcal{D}'$.

We consider ${}^*\mathcal{E}$ as a differential algebra over ${}^*\mathbb{C}$ with respect to pointwise addition, multiplication and internal partial differentiation. As we already mentioned, for integration over ${}^*\mathbb{R}^d$ we use the $*$-Lebesgue integral (Tom Lindstrøm [8], Ch. I).

(2.1) Definition (Equivalence Relation in ${}^*\mathcal{E}$): We write $f \cong g$ for $f, g \in {}^*\mathcal{E}$, if

$$\int\limits_{{}^*\mathbb{R}^d} f(x)\, {}^*\varphi(x)dx = \int\limits_{{}^*\mathbb{R}^d} g(x)\, {}^*\varphi(x)dx$$

for all φ in $\mathcal{D}$, where ${}^*\varphi$ is the nonstandard extension of φ.

(2.2) Lemma:
(i) *The equivalence relation $\cong$ in ${}^*\mathcal{E}$ preserves addition, multiplication by scalars in ${}^*\mathbb{C}$, partial differentiation of any standard order and multiplication by a standard function ${}^*\psi$, where $\psi \in \mathcal{E}$.*
(ii) *Let $f, g \in {}^*\mathcal{E}$, let T be a Schwartz distribution with compact support and ${}^*T$ be its nonstandard extension. Then $f \cong g$ implies ${}^*T * f \cong {}^*T * g$.*

Proof: (i) The preservation of addition and scalar multiplication is obvious. The preservation of the partial differentiation follows easily by integration by parts (and the transfer principle):

$$\int\limits_{{}^*\mathbb{R}^d} \partial^\alpha f(x)\, {}^*\varphi(x)dx = (-1)^{|\alpha|} \int\limits_{{}^*\mathbb{R}^d} f(x)\, {}^*(\partial^\alpha\varphi)(x)dx =$$
$$= (-1)^{|\alpha|} \int\limits_{{}^*\mathbb{R}^d} g(x)\, {}^*(\partial^\alpha\varphi)(x)dx = \int\limits_{{}^*\mathbb{R}^d} \partial^\alpha g(x)\, {}^*\varphi(x)dx,$$

i.e. $\partial^\alpha f \cong \partial^\alpha g$. The preservation of the multiplication by a standard function follows simply from the fact that ${}^*(\psi\varphi) = {}^*\psi\, {}^*\varphi$.

(ii) Suppose $\varphi \in \mathcal{D}$. Since $T * \check{\varphi} \in \mathcal{D}$, we have

$$\int_{^*\mathbb{R}^d} (^*T * f)(x)\, {}^*\varphi(x)dx = \int_{^*\mathbb{R}^d} f(x)\, {}^*(T * \check{\varphi})(x)dx = \int_{^*\mathbb{R}^d} g(x)\, {}^*(T * \check{\varphi})(x)dx$$
$$= \int_{^*\mathbb{R}^d} (^*T * g)(x)\, {}^*\varphi(x)dx,$$

as required. The proof is complete. ■

The above lemma suggests the following definition.

(2.3) Defininition (Generalized Distributions): (i) The elements of the factor space $^*\mathcal{E}(\mathbb{R}^d)/\cong$, denoted for short by $\widehat{\mathcal{E}}$, will be called "generalized distributions on $\mathbb{R}^d$". We define the inclusion $\mathcal{E} \subset \widehat{\mathcal{E}}$ by the injection $f \to q(^*f)$, where $q : {}^*\mathcal{E} \to \widehat{\mathcal{E}}$ is the quotient mapping. If $\Phi \subseteq {}^*\mathcal{E}$, we write $\widehat{\Phi} = q[\Phi]$. In particular, when $\Phi = {}^*S$ for some $S \subseteq \mathcal{E}$, we write $\widehat{S}$ (instead of the more precise $\widehat{^*S}$).

(ii) We supply $\widehat{\mathcal{E}}$ with addition, multiplication by scalars in $^*\mathbb{C}$ and partial differentiation of any standard order inherited from $^*\mathcal{E}$.

(iii) We define a multiplication by a function in $\mathcal{E}$ by $\psi q(f) = q(^*\psi f)$ for $\psi \in \mathcal{E}$ and $f \in {}^*\mathcal{E}$, where $^*\psi$ is the nonstandard extension of ψ.

(iv) We define a convolution with a Schwartz distribution with compact support by $T * q(f) = q(^*T * f)$ for $T \in \mathcal{E}'$ and $f \in {}^*\mathcal{E}$, where $^*T$ is the nonstandard extension of T.

(v) We define the pairing between $\widehat{\mathcal{E}}$ and $\mathcal{D}$ with values in $^*\mathbb{C}$, by

$$\langle q(f), \varphi \rangle = \int_{^*\mathbb{R}^d} f(x)\, {}^*\varphi(x)dx \tag{2.4}$$

for $f \in {}^*\mathcal{E}$ and $\varphi \in \mathcal{D}$, where $^*\varphi$ is the nonstandard extension of φ.

The correctness of the above definitions follows immediately from Lemma (2.2). We shall show only that the mapping $f \to q(^*f)$ from $\mathcal{E}$ into $\widehat{\mathcal{E}}$ is injective: Assume that $q(^*f) = 0$ in $\widehat{\mathcal{E}}$, i.e.

$$\int_{^*\mathbb{R}^d} {}^*f(x)\, {}^*\varphi(x)dx = 0$$

for all $\varphi \in \mathcal{D}$. It follows that $f = 0$ in $\mathcal{E}$, as required since

$$\int_{^*\mathbb{R}^d} {}^*f(x)\, {}^*\varphi(x)dx = \int_{\mathbb{R}^d} f(x)\varphi(x)dx, \tag{2.5}$$

by the transfer principle (Tom Lindstrøm [8], Chapter IV).

We should mention that the pointwise multiplication in ${}^*\mathcal{E}$ can not be transferred to $\widehat{\mathcal{E}}$ since the equivalence relation $\cong$ does not, in general, preserve the multiplication in ${}^*\mathcal{E}$.

(2.6) Proposition:
(i) *$\widehat{\mathcal{E}}$ is a differential linear space over ${}^*\mathbb{C}$ and $\mathcal{E}$ is a differential linear subspace of $\widehat{\mathcal{E}}$ over the standard scalars $\mathbb{C}$.*
(ii) *$\widehat{\mathcal{E}}$ is a module over $\mathcal{E}$ and the mapping $f \to q({}^*f)$ from $\mathcal{E}$ into $\widehat{\mathcal{E}}$ is an $\mathcal{E}$ - module homomorphism. (Or, equivalently, the multiplication by a function in $\mathcal{E}$ generalizes the usual multiplication in $\mathcal{E}$ in the sense that if $\psi, f \in \mathcal{E}$, then $\psi q({}^*f) = q({}^*(\psi f))$).*
(iii) *If $F \in \widehat{\mathcal{E}}$, then $\langle F, \ \varphi\rangle = 0$ for all $\varphi \in \mathcal{D}$ implies $F = 0$ in $\widehat{\mathcal{E}}$. Moreover, the pairing between $\widehat{\mathcal{E}}$ and $\mathcal{D}$ is an extension of the usual pairing:*

$$\langle f, \varphi\rangle = \int_{\mathbb{R}^d} f(x)\varphi(x)dx$$

between $\mathcal{E}$ and $\mathcal{D}$ in the sense that $\langle q({}^*f), \ \varphi\rangle = \langle f, \ \varphi\rangle$ for all f in $\mathcal{E}$ and φ in $\mathcal{D}$.

The statements follow immediately from Lemma (2.2). In particular, $\langle q({}^*f), \varphi\rangle = \langle f, \varphi\rangle$ follows directly from (2.5).

Let F be a generalized distribution, Ω be an open set of $\mathbb{R}^d$ and $\mathcal{D}(\Omega)$ be the space of test functions on Ω. We say that "F is zero on Ω" and write "$F = 0$ on Ω", if $\langle F, \varphi\rangle = 0$ for all φ in $\mathcal{D}(\Omega)$. Exactly as in the case of the Schwartz distributions, we have $F = 0$ on Ω if and only if $F = 0$ on some open neighborhood of every point in Ω. Hence we can define the concept of "support": Let $\mathcal{O}_F$ be the union of all open sets of $\mathbb{R}^d$ on which F is zero. Then the support of F in $\mathbb{R}^d$ is the complement of $\mathcal{O}_F$ to $\mathbb{R}^d$, i.e. supp $F = \mathbb{R}^d - \mathcal{O}_F$.

Our next goal is to show that the space of Schwartz distributions $\mathcal{D}'$ "lies between $\mathcal{E}$ and $\widehat{\mathcal{E}}$", i.e. that $\mathcal{E} \subset \mathcal{D}' \subset \widehat{\mathcal{E}}$. We shall use some results of [12] and [13].

Let $\Delta \in {}^*\mathcal{D}$ be a density (kernel) of the Dirac distribution δ in the sense that

$$\int_{{}^*\mathbb{R}^d} \Delta(x)\,{}^*\varphi(x)dx = \varphi(0) \tag{2.7}$$

for all φ in $\mathcal{D}$ where ${}^*\varphi$ is the nonstandard extension of φ. A proof of the existence of a nonstandard function Δ with these properties can be found in [12].

(2.8) Definition: We define the inclusion $\mathcal{D}' \subset \widehat{\mathcal{E}}$ by the mapping $T \to q({}^*T * \Delta)$ where ${}^*T$ is the nonstandard extension of T.

(2.9) Remark (Justification): In [15] it is shown that if T is a Schwartz distribution, then the nonstandard function ${}^*T * \Delta \in {}^*\mathcal{E}$ is a density (kernel) of T in the sense that

$$\int_{{}^*\mathbb{R}^d} ({}^*T * \Delta)(x)\,{}^*\varphi(x)dx = \langle T, \ \varphi\rangle \tag{2.10}$$

for all φ in $\mathcal{D}$. Hence the generalized distribution $q({}^*T * \Delta)$ does not depend on the choice of Δ which justifies the above definition.

(2.11) Lemma: *Let $T \in \mathcal{D}'$ and $\psi \in \mathcal{E}$ and let ψT be the product of ψ and T in $\mathcal{D}'$ in the sense of distribution theory and ${}^*\psi({}^*T * \Delta)$ be the pointwise product of ${}^*\psi$ and ${}^*T * \Delta$ in ${}^*\mathcal{E}$. Then:*

$$ {}^*(\psi T) * \Delta \cong {}^*\psi({}^*T * \Delta). $$

Proof: In view of (2.10), we obtain:

$$ \int_{{}^*\mathbb{R}^d} ({}^*(\psi T) * \Delta)(x)\, {}^*\varphi(x)dx = \langle \psi T, \varphi \rangle = \langle T, \psi\varphi \rangle = $$

$$ = \int_{{}^*\mathbb{R}^d} {}^*\psi(x)({}^*T * \Delta)(x)\, {}^*\varphi(x)dx \quad , $$

for all φ in $\mathcal{D}$, as required. ∎

Notice that according to the above lemma, the nonstandard function ${}^*\psi({}^*T * \Delta)$ is a density (kernel) of the Schwartz distribution ψT.

(2.12) Lemma ${}^*f * \Delta \cong {}^*f$ *for all f in $\mathcal{E}$, where ${}^*f$ is the nonstandard extension of f.*

Proof: Applying (2.10) for $T = f$ (in the sense of the distribution theory) we have

$$ \int_{\mathbb{R}^d} f(x)\varphi(x)dx = \int_{\mathbb{R}^d} ({}^*f * \Delta)(x)\, {}^*\varphi(x)dx $$

for all φ in $\mathcal{D}$ and the result follows in view of (2.5). The proof is complete. ∎

(2.13) Proposition:

(i) *The mapping $T \to q({}^*T * \Delta)$ from $\mathcal{D}'$ into $\widehat{\mathcal{E}}$ is injective and preserves the addition, multiplication by a complex (standard) number, partial differentiation of any standard order and multiplication by a standard smooth function in $\mathcal{E}$.*

(ii) *The mapping $T \to q({}^*T * \Delta)$ from $\mathcal{D}'$ into $\widehat{\mathcal{E}}$ preserves the pairing between $\mathcal{D}'$ and $\mathcal{D}$ in the sense that if $T \in \mathcal{D}'$ and $\varphi \in \mathcal{D}$, then*

$$ \langle T, \varphi \rangle = \langle q({}^*T * \Delta), \varphi \rangle. $$

(iii) *The mapping $T \to q({}^*T * \Delta)$ from $\mathcal{D}'$ into $\widehat{\mathcal{E}}$ is an extension of the mapping $f \to q({}^*f)$, from $\mathcal{E}$ into $\widehat{\mathcal{E}}$ (recall that $\mathcal{E} \subset \mathcal{D}'$ in the sense of distribution theory).*

(iv) *The mapping $T \to q({}^*T * \Delta)$ from $\mathcal{D}'$ into $\widehat{\mathcal{E}}$ preserves the "support" in $\mathcal{D}'$, in the sense that* $\operatorname{supp} T = \operatorname{supp} q({}^*T * \Delta)$, *where* $\operatorname{supp} T$ *is in the sense of $\mathcal{D}'$ and* $\operatorname{supp} q({}^*T * \Delta)$ *is in the sense of $\widehat{\mathcal{E}}$.*

Proof: (i) Assume that $q({}^*T * \Delta) = 0$ in $\widehat{\mathcal{E}}$ for some T. That is

$$\int_{{}^*\mathbb{R}^d} ({}^*T * \Delta)(x)\, {}^*\varphi(x)dx = 0$$

for all $\varphi \in \mathcal{D}$, which implies $T = 0$ in $\mathcal{D}'$, by (2.10). The preservation of the addition, scalar multiplication and partial differentiation follows from the fact that the mapping $T \to {}^*T * \Delta$ from $\mathcal{D}'$ into ${}^*\mathcal{E}$ preserves those operations [13]. The preservation of the multiplication by a standard function ψ follows directly from Lemma (2.11). (ii) follows directly from (2.10) and (iii) follows directly from Lemma (2.12). (iv) follows directly from (ii), since $\langle T, \varphi\rangle$ and $\langle q({}^*T * \Delta), \varphi\rangle$ are both zero or non-zero. The proof is complete. ∎

(2.14) Examples
1. $\delta \to q(\Delta)$ since we have ${}^*\delta * \Delta = \Delta$, by the transfer principle. Hence $q(\Delta)$ represents the Dirac distribution δ in $\hat{\mathcal{E}}$. 2. The generalized distributions $q(\Delta^2)$, $q(\exp(i\Delta))$ and $q(\exp(\Delta))$ all lie in $\widehat{\mathcal{E}} - \mathcal{D}'$.

(2.15) Definition: Let $P(x, \partial) : \mathcal{E} \to \mathcal{E}$ be a linear partial differential operator with C^∞-coefficients given by the symbol (0.2) (see the beginning of the Introduction). We define $\widehat{P}(x, \partial) : \widehat{\mathcal{E}} \to \widehat{\mathcal{E}}$ by the same symbol (0.2):

$$\widehat{P}(x, \partial) = \sum_{|\alpha|\le m} c_\alpha(x)\partial^\alpha,$$

where all operations are in the sense of $\widehat{\mathcal{E}}$ (Definition (2.3)).

(2.16) Lemma: *Let $P(x, \partial) : \mathcal{E} \to \mathcal{E}$ be a linear partial differential operator with C^∞-coefficients, $P'(x, \partial) : \mathcal{D}' \to \mathcal{D}'$ be its extension on $\mathcal{D}'$ in the sense of the distribution theory and ${}^*P(x, \partial) : {}^*\mathcal{E} \to {}^*\mathcal{E}$ be its nonstandard extension. Then:*
(i) *For any $f \in {}^*\mathcal{E}$ we have*

$$\widehat{P}(x, \partial)[q(f)] = q({}^*P(x, \partial)f).$$

(ii) *The operator $\widehat{P}(x, \partial)$ is an extension of $P(x, \partial)$ in the sense that for any $f \in \mathcal{E}$ we have:*

$$\widehat{P}(x, \partial)[q({}^*f)] = q({}^*(P(x, \partial)f)).$$

(iii) *The operator $\widehat{P}(x, \partial)$ is an extension of $P'(x, \partial)$ in the sense that for any $T \in \mathcal{D}'$ we have:*

$$\widehat{P}(x, \partial)q({}^*T * \Delta) = q({}^*(P'(x, \partial)T) * \Delta).$$

Proof: (i) follows directly from the definition of the operations in $\widehat{\mathcal{E}}$ (Definition (2.3)) and (ii) follows simply from the fact that ${}^*(P(x, \partial)f) = {}^*P(x, \partial)\, {}^*f$ (by the transfer principle).
(iii) We have to show that

$${}^*P(x, \partial)({}^*T * \Delta) \cong {}^*(P'(x, \partial)T) * \Delta.$$

Using (2.10) twice (and also "integrating by parts" along with the transfer principle), we get

$$\int_{{}^*\mathbb{R}^d} {}^*P(x,\partial)({}^*T * \Delta)(x)\,{}^*\varphi(x)dx = \int_{{}^*\mathbb{R}^d} ({}^*T * \Delta)(x)\,{}^*(P^t(x,\partial)\varphi(x))dx$$

$$= \langle T, P^t(x,\partial)\varphi(x)\rangle = \langle P'(x,\partial)T, \varphi\rangle = \int_{{}^*\mathbb{R}^d} ({}^*(P'(x,\partial)T) * \Delta)(x)\,{}^*\varphi(x)dx,$$

for all φ in $\mathcal{D}$, as required. ∎

(2.17) Simpler Notation:
(i) We shall often write simply $P(x,\partial)$ instead of the more precise $\widehat{P}(x,\partial)$ when no confusion could arise, e.g. the action of $P(x,\partial)$ in $\widehat{\mathcal{E}}$ will be written as:

$$P(x,\partial)[q(f)] = q({}^*P(x,\partial)f), \tag{2.18}$$

for $f \in {}^*\mathcal{E}$.
(ii) We shall sometimes use the same notation T for the original distribution T and its image $q({}^*T * \Delta)$ in $\widehat{\mathcal{E}}$ writing sometimes $T \in \mathcal{D}' \subset \widehat{\mathcal{E}}$ to stress the fact that T is considered as an element of $\widehat{\mathcal{E}}$. For example, $\delta \in \mathcal{D}' \subset \widehat{\mathcal{E}}$ will denote the image $q(\Delta)$ in the class $\widehat{\mathcal{E}}$ of the original Dirac distribution $\delta \in \mathcal{D}'$.

3 An existence result

We obtain an existence result for a class of partial differential equations with C^∞-coefficients in the space of generalized distributions $\widehat{\mathcal{E}}$. In particular, we prove that H. Lewy's [7] equation has a solution for any choice of the right hand side a classical smooth function or Schwartz distribution. Recall that $\widehat{\mathcal{D}} = \{q(f) : f \in {}^*\mathcal{D}\}$ is a subspace of $\widehat{\mathcal{E}}$ (Definition (2.3) - (i)).

(3.1) Theorem: *Let $P(x,\partial)$ be a regular linear partial differential operator with C^∞-coefficients (Section 1). Then the equation:*

$$P(x,\partial)U = F, \tag{3.2}$$

has a solution U in $\widehat{\mathcal{D}}$ for any choice of F in $\widehat{\mathcal{E}}$ (in particular, for any choice of F in $\mathcal{D}'$).

Proof: Suppose that $F = q(f)$ for some f in ${}^*\mathcal{E}$. We show first that there exists u in ${}^*\mathcal{D}$ such that

$${}^*P(x,\partial)u \cong f$$

in ${}^*\mathcal{E}$. Or, equivalently, we have to show that

$${}^*P_\varphi(u) = \int_{{}^*\mathbb{R}^d} f(x)\,{}^*\varphi(x)dx,$$

for all φ in $\mathcal{D}$, where $^*P_\varphi$ is the nonstandard extension of P_φ (Section 1). Define the family of internal sets $\mathcal{A}_\varphi$, $\varphi \in \mathcal{D}$, by

$$\mathcal{A}_\varphi = \{u \in {}^*\mathcal{D} : {}^*P_\varphi(u) = \int_{{}^*\mathbb{R}^d} f(x)\,{}^*\varphi(x)dx\ \},$$

and observe that, by Lemma (1.7), it has the finite intersection property. Hence, by the saturation principle (Tom Lindstrøm [8], Chapter III), the intersection

$$\mathcal{A} = \bigcap_{\varphi \in \mathcal{D}} \mathcal{A}_\varphi$$

is non-empty. Hence, every u in $\mathcal{A}$ has the desired property. Now the generalized distribution $U = q(u)$ satisfies (3.2) since, in view of (2.18), we have

$$P(x, \partial)U = P(x, \partial)q(u) = q({}^*P(x, \partial)u) = q(f) = F,$$

as required. The proof is complete. ∎

(3.3) Corollary *Every regular linear partial differential operator with C^∞-coefficients $P(x, \partial)$ has a fundamental solution E in $\widehat{\mathcal{D}}$, i.e.*

$$P(x, \partial)E = \delta,$$

where δ is the Dirac delta distribution in $\widehat{\mathcal{E}}$.

Proof: Let $F = \delta$ in Theorem (3.1). ∎

(3.4) Corollary: *H. Lewy's equation:*

$$\frac{\partial U}{\partial x_1} + i\frac{\partial U}{\partial x_2} - 2i(x_1 + ix_2)\frac{\partial U}{\partial x_3} = F \tag{3.5}$$

has a solution U in $\widehat{\mathcal{D}}(\mathbb{R}^3)$ for any choice of F in $\widehat{\mathcal{E}}(\mathbb{R}^3)$ (in particular, for any choice of F in $\mathcal{D}'(\mathbb{R}^3)$).

Proof: The existence of U follows immediately from Theorem (3.1), since the operator

$$L(x, \partial) = \frac{\partial}{\partial x_1} + i\frac{\partial}{\partial x_2} - 2i(x_1 + ix_2)\frac{\partial}{\partial x_3}$$

is regular (Section 1). ∎

Recall that H. Lewy's equation does not have solutions for a large choice of the right hand side F even in $\mathcal{D}$: neither in the class of Schwartz distributions (H. Lewy [7]), nor in the class of hyperfunctions (P. Schapira [10]). If the right hand side F is in $\mathcal{E}'$ (in particular, in $\mathcal{D}$), H. Lewy's equation (3.5) does have local solutions in the class of the analytic functionals (F. Treves [14], [15]). However, this result does not include the case when F is in $\mathcal{D}' - \mathcal{E}'$ (or in $\mathcal{E} - \mathcal{D}$).

The next result shows that the class of generalized distributions $\widehat{\mathcal{E}}$ preserves the solvability of the linear partial differential equations with constant coefficients.

(3.6) Corollary: *Let $P(\partial)$ be a linear partial differential operator with constant coefficients. Then:*
(i) The equation

$$P(\partial)U = F, \tag{3.7}$$

has a solution U in $\widehat{\mathcal{D}}$ for any choice of F in $\widehat{\mathcal{E}}$.
(ii) $P(\partial)$ has a fundamental solution E in $\widehat{\mathcal{D}}$, i.e.

$$P(\partial)E = \delta \tag{3.8}$$

(iii) If F is a Schwartz distribution with compact support, then the generalized distribution given by the formula:

$$U = F * E, \tag{3.9}$$

is a solution of the equation (3.7) for any choice of the fundamental solution E of the operator $P(\partial)$ in $\widehat{\mathcal{E}}$ (even when E is outside $\mathcal{D}'$).
(iv) The equality (3.9) holds for any solution U of (3.7) in $\mathcal{E}'$ and any fundamental solution E of the operator $P(\partial)$ in $\widehat{\mathcal{E}}$ (even when E is outside $\mathcal{D}'$).

Proof: (i) follows directly from Theorem (3.1), and (ii) follows directly from Corollary (3.3), since $P(\partial)$ is regular (Section 1).
(iii) Suppose that $E = q(v)$ for some $v \in {}^*\mathcal{E}$, i.e. ${}^*P(\partial)v \cong \Delta$. In view of (2.18), we have

$$\begin{aligned} P(\partial)U = P(\partial)(F * q(v)) = P(\partial)q({}^*F * v) = q({}^*P(\partial)({}^*F * v)) = \\ = q({}^*F * {}^*P(\partial)v) = q({}^*F * \Delta) = F, \end{aligned}$$

as desired, since ${}^*F * ({}^*P(\partial)v) \cong {}^*F * \Delta$, by Lemma (2.2) - (ii), and $q({}^*F * \Delta) = F$ (Notation (2.17) - (ii)). (iv) $U \in \mathcal{E}'$ implies $F \in \mathcal{E}'$ and $P(\partial)U = F$ in $\mathcal{D}'$. Suppose that $E = q(v)$ for some $v \in {}^*\mathcal{E}$, i.e. ${}^*P(\partial)v \cong \Delta$. By Lemma (2.2) -(ii), we get

$${}^*U * \Delta \cong {}^*U * ({}^*P(\partial)v) = {}^*(P(\partial)U) * v = {}^*F * v$$

in ${}^*\mathcal{E}$. Hence it follows $q({}^*U * \Delta) = F * E$ in $\widehat{\mathcal{E}}$ which coincides with (3.9) in view of Notation (2.17) - (ii). The proof is complete. ∎

A Concluding Remark: We would like to stress that in our view the solvability of H. Lewy's equation in $\widehat{\mathcal{E}}$ is due to the simple fact that $\widehat{\mathcal{E}}$ is larger than $\mathcal{D}'$ rather than to some unique properties of the class $\widehat{\mathcal{E}}$. In contrast to this, in the case of F. Treves' solutions in the class of analytic functionals, the property "to be solvable" for a given equation is connected with the property "to be localizable" for the corresponding solutions: "... what lies at the root of nonsolvability of PDEs is our insistence in having solutions which are localizable generalized functions" (F. Treves [16], p. 571). Here it is meant that the Schwartz distributions are "localizable" but the analytic functionals are not. Notice that the generalized distributions in $\widehat{\mathcal{E}}$ are also "localizable" in exactly the same sense as the Schwartz distributions are. This is a simple consequence of the existence of a pairing between $\widehat{\mathcal{E}}$ and $\mathcal{D}$ (Proposition (2.6), (iii)).

Acknowledgement: The author would like to thank Thomas K. Boehme, Ed Glassco, Goro Kato and Kent Morrison for the friendly critical discussion of the manuscript and also to Gerald B. Folland, Michael Oberguggenberger, Jean-Pierre Rosay and Francois Treves for their fruitful correspondence.

References

[1] J.F. Colombeau, "New Generalized Functions and Multiplication of Distributions", North-Holland Math. Studies 84, 1984.

[2] J.F. Colombeau, "New General Existence Result for Partial Differential Equations in the C^∞ Case", Preprint of the Universite de Bordeaux, 1984.

[3] J.F. Colombeau, A. Heibig, M. Oberguggenberger, "Generalized solutions to partial differential equations of evolution type", Preprint of Ecole Normale Superieure de Lyon, France, 1991.

[4] N. Dunford and J.T. Schwartz, "Linear Operators, Part I: General Theory", Interscience Publishers, Inc. New York, 1958.

[5] L. Ehrenpreis, "Solutions of some problems of division III", Am. J. Math., 78(1956), p. 685.

[6] L. Hörmander, "Linear partial differential operators", Grund. Math. Wis., 116(1963), Springer Verlag.

[7] H. Lewy, "An example of a smooth linear partial differential equation without solution", Annals of Mathematics, Vol. 66, No. 1, July, 1957.

[8] T. Lindstrøm, "An Invitation to Nonstandard Analysis" in "Nonstandard Analysis and its Applications", edited by Nigel Cutland, London Mathematical Society Student Texts 10, Cambridge University Press, 1988.

[9] B. Malgrange, "Sur la propagation de la regularite des solutions des equations a coefficients constants", Bull. Math. Soc. Sci. Math. Phys. R.P. Romuanie 3 (53) (1959), 433-440.

[10] P. Schapira, Une equation aux derivees partielles sans solutions dans l'espace des hyperfonctions, C.R. Acad. Sci. Paris Ser. A-B 265 (1967), A665-A667. MR 36, #4112.

[11] L. Schwartz, "Theorie des Distributions I, II", Herman, Paris 1950, 1951.

[12] T. Todorov, "Nonstandard Delta Function", Proceedings of the American Mathematical Society, Volume 110, N_o 4, December 1990, 1143–1144.

[13] T. Todorov, "Pointwise Kernels of Schwartz Distributions", Proceedings of the American Mathematical Society, Vol. 114, N_o 3, March 1992, 817-819.

[14] F. Treves, "On Local Solvability of Linear Partial Differential Equations", Bulletin of the American Mathematical Society, Vol. 76, N_o 3, May, 1970, 552-571.

[15] F. Treves, "Ovcyannikov theorem and hyperdifferential operators", Notas de Matematica, no. 46. Inst. Mat. Pura. Apl. con. Nac. Pesquisas, Rio de Janeiro, 1968.

ON THE GENERATION OF TOPOLOGY BY EXTERNAL EQUIVALENCE–RELATIONS

Bernd Wietschorke
Fachbereich 7 – Mathematik
Berg. Univ. Wuppertal, Gau

str. 20
D–42119 Wuppertal

Abstract

Within the framework of Nelson's Internal Set Theory and the theory of external equivalence relations developed by Reeken we consider such relations in a vectorspace. In general the external set of infinitesimals is not a monad and therefore doesn't define a topology. We give a necessary and sufficient condition for the smallest monad containing this set to be the monad of a (not necessarily linear) topology. Furthermore a criterion is deduced that depends only on the defining sets of the given relation and makes no use of intrinsic information about this monad. Finally an application to a certain class of locally convex spaces is given.

Preliminary remark: All results presented in this article will become part of my thesis which will be finished in 1993. All statements or comments without proof will appear there.

The main context of this work is Nelson's Internal Set Theory (IST) [3]. External sets in the sense of IST are underlined or carry an index like σ (for the external set of all standard elements in a given set), l and so on. In general external sets are pointed out.
Reeken has shown in [5] how in IST external quotients of standard spaces with respect to external equivalence relations define new standard spaces. The construction encompasses topological completions but is more general and transcends the topological framework.
Without loss of generality we assume that the relation on a standard set X has the following form:

$$x \simeq y \Leftrightarrow \exists^{st} u \in U \forall^{st} v \in V : Q(x, y, u, v)$$

Here U and V are standard directed sets and Q is a standard formula, that is directed in U and codirected in V (see [5]).
Of course, the relation is assumed to be symmetric and transitve, and we define

$$X_{\simeq} := \{x \in X | x \simeq x\} \qquad \textit{and} \qquad X^{\simeq} := \{[x] \mid x \in X_{\simeq}\},$$

which we call the domain respectively the classes of the relation. Both are external sets in the sense of IST.
Furthermore we assume two additional properties of the relation:

(i) ${}^S\simeq = id_{X\times X}$, i.e. the standardization of the relation is the equality

(ii) the density condition: $\forall x \in X_{\simeq} \exists^{st} u \forall^{st} v \exists^{st} y \in X : Q(x, y, u, v)$

If these conditions are satisfied, the main result of Reeken is:
There exists a standard set $\mathbf{X}^{\approx}$, called the shadowspace, and an external bijection

$$\underline{k} : X^{\simeq} \to \mathbf{X}^{\approx}_{\sigma},$$

where the index σ denotes the external set of all standard elements of the given set. So the classes of the relation represent the standard elements of $\mathbf{X}^{\approx}$.

In the following we consider the problem, how such an external relation can be used to construct a topology and how this topology can be described. This can be done in a rather general frame but within the limitations of this article we restrict to a special situation.
Let X be a vectorspace and $\simeq$ an external relation on X which is compatible with the vectorspace-operations:

$$\begin{array}{l} x \simeq x' \wedge y \simeq y' \quad \Rightarrow x + y \simeq x' + y' \\ x \simeq x' \wedge \lambda \simeq_{\mathbf{K}} \lambda' \Rightarrow \lambda x \simeq \lambda' x' \end{array}$$

Here $\simeq_{\mathbf{K}}$ denotes the usual relation in the field $\mathbf{K}$ ($\mathbf{R}$ or $\mathbf{C}$), i.e. both scalars are limited with infinitesimal distance.
Then the class of a standard element x is given by $[x] = x + [0]$. For abbreviation let

$$\underline{\mu} := [0] = \{x \in X | x \simeq 0\}.$$

It is well known, that in case $\underline{\mu}$ is a monad it directly induces a vectorspace-topology on X. In general $\underline{\mu}$ is an external collection, which can be represented as an external union of monads. If we denote

$$Q_{uv} := \{x \in X | Q(x, 0, u, v)\},$$

and for $u \in U_\sigma$

$$\underline{\mu}_u := \bigcap_{v \in V_\sigma} Q_{uv},$$

$\underline{\mu}$ has the form:

$$\underline{\mu} = \bigcup_{u \in U_\sigma} \underline{\mu}_u$$

In order to construct a topology $\underline{\mu}$ has to become a monad, and the first attempt leads to the smallest monad that contains $\underline{\mu}$, the so called discrete monad $\underline{\delta}(\underline{\mu})$: if $\mathcal{F}$ is defined as

$$\mathcal{F} :=^S \{F \subseteq X | \underline{\mu} \subset F\},$$

then $\mathcal{F}$ is a filter and $\underline{\delta}(\underline{\mu})$ the corresponding filter-monad:

$$\underline{\delta}(\underline{\mu}) = \bigcap_{F \in \mathcal{F}_\sigma} F$$

Benninghofen, Richter and Stroyan [1] have discussed this monad with the aid of π-monads. They have shown, that $\underline{\delta}(\underline{\mu})$ in general does not define a topology, and they gave a necessary and sufficient condition when $\underline{\delta}(\underline{\mu})$ does.
In the following we present an explicit construction of the finest topology a given relation gives rise to, and we deduce a criterion which states when $\underline{\delta}(\underline{\mu})$ is the monad of a topology. This criterion is not an intrinsic condition for $\underline{\delta}(\underline{\mu})$ like in the mentioned article but depends only on the defining sets U and V of the relation.
At first it is necessary to deal with the relationship between X and the shadowspace $\mathbf{X}^{\approx}$.
As already mentioned there exists an external bijection $\underline{k} : X^{\simeq} \to \mathbf{X}^{\approx}_\sigma$. If $\underline{r} : X_{\simeq} \to X^{\simeq}$ is the external map which sends each element in $X_{\simeq}$ to its class, we get by standardization of $\underline{k} \circ \underline{r}$ a map

$$i : X \to \mathbf{X}^{\approx},$$

which is injective by (i) from above and in this special context also linear.
For a standard set $M \subseteq X$ we define

$$^{\mathbf{S}_\mathbf{X}}M := {}^S\{z \in \mathbf{X}^{\approx} | \exists x \in M \cap X_{\simeq} : z = \underline{k} \circ \underline{r}(x)\}$$

So $^{\mathbf{S}_\mathbf{X}}M$ is the standard set in the shadowspace the standard elements of which are represented by classes in M. The operation $^{\mathbf{S}_\mathbf{X}}$ will play the central role in the further context.

The method we use works with what we call the Robinson-Principle.
The well known lemma of Robinson states, that for example an internal sequence, which is infinitesimal in all standard components, must stay infinitesimal up to a nonstandard component. Essential for this property is, that "infinitesimal" is described by a monad. For an arbitrary external relation there is at first no hope, that this theorem remains valid. But in some part it can be generalized and turns out to be useful.
For abbreviation we put

$$\underline{\mathcal{A}}(J) := \{M \subseteq J | \text{M finite and } J_\sigma \subseteq M\}$$

Definition 1 If J is a standard directed set, $\simeq$ is said to satisfy the Robinson-Principle with respect to J if the following holds:

$$(y_\iota)_{\iota\in J} \subset X \ \wedge \ \forall^{st}\iota : y_\iota \simeq 0 \ \Rightarrow \ \exists M \in \underline{A}(J)\forall\iota \in M : y_\iota \simeq 0$$

Definition 2 If J and K are directed sets, J is called larger than K if the following is true:

$$\forall\phi : K \to J \,\exists\, \iota \in J \,\forall\, \kappa \in K : \phi(\kappa) \leq \iota$$

This definition may appear somewhat curious, but later on an example is given. We say that the formula Q is monadically directed in U iff:

$$u \leq u' \Rightarrow \underline{\mu}_u \subseteq \underline{\mu}_{u'}$$

Then a first result is:

Proposition 1 *If U is larger than J and Q is monadically directed in U, then $\simeq$ satisfies the Robinson-Principle with respect to J.*

Proof Let $(y_\iota)_{\iota\in J}$ be a directed net with the property: $\forall^{st}\iota : y_\iota \simeq 0$. This means explicitly:

$$\forall^{st}\iota\exists^{st}u\forall^{st}v : y_\iota \in Q_{uv}$$

Commutation of the first two quantifiers yields:

$$\exists^{st}\phi \in U^J\forall^{st}\iota\forall^{st}v : y_\iota \in Q_{\phi(\iota)v}$$

U being larger than J there is (by transfer) an $u \in U_\sigma$ with $\phi(\iota) \leq u$ for all $\iota \in J$. Because Q is monadically directed we get:

$$\forall^{st}\iota : y_\iota \in \underline{\mu}_{\phi(\iota)} \subseteq \underline{\mu}_u$$

Now the proposition follows by an overspill argument according to the usual Robinson-Lemma. ∎

Remark: The reverse statement is also true if the relation satisfies some additional properties.

For the next definition let for a directed set J

$$J_l := \{\iota \in J | \forall^{st}\iota' : \iota' \leq \iota\}.$$

The elements of J_l are called large.

Definition 3 Let $(x_\iota)_{\iota\in J} \subset X$ be a standard net. (x_ι) is called $\underline{\mu}$-convergent with $\underline{\mu}$-limit $x \in X_\sigma$ if and only if:

$$\forall\iota \in J_l : x_\iota \simeq x$$

We denote this convergence by $x_\iota \underset{\mu}{\to} x$.
By standardization $\underset{\mu}{\to}$ defines a convergence structure on X which is compatible

with the vectorspace operations. Sometimes this is also called a limit vectorspace.

Definition 4 A standard set $A \subseteq X$ is called closed if the following holds:

$$(x_\iota) \subseteq A \ \wedge \ x_\iota \underset{\mu}{\rightarrow} x \ \Rightarrow \ x \in A$$

A standard set $O \subseteq X$ is called open, if $X \setminus O$ is closed.

Proposition 2 *Let $A \subseteq X$ be standard and $y \in A \cap X_{\simeq}$. Then there is a standard net $(y_v)_{v \in V} \subset A$ and a set $M \in \underline{A}(V)$ such that:*

$$\forall v \in M_l : y_v \simeq y$$

Proof The density condition for y implies:

$$\exists^{st} u \forall^{st} v \exists^{st} x \in X : Q(x, y, u, v)$$

Commutation yields:

$$\exists^{st} u \exists^{st} \phi \in X^V \forall^{st} v : Q(\phi(v), y, u, v)$$

By overspill we can find some $M \in \underline{A}(V)$ with

$$\forall v \in M_l : \phi(v) \simeq y. \qquad (*)$$

For $v \in V_\sigma$ arbitrary we have $\{z \in A | Q(\phi(v), z, u, v)\} \neq \emptyset$, because this set contains y. Being standard, we can find some standard z in it and get by transfer:

$$\exists^{st} \psi \in A^V \forall v : Q(\phi(v), \psi(v), u, v)$$

For $v \in M_l$ it then follows from $(*)$: $\psi(v) \simeq \phi(v) \simeq y$ and the proof is finished. ∎

Proposition 3

(i) A standard set A is closed if and only if: $(y \in A, \, x \in X_\sigma \wedge y \simeq x) \Rightarrow x \in A$.

(ii) A standard set O is open if and only if: $\forall^{st} x \in O : x + \underline{\mu} \subset O$

(iii) The open sets in the sense of definition 4 satisfy the axioms of a topology. This topology is translation-invariant and X equipped with this topology is a T_1-space.

Proof

(i) Let A be closed, $y \in A$, $x \in X_\sigma$ with $x \simeq y$. By proposition 2 there is a standard net $(y_v) \subset A$ and a $M \in \underline{A}(V)$ such that $y_v \simeq y$ for $v \in M_l$. It follows $y_v \underset{\mu}{\rightarrow} x$, (what will not be shown here because it requires a lot of quantifier manipulation) and thus $x \in A$ because A is closed.
The reverse statement is trivial.

(ii) Follows immediately from (i)

(iii) That the axioms of a topology are satisfied follows immediately from (ii). Having (ii) in mind it is easy to see that O is an open neighborhood of $x \in X_\sigma$ if and only if $x - O$ is an open neighborhood of 0. So the topology is translation invariant. For $x \in X_\sigma$ $O := X \setminus \{x\}$ is open not containing x. This shows that X is a T_1-space. ∎

The last proposition shows that $\underline{\mu}$ and $\simeq$ characterize open and closed standard sets in the usual nonstandard way though $\underline{\mu}$ is not a monad.
The now established topology is called the associated topology with respect to the relation. Property (iii) of the last theorem shows that the associated topology is determined by the neighborhood-monad of 0. Let us denote

$$\underline{\delta}_{\mathcal{O}}(\underline{\mu}) := \bigcap_{O \in \mathcal{O}_\sigma} O \qquad with \qquad \mathcal{O} :=^S \{O \subseteq X | \text{O open and } 0 \in O\}.$$

The question, when $\underline{\delta}(\underline{\mu})$ defines a topology can now be transformed in the question: under what conditions does $\underline{\delta}_{\mathcal{O}}(\underline{\mu}) = \underline{\delta}(\underline{\mu})$ hold? Before answering this question we establish some properties of closed sets in that topology.

Proposition 4 *For a standard set $A \subseteq X$ the following statements are equivalent:*

(i) A is closed

(ii) $(x_v)_{v \in V} \subseteq A \wedge x_v \underset{\mu}{\to} x \Rightarrow x \in A$

(iii) $(i^{-1} \circ {}^{\mathbf{S}}\mathbf{x})(A) = A$

Proof (i)⇒ (ii) If A is closed (ii) holds trivially by definition of "closed".
(ii)⇒ (iii) The inclusion $A \subseteq (i^{-1} \circ {}^{\mathbf{S}}\mathbf{x})(A)$ is always true what can easily be checked. So let $x \in (i^{-1} \circ {}^{\mathbf{S}}\mathbf{x})(A)_\sigma$. There is some $y \in A$ such that $y \simeq x$. By proposition 2 we find some standard net $(y_v) \subset A$ $\underline{\mu}$-converging to x. Applying *(ii)* we get $x \in A$.
(iii) ⇒ (i) Let $x \in X_\sigma$, $y \in A$ with $y \simeq x$. Clearly $x \in (i^{-1} \circ {}^{\mathbf{S}}\mathbf{x})(A)$ and therefore $x \in A$. By proposition 3 A is closed. ∎

If V is a countable set, (i) says that closed and sequentially closed are equivalent notions in this topology. A topological space satisfying this property is usually called sequential.
The next theorem is the essential result and answers the before mentioned question.

Theorem 1 *The identity $\underline{\delta}_{\mathcal{O}}(\underline{\mu}) = \underline{\delta}(\underline{\mu})$ holds if and only if*

$${}^{S}(i^{-1} \circ {}^{\mathbf{S}}\mathbf{x}) : \mathcal{P}(X) \to \mathcal{P}(X)$$

is the topological closure-operator. This is true if and only if:

$$\forall^{st} M \subseteq X : (i^{-1} \circ {}^{\mathbf{S}}\mathbf{x})^2(M) = (i^{-1} \circ {}^{\mathbf{S}}\mathbf{x})(M) \qquad (*)$$

Proof For arbitrary standard sets A, B

$$A \subseteq (i^{-1}\circ^{\mathbf{Sx}})(A)$$

and

$$(i^{-1}\circ^{\mathbf{Sx}})(A \cup B) = (i^{-1}\circ^{\mathbf{Sx}})(A) \cup (i^{-1}\circ^{\mathbf{Sx}})(B)$$

is always true, so $i^{-1}\circ^{\mathbf{Sx}}$ is a topological closure operator if and only if $(*)$ holds. Let $(*)$ be true. Let $\mathcal{F} :=^{S} \{F \subseteq X | \underline{\mu} \subset F\}$. Then by definition:

$$\underline{\delta}(\underline{\mu}) = \bigcap_{F \in \mathcal{F}_\sigma} F$$

For $F \in \mathcal{F}_\sigma$ let

$$O_F := F \setminus (i^{-1}\circ^{\mathbf{Sx}})(X \setminus F).$$

We show that O_F is open and contains 0. For this let $x \in (O_F)_\sigma$ and $y \in \underline{\mu}$. If $x+y \notin F$ we get $x+y \in X \setminus F$ and thus because of $x \simeq x+y$ $x \in (i^{-1}\circ^{\mathbf{Sx}})(X \setminus F)$ in contradiction to $x \in O_F$. So we have $x+y \in F$. The assumption $x+y \in (i^{-1}\circ^{\mathbf{Sx}})(X \setminus F)$ yields due to $x \simeq x+y$

$$x \in (i^{-1}\circ^{\mathbf{Sx}})^2(X \setminus F) = (i^{-1}\circ^{\mathbf{Sx}})(X \setminus F)$$

what causes the same contradiction. Altogether we have shown $x+y \in O_F$, and by proposition 3 O_F is open.
Furthermore we have $\underline{\mu} \subset F$, i.e. $\underline{\mu} \cap (X \setminus F) = \emptyset$ what implies $0 \notin (i^{-1}\circ^{\mathbf{Sx}})(X \setminus F)$. So we conclude $0 \in O_F$. Now the first part of the theorem follows from

$$\underline{\delta}_{\mathcal{O}}(\underline{\mu}) \subseteq \bigcap_{F \in \mathcal{F}_\sigma} O_F \subseteq \bigcap_{F \in \mathcal{F}_\sigma} F = \underline{\delta}(\underline{\mu}) \subseteq \underline{\delta}_{\mathcal{O}}(\underline{\mu})$$

Now let $\underline{\delta}_{\mathcal{O}}(\underline{\mu}) = \underline{\delta}(\underline{\mu})$ and $A \subseteq X$ be standard. Let x be a standard element of $X \setminus (i^{-1}\circ^{\mathbf{Sx}})(A)$. Then $(x + \underline{\mu}) \cap A = \emptyset$. Therefore we can find some $F \in \mathcal{F}_\sigma$ such that $(x+F) \cap A = \emptyset$. The assumption gives us some open set $O \in \mathcal{O}_\sigma$ with $O \subseteq F$, consequently $(x+O) \cap A = \emptyset$. Since x was arbitrarily chosen we get

$$(X \setminus (i^{-1}\circ^{\mathbf{Sx}})(A)) \cap \overline{A} = \emptyset,$$

i.e. $\overline{A} \subseteq (i^{-1}\circ^{\mathbf{Sx}})(A)$. The reverse inclusion always being true (which is not difficult to show) the proof is complete. ∎

The next proposition links our former results with the last one.

Proposition 5 *If $\simeq$ satisfies the Robinson-Principle with respect to V then $(*)$ holds.*

Proof Let $z \in (i^{-1}\circ^{\mathbf{Sx}})^2(A)_\sigma$ for an arbitrary standard set A. We have to show $z \in (i^{-1}\circ^{\mathbf{Sx}})(A)$. To start with there is some $y \in (i^{-1}\circ^{\mathbf{Sx}})(A) \cap X_{\simeq}$ such that $y \simeq z$.

Choose a standard net $(y_v) \subset (i^{-1} \circ \mathbf{S_X})(A)$ due to proposition 2 suitable for y. Then we get:

$$\forall^{st} v \exists x \in A : x \simeq y_v$$

Therefore we get some internal net $(x_v) \subset A$ with the property: $\forall^{st} v : x_v \simeq y_v$. Now $\simeq$ satisfies the Robinson-Principle with respect to V, and so there is a $v \in V_l$ such that $x_v \simeq y_v$. v can be so chosen that in addition $y_v \simeq y$ holds. So we get $x_v \simeq y$ and from this $z \simeq x_v$, i.e. $z \in (i^{-1} \circ \mathbf{S_X})(A)$. This completes the proof. ■

If we combine this result with proposition 1 we see, that a certain relationship between U and V implies $\underline{\delta}_{\mathcal{O}}(\underline{\mu}) = \underline{\delta}(\underline{\mu})$, i.e. we don't have to know the special form of $\underline{\delta}(\underline{\mu})$.

Corollary 1 If U is larger than V $\underline{\delta}_{\mathcal{O}}(\underline{\mu}) = \underline{\delta}(\underline{\mu})$ holds.

In the following we deduce some remarkable properties of the topology induced by $\underline{\delta}_{\mathcal{O}}(\underline{\mu})$ in case $\underline{\delta}_{\mathcal{O}}(\underline{\mu}) = \underline{\delta}(\underline{\mu})$.
Let $(x_\iota)_{\iota \in J}$ be a standard net converging to the standard element x with respect to the topology. Let $A := \{x_\iota | \iota \in J\}$. By Theorem 1 we have $\overline{A} = (i^{-1} \circ \mathbf{S_X})(A)$. Since $x \in \overline{A}_\sigma$ there is some $y \in A \cap X_\simeq$ with $x \simeq y$. Proposition 2 then yields a standard net $(y_v)_{v \in V}$ with $y_v \underset{\mu}{\rightarrow} x$. Therefore we have proved that the topological closure of an arbitrary set B is determined by the $\underline{\mu}$-limits of nets in B with index set V. If V is countable, topological spaces possessing this property are usually called Fréchet spaces.
Now let V be a countable set and (x_n), (y_n) two standard sequences converging to 0 with respect to the topology. We assume that $(x_n + y_n)$ does not converge to 0, i.e. there is some $O \in \mathcal{O}_\sigma$ and a subsequence $(x_{n_k} + y_{n_k})$ such that:

$$\forall k : x_{n_k} + y_{n_k} \notin O \qquad (*)$$

(x_{n_k}) being a subsequence of (x_n) converges to 0 with respect to the topology. By what we have shown above there is a subsequence $(x_{n_{k_j}})$ with $x_{n_{k_j}} \underset{\mu}{\rightarrow} 0$. Then, of course, $y_{n_{k_j}}$ tends to 0 in the topology and a further application gives a (last) subsequence $(y_{n_{k_{j_i}}})$ with $y_{n_{k_{j_i}}} \underset{\mu}{\rightarrow} 0$. But $\underset{\mu}{\rightarrow}$ is compatible with addition and we get from $x_{n_{k_{j_i}}} + y_{n_{k_{j_i}}} \underset{\mu}{\rightarrow} 0$ a contradiction to $(*)$.
What we have shown is that convergent sequences form a vectorspace in this topology though it is not clear whether the topology is linear or not.
We summarize:

Corollary 2 Let $\underline{\delta}_{\mathcal{O}}(\underline{\mu}) = \underline{\delta}(\underline{\mu})$. Then the closure of an arbitrary set $A \subset X$ arises from taking $\underline{\mu}$-limits of arbitrary $\underline{\mu}$-convergent nets $(x_v)_{v \in V} \subset A$.
In case V is countable the set of all sequences convergent in the topology is a vectorspace.

Finally we give an example which shows how the methods developed so far can be applied.

Let X be a locally convex space with neighborhood base $\mathcal{U}$ and X' its topological dual. The space X is said to satisfy the countable neighborhood property (short c.n.p.), if the following holds: For any set of countably many neighborhoods $U_1, U_2, \ldots$ exist positive scalars $\rho_1, \rho_2, \ldots$ such that

$$U := \bigcap_{n \in \mathbf{N}} \rho_n U_n$$

is again a neighborhood of 0.
Consider the following external equivalence relation on X':

$$f \simeq g \Longleftrightarrow \exists^{st} U \in \mathcal{U} \forall^{st} n \in \mathbf{N} \exists^{st} \phi \in X' : |(f-\phi)_{|U}| \leq \frac{1}{n} \wedge |(\phi - g)_{|U}| \leq \frac{1}{n}$$

The required properties for $\simeq$ are easily checked. The class of 0, i.e. $\underline{\mu}$, is given by:

$$f \in \underline{\mu} \Longleftrightarrow \exists^{st} U \forall^{st} n : |f_{|U}| \leq \frac{1}{n} \qquad (\dagger)$$

Proposition 6 *Let X satisfy the c.n.p. Then the associated topology to $\simeq$ on X' is given by $\underline{\delta}(\underline{\mu})$. X' equipped with this topology has therefore all properties mentioned in corollary 2.*

Proof We want to apply corollary 1, so we have to show that $\mathcal{U}$ is larger than $\mathbf{N}$ and that the formula in $(\dagger)$ is monadically directed in $\mathcal{U}$. So at first we have the freedom of defining a suitable quasiordering (in the sense of Reeken [5]) that makes $\mathcal{U}$ larger than $\mathbf{N}$. The demand for monadical direction restricts this freedom and makes corollary 1 meaningful.
We define

$$U \leq V :\Longleftrightarrow \exists n \in \mathbf{N} : V \subset nU.$$

"$\leq$" is a quasiordering (i.e. a partial ordering that lacks the property of antisymmetry). We show $(\mathcal{U}, \leq)$ is larger than $\mathbf{N}$: For this let $\phi : \mathbf{N} \to \mathcal{U}$ be arbitrarily given. Due to the c.n.p. we find scalars (ρ_n) such that

$$U := \bigcap_{n \in \mathbf{N}} \rho_n \phi(n)$$

is a neighborhood of 0. For a given $n \in \mathbf{N}$ let a_n be an integer with $\rho_n \leq a_n$. Then it follows $U \subseteq a_n \phi(n)$ which means $\phi(n) \leq U$ by definition of "$\leq$". Since n and ϕ were arbitrary $(\mathcal{U}, \leq)$ is larger than $\mathbf{N}$.
Now let $U \leq V$, both standard. We have to show $\underline{\mu}_U \subseteq \underline{\mu}_V$. $f \in \underline{\mu}_U$ means: $\forall^{st} n : |f_{|U}| \leq \frac{1}{n}$. There is a standard integer m with $V \subseteq mU$ and this yields immediately $\forall^{st} n : |f_V| \leq \frac{1}{n}$, i.e. $f \in \underline{\mu}_V$. ∎

Remark: Denote $X'_U := \{f \in X' | \ f \text{ bounded on } U\}$. It turns out that $\underline{\delta}_{\mathcal{O}}(\underline{\mu})$ is the finest topology on X' that coincides with the canonical norm topology on each X'_U. This topology has been studied for example in Jarchow [2], where it is denoted by

$\eta^t(X', X)$. One class of spaces satisfying the c.n.p. are the so called gDF-spaces (see Jarchow [2]). It can be shown that for these spaces (in fact even for a weaker class, the so called df-spaces) $\underline{\delta}_{\mathcal{O}}(\underline{\mu})$ is the monad of a locally convex topology. It seems to be an open question if this is true for all spaces with the c.n.p. An example of a space with c.n.p. which is not a gDF-space can be found in Perez Carreras/ Bonet [4].

References

[1] Benninghofen B., Richter M.M., Stroyan K.D.: *Superinfinitesimals in topology and functional analysis,* London Math. Soc. Proc. 59, 1989, p.153-181

[2] Jarchow H.: *Locally convex spaces,* B.G.Teubner,1981

[3] Nelson E.: *Internal Set Theory,* Bull. Math. Soc., Vol.83, Nr.6, Nov.77, p.1165-1198

[4] Perez Carreras P., Bonet J.: *Barrelled Locally Convex Spaces,* North Holland 131

[5] Reeken M.: *On external constructions in Internal Set Theory,* Expositiones Math. 10 (1992), p.193-247

NONSTANDARD HULLS OF LEBESGUE–BOCHNER SPACES

G. Beate Zimmer
Dept. Mathematics, University of Illinois at Urbana-Champaign
Urbana, Il. 61 801, USA

Abstract

The problem of finding a representation of the nonstandard hull of $L_p(\mu, E)$ for $1 \leq p < \infty$ (E a Banach space) was posed by Henson and Moore in 1983 (see [3]). A related question was, whether this nonstandard hull can be described in some smooth way in terms of only $\widehat{L_p}(\mu)$ and $\widehat{E}$. We show that the answer to this question is negative.
The nonstandard hull $\widehat{L_p}(\mu, E)$ contains $L_p(\widehat{\mu}, \widehat{E})$ isometrically (where $\widehat{\mu}$ is the Loeb-measure obtained from μ). Between $L_p(\widehat{\mu}, \widehat{E})$ and $\widehat{L_p}(\mu, E)$ there is a space isometric to the space $M_p(\widehat{\mu}, \widehat{E})$ of $\widehat{E}$-valued, not necessarily measurable functions that have an SL^p-integrable μ-lifting. The Bochner integral extends in a natural way to $M_p(\widehat{\mu}, \widehat{E})$ and $M_p(\widehat{\mu}, \widehat{E})$ is a Banach space under the p-norm. For an infinite dimensional Banach space E, the inclusions $L_p(\widehat{\mu}, \widehat{E}) \subseteq M_p(\widehat{\mu}, \widehat{E}) \subseteq \widehat{L_p}(\mu, E)$ are both proper.
All of these observations are based on working with simple functions: the nonstandard hull of $L_p(\mu, E)$ equals the nonstandard hull of the space of simple E-valued functions.

Introduction

We use a superstructure model of nonstandard analysis and assume that the nonstandard model is at least $\aleph$-saturated, where $\aleph$ is an uncountable cardinal number. The nonstandard extension of an entity S is denoted by ${}^*S$. We use $\approx$ to denote "infinitesimally close". We write $a < \infty$ when a is nonnegative and finite in ${}^*\mathbb{R}$; if a is nonnegative and infinite, we set ${}^\circ a = +\infty$.

The basic setup is a standard finite nonatomic measure space $(X, \mathcal{A}, \mu)$ (we can assume it to be the unit interval with Lebesgue measure). From the nonstandard extension $({}^*X, {}^*\mathcal{A}, {}^*\mu)$ we obtain the corresponding Loeb space $({}^*X, L_\mu({}^*\mathcal{A}), \widehat{\mu})$ by composing ${}^*\mu$ with the standard part map, extending the resulting measure to the σ-algebra generated by ${}^*\mathcal{A}$, and then taking the completion. The Loeb space is again a standard finite measure space (see [4]). The nonstandard hull $\widehat{S}$ of a normed space S is obtained by taking the elements of finite norm in $^*S$ (denoted by $fin(^*S)$) and dividing out by those of infinitesimal norm. This construction works in settings as general as a uniform space, see [2]. We write π for the usual mapping of $fin(^*S)$ onto $\widehat{S}$; we call π the standard part mapping.

Throughout this paper E is an arbitrary Banach space. A function $g : X \to E$ is *simple* if it is a finite linear combination of characteristic functions of sets in $\mathcal{A}$ with coefficients in E, i.e. if it can be written as

$$g(x) = \sum_{i=1}^{n} e_i \, 1_{A_i}(x)$$

for $n \in \mathbb{N}$, $e_1, \ldots, e_n \in E$ and $A_1, \ldots, A_n$ a collection of pairwise disjoint sets in $\mathcal{A}$.

The standard concepts "μ-measurable", "Bochner integrable" and the definition of the Lebesgue-Bochner spaces $L_p(\mu, E)$ are used in the usual sense; see, for example, [1]. We denote by $\widehat{L_p}(\mu, E)$ the nonstandard hull of the Lebesgue-Bochner space $L_p(\mu, E)$.

1 The nonstandard extension of the space of simple functions

On the space of simple functions the "p-norm" $\|f\|_p = \left(\int_X \|f\|^p \, d\mu\right)^{\frac{1}{p}}$ is actually a seminorm. Identifying functions for which the difference has norm zero (or, equivalently, which agree μ-almost everywhere), we get a normed vector space which we will denote by $S_p(\mu, E)$. We usually don't distinguish between a function and its equivalence class.

1. Theorem: *Assume $1 \leq p < \infty$. Then the nonstandard hull of the space $S_p(\mu, E)$ of simple E-valued functions with the p-norm is $\widehat{L_p}(\mu, E)$.*

Proof: First note that if C is a normed space and D is a dense subspace of C then $\widehat{C} = \widehat{D}$. In fact, the condition that D is dense in C can be written as

$$\forall \epsilon \in \mathbb{R}^+ \; \forall c \in C \; \exists d \in D \;\text{ such that }\; \|d - c\| < \epsilon.$$

The $*$-transform of this formula asserts that for any element of $^*C$ there is an element of $^*D$ infinitesimally close. Identifying elements which are infinitesimally close in $fin(^*C)$ we see that each equivalence class contains an element of $fin(^*D)$, hence $\widehat{C} \subseteq \widehat{D}$. The reverse inclusion is obvious.
It is well-known that the (equivalence classes of) simple functions $S_p(\mu, E)$ form a dense subspace D of $C = L_p(\mu, E)$ for $1 \leq p < \infty$, and so the previous density argument applies. ∎

The nonstandard extension of $S_p(\mu, E)$ consists of all internal, $^*\mathcal{A}$-simple functions

$\varphi : {}^*X \to {}^*E$. Elements of ${}^*S_p(\mu, E)$ can be written as

$$\varphi(x) = \sum_{i=1}^{H} e_i \, 1_{A_i}(x)$$

with $H \in {}^*\mathbb{N}$, $e_1, \ldots, e_H \in {}^*E$ and $A_1, \ldots, A_H$ pairwise disjoint sets in ${}^*\mathcal{A}$.

2 SL^p-integrability

Definition: A function $\varphi : {}^*X \to {}^*E$ is called SL^p-integrable if for all $H \in {}^*\mathbb{N} \setminus \mathbb{N}$

$$\int_{\{\|\varphi\|^p \geq H\}} \|\varphi\|^p \, d\mu \approx 0.$$

If φ is SL^p-integrable then (by Theorem 3 of [4]) it follows that

$$\int_{{}^*X} {}^\circ\|\varphi\|^p \, d\widehat{\mu} = {}^\circ\left(\int_{{}^*X} \|\varphi\|^p \, d\mu\right) < \infty.$$

Actually in our case this condition is equivalent to SL^p-integrability because we assume that $(X, \mu, \mathcal{A})$ is a finite measure space.

If φ is SL^p-integrable and $p > q \geq 1$ then φ is SL^q-integrable (apply the defintion), but the converse does not hold:

Example 1: Fix $H \in {}^*\mathbb{N} \setminus \mathbb{N}$ and assume that p and q are real numbers with $p > q \geq 1$. Then any function which has constant norm $H^{\frac{1}{p}}$ on a set of measure $1/H$ and is zero elsewhere is SL^q-integrable, but not SL^p-integrable since $H^{\frac{q}{p}-1} \approx 0$ but $H^{\frac{p}{p}-1} = 1$.

It follows from Example 1 that $\int_{{}^*X} \|\varphi\|^p < \infty$ does not imply that φ is SL^p-integrable. This does, however, imply that φ is SL^q-integrable for $1 \leq q < p$, as we now show.

2. Proposition: *Let $p > q \geq 1$ be real numbers. Assume that $\varphi \in {}^*S_p(\mu, E)$ (or at least that $\|\varphi\|$ is an internally measurable function) with $\int_{{}^*X} \|\varphi\|^p \, d\mu < \infty$. Then φ is SL^q-integrable.*

Proof: We want to show that for any $M \in {}^*\mathbb{N} \setminus \mathbb{N}$

$$\int_{\{\|\varphi\|^q \geq M\}} \|\varphi\|^q \, d\mu \approx 0.$$

Now on $\{\|\varphi\|^q \geq M\}$ we have $M^{\frac{p-q}{q}} \|\varphi\|^q \leq \|\varphi\|^p$, and integrating on $\{\|\varphi\| \geq M^{\frac{1}{q}}\}$ we get

$$M^{\frac{p-q}{q}} \int_{\{\|\varphi\|^q \geq M\}} \|\varphi\|^q \, d\mu \leq \int_{{}^*X} \|\varphi\|^p \, d\mu < \infty.$$

Since by assumption $\frac{p-q}{q}$ is positive and real and $M \in {}^*\mathbb{N} \setminus \mathbb{N}$, we conclude that $M^{\frac{p-q}{q}}$ is not finite. Since the product at the left hand side of the inequality is finite, $\int_{\{\|\varphi\|^q \geq M\}} \|\varphi\|^q \, d\mu$ must be infinitesimal. ∎

3. Corollary: *For $p > q \geq 1$, the following implications hold for a function whose norm is internally measurable:*

$$\varphi \textit{ is } SL^p - \textit{integrable} \;\Rightarrow\; \int_{{}^*X} \|\varphi\|^p \, d\mu < \infty \;\Rightarrow\; \varphi \textit{ is } SL^q - \textit{integrable}.$$

Example 1 shows that the first implication of the Corollary can not be reversed, and the following example shows that the second implication can not be reversed:

Example 2: Take any (real valued) function $f \in L_q(\widehat{\mu}) \setminus L_p(\widehat{\mu})$. By the lifting theorem in [4] f has a SL^q-integrable μ-lifting $\varphi \in {}^*S_q(\mu, \mathbb{R})$. Assume $\int_{{}^*X} \|\varphi\|^p \, d\mu$ is finite. It follows (by a truncation argument) that $\int_{{}^*X} {}^\circ\|\varphi\|^p \, d\widehat{\mu}$ is finite. By the definition of a μ-lifting ${}^\circ\|\varphi\| = \|f\|$ $\widehat{\mu}$-almost everywhere, hence $\int_{{}^*X} \|f\|^p \, d\widehat{\mu} < \infty$, which contradicts $f \notin L_p(\widehat{\mu})$. Hence $\int_{{}^*X} \|\varphi\|^p \, d\mu$ is not finite. Multiplying any unit vector in ${}^*E$ by φ produces an SL^q-integrable function whose p-norm is infinite.

We will use Proposition 2 to describe how the nonstandard hull $\widehat{L_p}(\mu, E)$ sits inside $\widehat{L_q}(\mu, E)$ for $p > q \geq 1$. Before we do that, we need to describe the nonstandard hull $\widehat{L_p}(\mu, E)$ and the Banach space of extended Bochner integrable functions contained in it.

3 The SL^p-integrable functions

For any SL^p-integrable function φ,

$$\int {}^\circ\|\varphi\|^p \, d\widehat{\mu} = {}^\circ\left(\int \|\varphi\|^p \, d\mu\right) < \infty. \tag{3.1}$$

In what follows, we show that Equation (3.1) allows one to view SL^p-integrable functions as $\widehat{E}$-valued functions by ignoring the set where $\|\varphi\|$ is infinite. It also follows from Equation (3.1) that any SL^p-integrable function in ${}^*S_p(\mu, E)$ is necessarily an element of $fin({}^*S_p(\mu, E))$.

From an SL^p-integrable function φ we can obtain a function $f : {}^*X \to \widehat{E}$ by composing φ with the standard part map wherever $\varphi(x) \in fin({}^*E)$ and setting $f(x)$ equal to zero elsewhere. This function f has φ as a μ-lifting, i.e. $\pi \circ \varphi = f$ $\widehat{\mu}$-almost everywhere. Equation (3.1) ensures that we change $\|\varphi\|_p$ by at most an infinitesimal amount by composing φ with the standard part map.

The case $p = 1$ is considered in [6]: the space of equivalence classes of SL^1-integrable functions is (isometrically isomorphic to) a Banach space of extended Bochner integrable functions with values in $\widehat{E}$; these functions are not necessarily measurable.

In [6] we construct a normed space from the SL^1-integrable functions in ${}^*S_1(\mu, E)$ by taking the quotient modulo elements of infinitesimal norm. This space is isometrically isomorphic to the space $M(\widehat{\mu}, \widehat{E})$ formed from the set of equivalence classes of functions $f : {}^*X \to \widehat{E}$ which have an SL^1-integrable μ-lifting $\varphi \in {}^*S_1(\mu, E)$ (equivalence is modulo equality $\widehat{\mu}$-almost everywhere). The space $M(\widehat{\mu}, \widehat{E})$ is a Banach space under the norm $\|f\|_1 = \int_{{}^*X} \|f\| \, d\widehat{\mu}$. On $M(\widehat{\mu}, \widehat{E})$ we define an integral which generalizes the Bochner integral in a natural way as follows: we take an internal ${}^*\mathcal{A}$-simple and SL^1-integrable μ-lifting, integrate the μ-lifting internally and then apply the standard part map. The integral defined in this way shares many of the basic properties of the Bochner integral, but fails, for example, the convexity property (see [6]). The space $L_1(\widehat{\mu}, \widehat{E})$ is in general a proper subspace of $M(\widehat{\mu}, \widehat{E})$, and on it the extended integral coincides with the Bochner integral. If E is a Banach lattice then the space $M(\widehat{\mu}, \widehat{E})$ is the space $L_1^1(\widehat{\mu})$ introduced by Loeb and Osswald in [5], and the integral is the Loeb-Osswald integral. That is, their integral is extended by the integral introduced in [6].

For $p > 1$, an analogous construction works: let $M_p(\widehat{\mu}, \widehat{E})$ be the space of (equivalence classes of) functions $f : {}^*X \to \widehat{E}$ which have an SL^p-integrable μ-lifting in ${}^*S_p(\mu, E)$. The SL^p-integrability ensures a one-to-one correspondence between equivalence classes of functions and equivalence classes of μ-liftings in $S_p(\mu, E)$. The integral for f is defined as for $p = 1$. Again we get a Banach space for the p-norm (this follows by adapting the proof of Proposition 6.6 in [5], where a lattice structure is assumed). Any function in $L_p(\widehat{\mu}, \widehat{E})$ has a μ-lifting and the integral defined above coincides with the Bochner integral. All the proofs are similar to the proofs for $p = 1$ as given in [6].

Example 3: Assume E is infinite dimensional and $X = [0, 1]$. Take an Auerbach basis $\{e_1, \ldots, e_H\}$ with coordinate functionals $\{f_1, \ldots, f_H\}$ for an internal subspace of ${}^*E$ with internal dimension H for some $H \in {}^*\mathbb{N} \setminus \mathbb{N}$. Thus each e_i and f_i has norm 1 and $f_i(e_j) = 0$ if $i \neq j$, while $f_i(e_i) = 1$. Such an Auerbach basis was used by Henson and Moore in [3] to show that for infinite dimensional spaces $E \neq \widehat{E}$. Set $A_i = [(i-1)/H, i/H)$ for $1 \leq i \leq H$. This partitions ${}^*[0, 1]$ into H sets of internal measure $1/H$. Define $f : {}^*[0, 1] \to \widehat{E}$ by

$$f(x) = \pi\left(\sum_{i=1}^{H} e_i \, 1_{A_i}(x)\right).$$

This function has the SL^p-integrable μ-lifting $\varphi(x) = \sum_{i=1}^{H} e_i \, 1_{A_i}(x)$, which is internal and ${}^*\mathcal{A}$-simple; therefore $f \in M_p(\widehat{\mu}, \widehat{E})$. As for any $i \neq j$ we have $\|\pi(e_i) - \pi(e_j)\| \geq 1$, f can not be separably valued outside a set of $\widehat{\mu}$-measure zero.

4. Theorem *If E is infinite dimensional, then for each $p \geq 1$ the space $M_p(\widehat{\mu}, \widehat{E})$ contains functions which are not $\widehat{\mu}$-measurable.*

Proof: By the Pettis Measurability Theorem (see Theorem II.1.2 in [1]) a function $f : {}^*X \to \widehat{E}$ is $\widehat{\mu}$-measurable if and only if it is weakly $\widehat{\mu}$-measurable and $\widehat{\mu}$-essentially separably valued. Example 3 constructs a function in $M_p(\widehat{\mu}, \widehat{E})$ which is not $\widehat{\mu}$-essentially separably valued. ∎

4 Functions which are not SL^p-integrable

In [4] Loeb points out "in converting nonstandard integrals to standard integrals, we need to consider the fact that the product of any infinite positive number a with $\frac{1}{a}$ is 1 while ${}^\circ a \cdot {}^\circ(\frac{1}{a}) = +\infty \cdot 0$." SL^p-integrability is the concept which avoids this problem. However, the space $fin({}^*S_p(\mu, E))$ contains functions which are not SL^p-integrable. For these functions

$$\int_{{}^*X} {}^\circ\|\varphi\|^p \, d\widehat{\mu} < {}^\circ\left(\int_{{}^*X} \|\varphi\|^p \, d\mu\right).$$

Because of this inequality, we can not view the functions in $fin({}^*S_p(\mu, E))$ which are not SL^p-integrable as $\widehat{E}$-valued functions.

For $p > 1$, any function in $fin({}^*S_p(\mu, E))$ is, by Proposition 2, SL^1-integrable and hence extended Bochner integrable in the sense of the previous section. This makes $\widehat{L_p}(\mu, E)$ a space of equivalence classes of extended Bochner integrable functions. The equivalence classes are determined by the p-norm, not by equality $\widehat{\mu}$-almost everywhere. The following example shows that functions in $fin({}^*S_p(\mu, E))$ can agree $\widehat{\mu}$-almost everywhere but differ in p-norm. Hence we can not describe functions in $\widehat{L_p}(\mu, E)$ in terms of $\widehat{E}$-valued functions with μ-liftings.

Example 4: Take an $e \in {}^*E$ of norm one and an $r \in fin\,{}^*\mathbb{R}$. Define

$$\varphi = He\ 1_{[0,H^{-p}]} \text{ and } \psi = rHe\ 1_{[0,H^{-p}]}.$$

Both functions are zero $\widehat{\mu}$-almost everywhere, but $\|\varphi\|_p = 1$ and $\|\psi\|_p = r$.

For $p = 1$, the space $fin({}^*S_1(\mu, E))$ contains functions which are not SL^1-integrable and hence not extended Bochner integrable. Nevertheless, we can still make use of the internal integral of internal simple functions.

5. Proposition: *If $\varphi \in fin({}^*S_1(\mu, E))$, then $\int_X \varphi \, d\mu \in fin({}^*E)$.*

Proof: This follows from the internal triangle inequality:

$$\|\int_X \sum_{i=1}^H e_i\ 1_{A_i}\, d\mu\| = \|\sum_{i=1}^H e_i\ \mu(A_i)\| \leq \sum_{i=1}^H \|e_i\|\ \mu(A_i) = \int_X \|\sum_{i=1}^H e_i\ 1_{A_i}\|\, d\mu.$$

Hence if $\|\varphi\|_1 < \infty$ then $\int_X \varphi \, d\mu$ is finite in ${}^*E$. ∎

Proposition 5 allows us to compose the integral with the standard part map to get an $\widehat{E}$-valued integral for functions in $fin({}^*S_1(\mu, E))$. If $\|\varphi - \psi\|_1 \approx 0$ then $\pi(\int_{{}^*X} \varphi \, d\mu) = \pi(\int_{{}^*X} \psi \, d\mu)$, hence the integral on $\widehat{S_1}(\mu, E)$ is well defined.

Summary of results

For a fixed $p \geq 1$, we have

$$L_p(\widehat{\mu}, \widehat{E}) \subsetneq M_p(\widehat{\mu}, \widehat{E}) \subsetneq \widehat{L_p}(\mu, E).$$

Both inclusions are strict (see Examples 3 and 4). $L_p(\widehat{\mu}, \widehat{E})$ is characterized by measurability and $M_p(\widehat{\mu}, \widehat{E})$ is characterized by SL^p-integrability.

For $p > 1$, the elements of the space $\widehat{L_p}(\mu, E)$ can be regarded as equivalence classes of extended Bochner integrable $\widehat{E}$-valued functions. Unlike the case for the Lebesgue-Bochner spaces, however, the equivalence is not equality $\widehat{\mu}$-almost everywhere, but it is given by the internal p-norm. Example 5 illustrates the difference.

On $\widehat{L_1}(\mu, E)$, a $\widehat{E}$-valued integral is defined by the composition of the internal integral with the standard part map. Due to the lack of SL^1-integrability, not all functions in $\widehat{L_1}(\mu, E)$ can be regarded as $\widehat{E}$-valued functions when integrating.

For $p > q \geq 1$, it follows from Lemma 2 that $\widehat{L_p}(\mu, E) \subsetneq M_q(\widehat{\mu}, \widehat{E})$ and the inclusion is proper by example 2.

Remark: All of our constructions and definitions could have been done without excluding the case $0 < p < 1$, but for the Lebesgue-Bochner spaces it is common to assume $p \geq 1$.

References

[1] J. Diestel, J.J. Uhl: *Vector Measures,* Mathematical Surveys No. 15, American Mathematical Society, Providence, Rhode Island, 1977

[2] C.W. Henson: *The nonstandard hulls of a uniform space,* Pacific. Jour. Math. 44, No.1, 115-137, 1972

[3] C.W. Henson, L. Moore: *Nonstandard Analysis and the theory of Banach spaces,* in:A. E. Hurd (ed.): Nonstandard Analysis - Recent Developments, Springer Lecture Notes in Mathematics 983, 27-112, 1983

[4] P.A. Loeb: *Conversion from nonstandard measure spaces to standard measure spaces and applications to probability theory,* Trans. Amer. Math. Soc. 211, 113 - 122, 1975

[5] P.A. Loeb, H. Osswald: *Nonstandard Integration Theory in Topological Vector Lattices*, preprint, 1992

[6] G.B. Zimmer: *An extension of the Bochner integral generalizing the Loeb-Osswald integral,* preprint, 1993

Part II:

PROBABILITY THEORY

A NONSTANDARD APPROACH TO DIFFUSIONS ON MANIFOLDS AND NONSTANDARD HEAT KERNELS

Hiroshi Akiyama
Department of Applied Mathematics, Faculty of Engineering, Shizuoka University, Hamamatsu 432, Japan

Abstract

Internal ordinary differential equations are used to construct diffusions on manifolds and nonstandard heat kernels are obtained for heat equations.

1 Introduction

This paper deals with nonstandard constructions of diffusions on manifolds and nonstandard heat kernels. We construct a Brownian motion on a compact C^∞ Riemannian manifold M in §2 following [Ak1], using an internal ordinary differential equation on the nonstandard extension of the bundle of orthonormal frames. Then the heat kernel of a heat equation involving the Laplacian is treated in §3 with the use of a "nonstandard delta function" on ${}^*M$; the result improves [Ak1, Theorem 4.3]. In §4, we present a nonstandard heat kernel of a heat equation for sections of a vector bundle.

2 A nonstandard construction of a Brownian motion on a compact Riemannian manifold

Let $0 < n \in \mathbb{N}, \eta = \eta_0!$ with $\eta_0 \in {}^*\mathbb{N}_\infty := {}^*\mathbb{N} - \mathbb{N}$, and let $(\Omega, L(\mathcal{A}), \mu_L)$ be the Loeb space associated with the hyperfinite probability space $(\Omega = (\{-1,1\}^\eta)^n, \mathcal{A}, \mu)$ where $\mathcal{A}$ is the internal algebra of all internal subsets of Ω and $\mu(A) = |A|/|\Omega| \in {}^*[0,1]$, $A \in \mathcal{A}$ (see [L], [AFHL]). Each $\omega \in \Omega$ is expressed as $\omega = (\omega_k^\alpha), \omega_k^\alpha = \pm 1$, $(k = 1,2,\ldots,\eta;\ \alpha = 1,2,\ldots,n)$. Let $\{e_1, e_2, \ldots, e_n\}$ be the canonical basis of $\mathbb{R}^n$ (and thus of ${}^*\mathbb{R}^n$). Denote by $w : {}^*[0,1] \times \Omega \ni (t,\omega) \mapsto w(t,\omega) \in {}^*\mathbb{R}^n$ Anderson's $*$-random walk

$$w(t,\omega) = \sum_{\alpha=1}^{n} w_t^\alpha(\omega) e_\alpha, \quad w_t^\alpha(\omega) := \frac{1}{\sqrt{\eta}} \left(\sum_{k=1}^{[\eta t]} \omega_k^\alpha + (\eta t - [\eta t]) \omega_{[\eta t]+1}^\alpha \right),$$

where $[\eta t] \in {}^*\mathbf{Z}, [\eta t] \leq \eta t < [\eta t]+1$. *Anderson's Brownian motion* is given by $b(t,\omega) = \sum_{\alpha=1}^n b_t^\alpha(\omega) e_\alpha (t \in [0,1])$ with $b_t^\alpha(\omega) = {}^\circ w_t^\alpha(\omega)$ (the standard part of $w_t^\alpha(\omega)$) (see [An]). Put $t_j = j\Delta t = j/\eta\ (j=0,1,\dots,\eta)$; $\Delta_j w^\alpha = w^\alpha_{(j+1)\Delta t} - w^\alpha_{j\Delta t}, T_j = \{t \in {}^*\mathbb{R} : t_j < t < t_{j+1}\}\ (j = 0,1,\dots,\eta-1)$.

Let (M,g) be a compact connected C^∞ Riemannian manifold of dimension n and $\pi : O(M) \to M$ the bundle of orthonormal frames over M. Then $O(M)$ is compact. We use the Riemannian connection on $O(M)$ (cf. [KN]). For $\xi \in \mathbb{R}^n, B(\xi)$ denotes the *basic vector field* on $O(M)$ corresponding to ξ; that is, for $r \in O(M), B(\xi)_r$ is the unique horizontal vector at r such that $(\pi_*)_r(B(\xi)_r) = r\xi\ (\in T_{\pi(r)}M)$, where $(\pi_*)_r$ is the differential of the map π at r. From the map $B : \mathbb{R}^n \ni \xi \mapsto B(\xi) \in \mathcal{X}(O(M))$(:= the space of C^∞ vector fields on $O(M)$), we obtain ${}^*B : {}^*\mathbb{R}^n \ni \xi \mapsto {}^*B(\xi) \in {}^*(\mathcal{X}(O(M)))$.

Now for each $\omega \in \Omega$, consider the *-continuous curve $r_t(r,\omega), t \in {}^*[0,1]$, in ${}^*(O(M))$ starting from a standard point r (identified with ${}^*r$) of $O(M)$ at time $t=0$ and satisfying the internal ordinar y differential equation

$$\frac{dr_t}{dt} = {}^*B\left(\frac{dw(t,\omega)}{dt}\right)_{r_t}.$$

(Notice that $dw(t,\omega)/dt = \sum_{\alpha=1}^n (dw_t^\alpha(\omega)/dt)e_\alpha = \sum_{\alpha=1}^n (\Delta_j w^\alpha/\Delta t)e_\alpha$ if $t \in T_j$.) Observe that ${}^*\pi(r_t(r,\omega)), t \in {}^*[0,1]$, is a *-broken *-geodesic in ${}^*M$ (cf. [KN]). Put $X_t(r,\omega) := {}^{\circ *}\pi(r_t(r,\omega)) = \pi({}^\circ r_t(r,\omega))$.

Proposition 2.1 ([Ak1]). (1) $({}^\circ r_t(r,\omega))$, $t \in [0,1]$, *satisfies the stochastic differential equation in the Stratonovich form*

$$d\,{}^\circ r_t = \sum_{\alpha=1}^n B(e_\alpha)_{{}^\circ r_t} \circ db_t^\alpha, \tag{2.1}$$

where $\circ db_t^\alpha$ *is the Stratonovich stochastic differential of* b_t^α.

(2) $(X_t(r,\omega))$, $t \in [0,1]$, *is a Brownian motion on* (M,g), *that is, a* $\frac{1}{2}\Delta_M$*-diffusion where* Δ_M *is the Laplacian acting on* C^∞ *functions on* M.

Proof. (1) Let $t \in [0,1]$. For every C^∞ function $F : O(M) \to \mathbb{R}$, we have

$${}^*F(r_t(r,\omega)) - F(r) \approx \sum_{j=0}^{[\eta t]-1} \{{}^*F(r_{t_{j+1}}(r,\omega)) - {}^*F(r_{t_j}(r,\omega))\}$$

$$\approx \sum_{\alpha=1}^n \sum_{j=0}^{[\eta t]-1} {}^*(B(e_\alpha))_{r_{t_j}(r,\omega)}{}^*F \cdot \Delta_j w^\alpha(\omega) + \sum_{j=0}^{[\eta t]-1} \left(\frac{1}{2}\sum_{\alpha=1}^n {}^*(B(e_\alpha))^2\right)_{r_{t_j}(r,\omega)} {}^*F\ \mathrm{dot}\,\Delta t,$$

since the sum of the terms including $\Delta_j w^\alpha(\omega)\Delta_j w^\beta(\omega)\ (j=0,1,\dots,[\eta t]-1)$ with $\alpha \neq \beta$ is infinitesimal (cf. the proof of [An, Theorem 37]). Hence

$$F({}^\circ r_t(r)) - F(r) = \sum_{\alpha=1}^n \int_0^t B(e_\alpha)_{{}^\circ r_s(r)}F\cdot db_s^\alpha + \int_0^t \left(\frac{1}{2}\sum_{\alpha=1}^n (B(e_\alpha))^2\right)_{{}^\circ r_s(r)} F\ ds,$$

where $\cdot db_s^\alpha$ is the Itô stochastic differential of b_s^α. The right-hand side equals $\sum_{\alpha=1}^n \int_0^t B(e_\alpha)_{\circ r_s(r)} F \circ db_s^\alpha$, proving (2.1). Now (1) implies (2) (cf. [IW]). ∎

Let $f : M \to \mathbb{R}$ be C^∞. By Proposition 2.1, the expectation $u(t,x) = \mathbf{E}[f(X_t(r))] (t \in [0,1], x \in M, r \in \pi^{-1}(x))$ with respect to ν_L is well-defined (independent of the choice of r in $\pi^{-1}(x)$) and it solves the heat equation

$$\frac{\partial u}{\partial t} = \frac{1}{2}\Delta_M u, \;\; u(0,\cdot) = f, \;\; (u : [0,1] \times M \ni (t,x) \longmapsto u(t,x) \; in \mathbb{R}). \qquad (2.2)$$

3 A nonstandard representation of the heat kernel of Eq. (2.2)

Let (M,g) be a C^∞ Riemannian manifold with $\dim M = n\,(0 < n \in \mathbb{N})$ and dv_g the Riemannian volume density. Let $C^\infty(M;\mathbf{K})$ be the space of $\mathbf{K}$-valued C^∞ functions on M ($\mathbf{K} = \mathbb{R}$ or $\mathbf{K} = \mathbb{C}$) and $C_0^\infty(M;\mathbf{K}) := \{f \in C^\infty(M;\mathbf{K}) :$ the support of f is compact$\}$.

Theorem 3.1 ([Ak2]). *There exists a* $^*C^\infty$ *function* $\Delta : {}^*M \times {}^*M \to {}^*\mathbb{R}$ [*namely,* $\Delta \in {}^*(C^\infty(M\times M;\mathbb{R}))$] *satisfying the followin g properties:*

(1) $\Delta(x,y) = \Delta(y,x) = \sum_{i=1}^\nu \varphi_i(x)\varphi_i(y)$, $(x,\, y \in {}^*M)$, *for some* $\nu \in {}^*\mathbb{N}_\infty, \varphi_i \in {}^*(C_0^\infty(M;\mathbb{R}))$ $(i = 1,2,\ldots,\nu)$.

(2) *For all* $f \in C_0^\infty(M;\mathbb{C})$ *and* $x \in {}^*M$,

$${}^*f(x) = \int_{^*M} \Delta(x,y)\, {}^*f(y)\, {}^*dv_g(y).$$

(3) *Let* $T : C_0^\infty(M;\mathbb{C}) \to \mathbb{C}$ *be a Schwartz distribution on* M, *and define* $\gamma_T \in {}^*(C^\infty(M;\mathbb{C}))$ *by* $\gamma_T(y) = {}^*T(\Delta(\cdot,y)) = \sum_{i=1}^\nu {}^*T(\varphi_i)\varphi_i(y)$ $(y \in {}^*M)$. *Then*

$$T(f) = \int_{^*M} {}^*f(y)\, \gamma_T(y)\, {}^*dv_g(y), \qquad f \in C_0^\infty(M;\mathbb{C}).$$

Proof. By the saturation principle, there exists a hyperfinite dimensional internal vector subspace V of the internal vector space ${}^*(C_0^\infty(M;\mathbb{R}))$ over ${}^*\mathbb{R}$ such that the external set $\{{}^*f : f \in C_0^\infty(M;\mathbb{R})\}$ is a subset of V. Let $\nu = {}^*\dim V (\in {}^*\mathbb{N}_\infty)$ and take an orthonormal basis φ_i $(i = 1,2,\ldots,\nu)$ of V with respect to the internal inner product $(f_1,f_2) = \int_{^*M} f_1{\cdot}f_2\, {}^*dv_g$, $(f_1,f_2 \in V)$, so that for every $f \in C_0^\infty(M;\mathbb{C}) = C_0^\infty(M;\mathbb{R}) + \sqrt{-1}\,C_0^\infty(M;\mathbb{R})$, ${}^*f$ is ex pressed as ${}^*f = \sum_{i=1}^\nu c_i({}^*f)\varphi_i$ with $c_i({}^*f) = \int_{^*M} {}^*f(y)\,\varphi_i(y)\,{}^*dv_g(y) \in {}^*\mathbb{C}$. Then $\Delta(x,y) := \sum_{i=1}^\nu \varphi_i(x)\varphi_i(y)$ satisfies (1), (2). Moreover, (3) also holds since $\int_{^*M} {}^*f\gamma_T\, {}^*dv_g = \sum_{i=1}^\nu {}^*T(\varphi_i)c_i({}^*f) = {}^*T\left(\sum_{i=1}^\nu c_i({}^*f)\varphi_i\right) = {}^*T({}^*f) = T(f)$. ∎

Remark. When $M = \mathbb{R}^n$ and g is the Euclidean metric, T. Todorov ([T1], [T2]) proved the existence of a nonstandard delta function and (nonstandard) pointwise kernels of Schwartz distributions in a different manner.

Now assume that M is compact and connected. Let $\mathbf{K} = \mathbb{R}$. Consider $\Phi : [0,1] \times O(M) \times C^\infty(M;\mathbb{R}) \ni (t,r,f) \mapsto \mathbf{E}[f(X_t(r))] \in \mathbb{R}$ (see §2). This Φ is standard since it is identified with its graph, a subset of $([0,1] \times O(M) \times C^\infty(M;\mathbb{R})) \times \mathbf{R}$. Since $\Delta(\cdot,y) \in {}^*(C^\infty(M;\mathbb{R}))$ for each $y \in {}^*M$, we obtain the internal map ${}^*[0,1] \times {}^*(O(M)) \times {}^*M \ni (t,r,y) \mapsto {}^*\Phi(t,r,\Delta(\cdot,y)) \in {}^*\mathbb{R}$.

Theorem 3.2. *Let $e(t,x,y)$ denote the heat kernel of* (2.2), *which is known to exist. Then ${}^*\Phi(t,r,\Delta(\cdot,y))$ coincides with $e(t,x,y)$ for $t \in (0,1]$, x, $y \in M$, and $r \in \pi^{-1}(x)$.*

Proof. As is well-known, $e(t,x,\cdot)$ is smooth for fixed (t,x). Moreover, $\Phi(t,r,f) = \mathbf{E}[f(X_t(r))] = \int_M f(y)e(t,x,y)\,dv_g(y)$. Therefore it holds that ${}^*\Phi(t,r,\Delta(\cdot,y)) = \int_{{}^*M} \Delta(z,y)\,{}^*e(t,x,z)\,{}^*dv_g(z) = {}^*e(t,x,y) = e(t,x,y)$. ∎

4 A nonstandard heat kernel of a heat equation for sections of a vector bundle

Let $P(M,G,\pi_P)$ be a C^∞ principal fiber bundle over a compact connected C^∞ Riemannian manifold M; in this section, we do not assume that $\dim M = n$. Assume that G (the structure group) and P are compact. Let $\pi_E : E \to M$ be a C^∞ real vector bundle of rank $\ell \neq 0$ associated with P through a representation of G into $GL(\ell,\mathbb{R})$. Let E^* be the dual bundle of E. The fiber $\pi_E^{-1}(x)$ of E over $x \in M$ is denoted by E_x. For $x \in {}^*M$, we put ${}^*E_x := ({}^*(\pi_E))^{-1}(x)$. Let $A_0, A_1, \dots, A_n$ be projectable C^∞ vector fields on P invariant under the right translation by any element of G. Let $w_t^\alpha(\omega)$ and $b_t^\alpha(\omega)$ be as in §2. For convenience, we put $w_t^0 := t, \Delta_j w^0 := \Delta t$.

For each $\omega \in \Omega$, consider the *-continuous curve $(p_t(p,\omega))_{t\in{}^*[0,1]}$ in ${}^*P$ with $p_0(p,\omega) = p \in P$ governed by the internal ordinary differential equation

$$\frac{dp_t}{dt} = \sum_{\lambda=0}^{n} \frac{dw_t^\lambda(\omega)}{dt}({}^*A_\lambda)_{p_t}.$$

We often write $p_t(p)$ for $p_t(p,\omega)$ by suppressing ω. Each $p \in P$ with $\pi_P(p) = x$ is regarded as the admissible map $\mathbb{R}^\ell \ni v \mapsto p(v) \in E_x \subseteq E$. Observe that $p_t(pa) = p_t(p)a$, $(p \in P, a \in G)$. Then for $x \in M$, we can define $\theta_t(x) := \pi_P(p_t(p))$ and $\rho_{t,x} := p_t(p) \circ {}^*p^{-1} : {}^*E_x \to {}^*E_{\theta_t(x)}$ independently of the choice of $p \in \pi_P^{-1}(x)$. Given $\sigma \in \Gamma^\infty(E)$ (:= the space of C^∞ sections of E), define $F_\sigma : P \to \mathbb{R}^\ell$ by $F_\sigma(p) = p^{-1}(\sigma(\pi_P(p)))$, and set $\rho_t^{-1}({}^*\sigma)(x) = \rho_{t,x}^{-1}({}^*\sigma(\theta_t(x)))$, $(x \in M)$. We can also define $L_\lambda : \Gamma^\infty(E) \ni \sigma \mapsto L_\lambda\sigma \in \Gamma^\infty(E)$ by $L_\lambda\sigma(x) = p((A_\lambda)_p F_\sigma)$, $(x \in M,\ p \in \pi_P^{-1}(x),\ \lambda = 0,1,\dots,n)$.

Proposition 4.1 (cf. [Ak1]). (1) $({}^\circ p_t(p))_{t\in[0,1]}$ *satisfies the stochastic differential equation in the Stratonovich form*

$$d\,{}^\circ p_t = \sum_{\alpha=1}^{n} (A_\alpha)_{{}^\circ p_t} \circ db_t^\alpha + (A_0)_{{}^\circ p_t}\, dt.$$

(2) *Put* $H = \frac{1}{2}\sum_{\alpha=1}^{n}(L_\alpha)^2 + L_0$. *Let* $\sigma \in \Gamma^\infty(E)$, $(t,x) \in [0,1] \times M$. *Then*

$$\begin{aligned} {}^\circ(\rho_t^{-1}({}^*\sigma)(x)) - \sigma(x) &= \sum_{\alpha=1}^{n} \int_0^t {}^\circ(\rho_s^{-1}({}^*(L_\alpha\sigma))(x)) \cdot db_s^\alpha \\ &\quad + \int_0^t {}^\circ(\rho_s^{-1}({}^*(H\sigma))(x))\, ds. \end{aligned} \tag{4.1}$$

(3) *For* $(t,x) \in [0,1] \times M, v_t(x) = \mathbf{E}[{}^\circ(\rho_t^{-1}({}^*\sigma)(x))]$ $(\sigma \in \Gamma^\infty(E))$ *s olves*

$$\frac{\partial v_t}{\partial t} = H v_t, \quad v_0 = \sigma. \tag{4.2}$$

Proof. Part (1) is proved in a similar way as in (1) of Proposition 2.1.
(2) Since ${}^*p(({}^*A_\lambda)_{p_t(p)}{}^*(F_\sigma)) = \rho_{t,x}^{-1}({}^*(L_\lambda\sigma)(\theta_t(x)))$, we have, for $t \in T_j$,

$$\frac{d}{dt}\rho_{t,x}^{-1}({}^*\sigma(\theta_t(x))) = {}^*p\left(\frac{d({}^*(F_\sigma)(p_t(p)))}{dt}\right) = \sum_{\lambda=0}^{n} \frac{\Delta_j w^\lambda}{\Delta t}\rho_{t,x}^{-1}({}^*(L_\lambda\sigma)(\theta_t(x))),$$

$$\left(\frac{d}{dt}\right)^2 \rho_{t,x}^{-1}({}^*\sigma(\theta_t(x))) = \sum_{\kappa,\lambda=0}^{n} \frac{\Delta_j w^\kappa}{it\Delta t}\frac{\Delta_j w^\lambda}{\Delta t}\rho_{t,x}^{-1}({}^*(L_\kappa L_\lambda\sigma)(\theta_t(x))),\ (p \in \pi_P^{-1}(x)).$$

Use these formulas to obtain (4.1) (cf. [Ak1]); notice that
${}^\circ(\rho_t^{-1}({}^*\sigma)(x)) = {}^\circ({}^*p({}^*(F_\sigma)(p_t(p)))) = p(F_\sigma({}^\circ p_t(p))) \in E_x$. Part (3) follows from (2). ∎

Let h be a C^∞ fiber metric in E. By the saturation principle, there exists a hyperfinite dimensional internal vector subspace $\tilde{V}$ of ${}^*(\Gamma^\infty(E))$ over ${}^*\mathbb{R}$ such that $\{{}^*\sigma : \sigma \in \Gamma^\infty(E)\} \subseteq \tilde{V}$. Let $\tilde{\nu} = {}^*\dim\tilde{V}$, and take an orthonormal basis ψ_i $(i = 1, 2, \ldots, \tilde{\nu})$ of $\tilde{V}$ with respect to the internal inner product $(\sigma_1, \sigma_2)_{\tilde{V}} = \int_{{}^*M} {}^*h(\sigma_1, \sigma_2)\, {}^*dv_g$, $(\sigma_1, \sigma_2 \in \tilde{V})$. Put $D(x,y) := \sum_{i=1}^{\tilde{\nu}} \psi_i(x) \otimes \psi_i^h(y)$, $(x, y \in {}^*M)$, where $\psi_i^h \in {}^*(\Gamma^\infty(E^*))$ is obtained by the evaluation map $({}^*E_x^* \odot {}^*E_x^*) \otimes {}^*E_x \to {}^*E_x^*, {}^*h(x) \otimes \psi_i(x) \mapsto \psi_i^h(x) = {}^*h(x) \cdot \psi_i(x)$, $(x \in {}^*M)$. (For the tensor product $\otimes$ and the symmetric tensor product $\odot$, we denote ${}^*\otimes$ and ${}^*\odot$ simply by $\otimes$ and $\odot$, respectively.)

Let $\Psi : [0,1] \times M \times \Gamma^\infty(E) \ni (t, x, \sigma) \mapsto \mathbf{E}[{}^\circ(\rho_t^{-1}({}^*\sigma)(x))] \in E_x \subseteq E$. Then Ψ is now standard since it is identified with its graph.

Theorem 4.2. *For* $t \in [0,1], x \in M$, *and* $y \in {}^*M$, *put* ${}^*\Psi(t,x,D(\cdot,y)) := \sum_{i=1}^{\tilde{\nu}} {}^*\Psi(t,x,\psi_i) \otimes \psi_i^h(y) \in {}^*E_x \otimes {}^tE_y^*$. *Then* ${}^*\Psi(t,x,D(\cdot,y))$ *is a "nonstandard heat kernel" of* (4.2) *as long as* $t \in [0,1]$, *in the sense that*

$$\mathbf{E}[{}^\circ(\rho_t^{-1}({}^*\sigma)(x))] = \int_{{}^*M} {}^*\Psi(t,x,D(\cdot,y)) \cdot {}^*\sigma(y)\, {}^tdv_g(y), \quad (t \in [0,1], x \in M),$$

regardless of whether there exists a standard (in the sense of nonstandard analysis) heat kernel of (4.2) *or not.*

Proof. By ${}^*\sigma = \sum_{i=1}^{\tilde{\nu}} \tilde{c}_i({}^*\sigma)\psi_i$ with $\tilde{c}_i({}^*\sigma) = ({}^*\sigma, \psi_i)_{\tilde{V}} = \int_{{}^*M} \psi_i^h \cdot {}^*\sigma\, {}^*dv_g$, we get $\mathbf{E}[{}^\circ(\rho_t^{-1}({}^*\sigma)(x))] = \Psi(t,x,\sigma) = {}^*\Psi(t,x,{}^*\sigma) = \sum_{i=1}^{\tilde{\nu}} \tilde{c}_i({}^*\sigma)\, {}^*\Psi(t,x,\psi_i) = \int_{{}^*M} {}^*\Psi(t,x,D(\cdot,y)) \cdot {}^*\sigma(y)\, {}^*dv_g(y)$. ∎

References

[Ak1] H. Akiyama, *Applications of nonstandard analysis to stochastic flows and heat kernels on manifolds*, in "*Geometry of Manifolds*" (K. Shiohama, ed.), Perspectives in Mathematics, Vol. 8, Academic Press, Boston, 1989, pp. 3–27.

[Ak2] H. Akiyama, *Nonstandard representations of generalized sections of vector bundles*, preprint.

[AFHL] S. Albeverio, J. E. Fenstad, R. Høegh-Krohn and T. Lindstrøm, *Nonstandard Methods in Stochas tic Analysis and Mathematical Physics*, Academic Press, New York, 1986.

[An] R. M. Anderson, *A non-standard representation for Brownian motion and Itô integration*, Israel J. Math. **25** (1976), 15–46.

[IW] N. Ikeda and S. Watanabe, *Stochastic Differential Equations and Diffusion Processes*, 2nd ed., North-Holland/K odansha, Amsterdam/Tokyo, 1989.

[KN] S. Kobayashi and K. Nomizu, *Foundations of Differential Geometry*, I, John Wiley & Sons (Interscience), New York, 1963.

[L] P. A. Loeb, *Conversion from nonstandard to standard measure spaces and applications in probability theory*, Trans . Amer. Math. Soc. **211** (1975), 113–122.

[T1] T. Todorov, *A nonstandard delta function*, Proc. Amer. Math. Soc. **110** (1990), 1143–1144.

[T2] T. Todorov, *Pointwise kernels of Schwartz distributions*, Proc. Amer. Math. Soc. **114** (1992), 817–81 9.

A NONSTANDARD APPROACH TO THE MALLIAVIN CALCULUS

Nigel J. Cutland and Siu-Ah Ng
Dept. of Pure Mathematics, University of Hull,
Cottingham Road, Hull HU 6 7RX, England

Abstract

We outline an intuitive approach to the Malliavin calculus for the classical Wiener space, showing that the basic operators of this calculus have natural descriptions as classical differential operators on a nonstandard space ${}^*\mathbb{R}^N$ for an infinite natural number N.

1 Introduction

Let W denote Wiener measure on the space $\mathcal{C} = C[0,1]$, and write $L^2(W)$ for the space $L^2(\mathcal{C}, W)$ of square integrable Wiener functionals. In its simplest form, the Malliavin calculus is a kind of partial differential calculus for functionals $\varphi(b)$ in $L^2(W)$, embodying the following intuition. Each fixed sample point $b \in \mathcal{C}$ can be thought of as given by the family of increments $x_t = \mathrm{d}b_t = b_{t+\mathrm{d}t} - b_t$ $(t \in [0,1])$. Under W these increments x_t vary independently with mean 0 and variance $\mathrm{d}t$. Thus, informally, we can think of $\varphi \in L^2(W)$ as $\varphi = \varphi(x)$, where $x = (x_t)_{t\in[0,1]}$. The *gradient* or *derivation* operator D is then given intuitively by

$$D\varphi(b,t) = \frac{\partial \varphi}{\partial x_t}(b).$$

We will outline below how this idea can be made precise in a nonstandard framework, which allows the basic properties of D to be derived rigorously in an intuitive way, by application of the rules of classical differential calculus.

In a similar way we treat the other basic operators of the Malliavin calculus - the *Skorohod integral* operator δ and the *Malliavin* operator L. These are given intuitively by

$$\delta u(b) = \int_0^1 u(b,t)\mathrm{d}b_t \ - \ \int_0^1 \frac{\partial u(b,t)}{\partial x_t}\mathrm{d}t$$

for integrands $u(b,t)$ that are not necessarily adapted. This is an extension of the Itô integral (since, informally, an adapted integrand $u(b,t)$ depends only on $(b_s)_{s\leq t}$, hence on $(x_s)_{s<t}$, and so $\partial u(b,t)/\partial x_t = 0$).

For L we have

$$L\varphi(b) = \delta D\varphi(b) = \int_0^1 \frac{\partial \varphi}{\partial x_t}(b)\mathrm{d}b_t \ - \ \int_0^1 \frac{\partial^2 \varphi}{\partial x_t^2}(b)\mathrm{d}t$$

which is a generalised Ornstein-Uhlenbeck operator. The nonstandard construction of Brownian motion/Wiener measure that we need for our treatment is the modification [4] of Anderson's construction [2], in which the infinitesimal random walk $\Delta B_t = \pm\sqrt{\Delta t}$, where Δt is a positive infinitesimal, is replaced by ΔB_t which is Gaussian distributed with mean 0 and variance Δt. Details are given in Sec.2, where we also sketch a nonstandard proof of the so-called Wiener-Itô chaos decompositon theorem for $L^2(W)$ which is crucial to the standard approach to the Malliavin calculus. In subsequent sections we give our treatment of the operators D, δ and L and their fundamental properties, and indicate some of the many applications of this theory.

In this short paper we have been selective in the topics we mention, confining ourselves mostly to basics of the theory. We omit or only sketch most proofs; those we include are chosen to indicate the flavour of the nonstandard approach to the Malliavin calculus. In a subsequent paper [7] we will provide full details together with a more comprehensive treatment.

2 Wiener Measure, Itô Integrals and Chaos

The following is the construction of Brownian motion/Wiener measure given in [4]; we restrict to the time interval [0,1] and to 1-dimension for convenience only - our approach extends easily to d-dimensions and time $t \in [0,\infty[$.

Fix an infinite $N \in {}^*\mathbb{N}$, let $\Delta t = N^{-1} \approx 0$ and let $\Omega = {}^*\mathbb{R}^T$, where T is the hyperfinite time line $T = \{0, \Delta t, \ldots, 1-\Delta t\}$. Elements of Ω are denoted by $x = (x_t)_{t\in T}$. The internal probability Γ is defined on Ω as the product of the normal distribution $\mathcal{N}(0,\Delta t)$ on each coordinate:

$$\Gamma(A) = (2\pi\Delta t)^{-\frac{N}{2}} \int_A \exp\Big(-\frac{1}{2\Delta t}\sum_{t\in T} x_t^2\Big)\mathrm{d}x$$

for $A \in {}^*\mathcal{B}$, the *Borel subsets of Ω, where $\mathrm{d}x$ denotes *Lebesgue measure on Ω. We denote by $\mathbf{\Omega}$ the complete Loeb space $\mathbf{\Omega} = (\Omega, \mathcal{F}, P)$ obtained from $(\Omega, {}^*\mathcal{B}, \Gamma)$.

Let $\bar{T} = T \cup \{1\}$. An internal process $(B_t)_{t\in\bar{T}}$ is defined on Ω by

$$B_t(x) = \sum_{s<t} x_s$$

i.e.

$$\begin{cases} B_0 &= 0 \\ \Delta B_t &= B_{t+\Delta t}(x) - B_t(x) = x_t \end{cases}$$

The following was proved in [4] (following the ideas of [2]):

Theorem 2.1 *For $P-$almost all $x \in \Omega$, $B(x)$ is S-continuous, and the process $b_t(x)$ defined on Ω by*

$$b_{^\circ t}(x) = {}^\circ B_t(x)$$

for $t \in T$ is Brownian motion, so Wiener measure is given on $\mathcal{C}$ by

$$W(A) = P(\{x \in \Omega : b(x) \in A\})$$

Itô integration with respect to b is developed exactly as in [2] or [8]. A standard right-continuous filtration $(\mathcal{F}_t)_{t\in[0,1]}$ is defined on Ω by $\mathcal{F}_t = \mathcal{N} \vee \sigma(\bigcup_{^\circ s=t} \mathcal{A}_s)$, where $\mathcal{N}$ denotes the P-null sets and $\mathcal{A}_s$ ($s \in T$) is the internal ${}^*\sigma$-algebra generated by $(x_u)_{u<s}$. We have

Theorem 2.2 *Let $f : \Omega \times [0,1] \to \mathbb{R}$ be an adapted process with $E(\int_0^1 f^2 dt) < \infty$. Then*

(1) *f has an SL^2 lifting $F : \Omega \times T \to {}^*\mathbb{R}$ that is nonanticipating (i.e. $F(\cdot, t)$ is $\mathcal{A}_t$-measurable for each $t \in T$);*

(2) *for any such lifting F, define*

$$G(x,t) = \sum_{s<t} F(x,s)\Delta B_s$$

and then $G(x,\cdot)$ is S-continuous a.s., and for a.a. x

$$\int_0^{^\circ t} f(x,s)\,db_s = {}^\circ G(x,t) \qquad \text{all } t \in T.$$

Wiener-Itô chaos expansion

Let us fix some notation. For $n \geq 1$ let $\Delta_n = \{(t_1, \ldots, t_n) : 0 \leq t_1 \leq \ldots \leq t_n \leq 1\}$. For $f \in L^2(\Delta_n)$, the *multiple Wiener integral* is defined by

$$\begin{aligned} \int_{\Delta_n} f \mathrm{d}b^{(n)} &= \int_{t_n=0}^1 \Big(\int_{t_{n-1}=0}^{t_n} \cdots \Big(\int_{t_1=0}^{t_2} f(t_1, \ldots, t_n) \mathrm{d}b_{t_1} \Big) \ldots \mathrm{d}b_{t_{n-1}} \Big) \mathrm{d}b_{t_n} \\ &= \int_{\Delta_n} f(t_1, \ldots, t_n) \mathrm{d}b_{t_1} \ldots \mathrm{d}b_{t_n} \end{aligned}$$

say. We will also write $\int f \mathrm{d}b^{(n)}$ and $I_n(f)$ to denote this integral. The following properties of $I_n(f)$ are routine:

Proposition 2.3 (a) $\mathrm{E}(I_n(f)) = 0$

(b) $\mathrm{E}(I_n(f)^2) = \|I_n(f)\|^2 = \|f\|^2_{\Delta_n}$

(c) *if* $f \in L^2(\Delta_n)$ *and* $g \in L^2(\Delta_m)$ *then*

$$E(I_n(f)I_m(g)) = \begin{cases} 0 & \text{if } n \neq m \\ \int_{\Delta_n} fg & \text{if } n = m \end{cases}$$

The nonstandard representation of the Itô integral gives the following representation for the multiple Wiener integral. The counterpart of Δ_n for the time set T is the set $\Delta_n^T = \{(t_1, \ldots, t_n) \in T^n : t_1 < t_2 < \ldots t_n\}$, and we have

Theorem 2.4 *Let* $f \in L^2(\Delta_n)$ *with* SL^2 *lifting* $F : \Delta_n^T \to {}^*\mathbb{R}$. *Then*

$$\begin{aligned} I_n(f) &= {}^\circ \sum_{t_1 < t_2 \ldots < t_n} F(t_1, \ldots, t_n) \Delta B_{t_1} \ldots \Delta B_{t_n} \\ &= {}^\circ \sum_{t_1 < t_2 \ldots < t_n} F(t_1, \ldots, t_n) x_{t_1} \ldots x_{t_n} \end{aligned}$$

and $\sum_{t_1 < t_2 \ldots < t_n} F(t_1, \ldots, t_n) x_{t_1} \ldots x_{t_n}$ *is* SL^2 .

It is convenient to write $I_n(F) = \sum_{t_1 < t_2 \ldots < t_n} F(t_1, \ldots, t_n) x_{t_1} \ldots x_{t_n}$. It is immediate that the counterpart of Proposition 2.3 holds for $I_n(F)$, and from this Proposition 2.3 itself can be proved.

The fundamental Wiener-Itô chaos decomposition result for $L^2(W)$ is now:

Theorem 2.5 *Let* $\varphi \in L^2(W)$. *Then* φ *has a unique expression as*

$$\varphi(b) = \varphi_0 + \sum_{n=1}^{\infty} I_n(f_n)$$

where $\varphi_0 = E(\varphi)$ *and* $f_n \in L^2(\Delta_n)$ *for* $n \geq 1$. *Moreover*

$$\|\varphi\|^2 = \varphi_0^2 + \sum_{n=1}^{\infty} \|f_n\|^2$$

where $\|f_n\|$ *means* $\|f_n\|_{\Delta_n}$.

We will sketch a proof of this below.

The closed subspace of $L^2(W)$ given by

$$Z_n = \{I_n(f) : f \in L^2(\Delta_n)\}$$

is called the nth *Wiener chaos;* so Theorem 2.5 says that

$$L^2(W) = \oplus_{n=0}^{\infty} Z_n$$

where we set $Z_0 = \mathbb{R}$. Let $Z^{(n)}$ denote $\oplus_{m=0}^{n} Z_m$. For $\varphi \in L^2(W)$ we write φ_n for the projection of φ onto Z_n and we write

$$\varphi^{(n)} = \sum_{m=0}^{n} \varphi_m$$

for the projection onto $Z^{(n)}$. There are a number of proofs of Theorem 2.5 in the literature - for example [9] or [17]. In the paper [6] we gave two elementary nonstandard proofs and we sketch one of them here.

Proof of Theorem 2.5 It is well known that $L^2(W)$ has a dense subset spanned by linear combinations of functions for the form

$$\varphi(b) = \exp\left(\int_0^1 f(t)\mathrm{d}b_t - \tfrac{1}{2}\int_0^1 f^2(t)\mathrm{d}t\right),$$

so it is sufficient to show that $\varphi = 1+\sum_{n=1}^{\infty} I_n(f_n) = \psi(b)$, say, where $f_n(t_1,\dots,t_n) = f(t_1)\dots f(t_n)$. Take an SL^2 lifting F of f and define an internal function Φ

$$\Phi(B) = \prod_{t\in T}(1 + F(t)\Delta B_t).$$

The proof is completed by showing that (i) $\Phi(B) \approx \varphi(b)$ a.s. and (ii) $\Phi(B) \approx \psi(\mathrm{b})$ a.s. The first follows by showing that a.s. we have

$$\begin{aligned}
\log\Phi(B) &= \sum_{t\in T}\log(1 + F(t)\Delta B_t) \\
&= \sum_{t\in T}(F(t)\Delta B_t - \tfrac{1}{2}F^2(t)\Delta B_t^2 + \varepsilon_t) \\
&\approx \sum_{t\in T}(F(t)\Delta B_t - \tfrac{1}{2}F^2(t)\Delta t) \\
&\approx \log\varphi(b)
\end{aligned}$$

For (ii) we have the exact expansion

$$\begin{aligned}
\Phi(B) &= 1 + \sum_{n=1}^{N}\sum_{\Delta_n^T} F(t_1)F(t_2)\dots F(t_n)\Delta B_{t_1}\dots\Delta B_{t_n} \\
&= 1 + \sum_{n=1}^{N} I_n(F_n)
\end{aligned}$$

where $F_n(t_1,\dots,t_n) = F(t_1)\dots F(t_n)$. We have $I_n(F_n) \approx I_n(f_n)$ a.s. for all finite n, by Theorem 2.4. The observation that

$$\mathrm{E}(I_n(F_n)^2) = \sum_{\Delta_n^T} F_n(t_1,\dots,t_n)^2\Delta t^n \le \frac{1}{n!}\Big(\sum_T F(t)^2\Delta t\Big)^n$$

which is infinitesimal for infinite n completes the proof. ■

Wiener thought of the integrands f_n in the chaos expansion of $\varphi \in L^2(W)$ as being given by the following 'recipe' (see [9]):

$$f_n(t_1, \ldots, t_n) = \mathrm{E}(\varphi(b)\dot{b}_{t_1}\dot{b}_{t_2}\ldots\dot{b}_{t_n}).$$

We can make sense of this as follows (see [6] for details):

Theorem 2.6 *Let $\varphi \in L^2(W)$ have chaos expansion given by Theorem 2.5, and let Φ be an SL^2 lifting of φ. Then for finite n the following function F_n is an SL^2 lifting of f_n:*

$$F_n(t_1, \ldots, t_n) = E(\Phi(B)\dot{B}_{t_1}\dot{B}_{t_2}\ldots\dot{B}_{t_n}) \tag{2.1}$$

where $\dot{B}_t = \Delta B_t/\Delta t$.

Proof (Sketch) Take SL^2 liftings $\bar{F}_n$ of f_n; then for sufficiently small infinite M, the function

$$\Psi = \varphi_0 + \sum_{n=1}^{M} I_n(\bar{F}_n)$$

is an SL^2 lifting of φ. It is routine to see that $\mathrm{E}(\Psi x_{t_1}x_{t_2}\ldots x_{t_n}) = \bar{F}_n(t_1, \ldots, t_n)\Delta t^n$ and hence that $\|\bar{F}_n - F_n\|^2 \le \mathrm{E}((\Psi - \Phi)^2) \approx 0$ for finite n. ■

Suppose we now take $\varphi \in L^2(W)$ with SL^2 lifting Φ and define for all $n \in^* \mathbb{N}$

$$\Phi_n = I_n(F_n)$$

where F_n is given by the recipe (2.1), and

$$\Phi^{(m)} = \sum_{n=0}^{m} \Phi_n = F_0 + \sum_{n=1}^{m} \sum_{t_1<t_2<\ldots t_n} F_n(t_1, \ldots, t_n)x_{t_1}\ldots x_{t_n}.$$

The above considerations show that

Theorem 2.7 *For all infinite M, $\Phi^{(M)}$ is an SL^2 lifting of φ.*

Each lifting $\Phi^{(M)}$ is a *monomial* lifting of φ; i.e. $\Phi^{(M)}$ is a (nonstandard) polynomial in $(x_t)_{t\in T}$ with no quadratic or higher powers of any x_t. The existence of such liftings is one way to understand the Wiener-Itô chaos decomposition, and to some extent our nonstandard development shows that the Malliavin calculus on $L^2(W)$ can be seen as nonstandard classical calculus of monomials in ${}^*L^2(\Omega)$. However, monomials form a somewhat restricted class with few closure properties, so we do move outside this class.

Another way to understand the chaos decomposition from the nonstandard viewpoint is using Hermite polynomials. By the transfer of classical theory, *any* $\Phi \in {}^*L^2(\Omega)$ has an expansion in terms of *Hermite polynomials (these form an orthonormal basis for ${}^*L^2(\Omega)$). For those Φ that are liftings the non-monomial terms make only infinitesimal contribution.

3 The Derivation Operator

The *derivation* or *gradient* operator D is a densely defined operator

$$D : L^2(W) \to L^2(W \times [0,1])$$

(where we take Lebesgue measure on [0,1]) with domain $\mathbb{D}^{2,1} \subseteq L^2(W)$, defined standardly as follows:

Definition 3.1

$$\mathbb{D}^{2,1} = \{\varphi \in L^2(W) : \sum_{n=1}^{\infty} n\|\varphi_n\|^2 < \infty\}$$

and for $\varphi \in \mathbb{D}^{2,1}$,

$$D_s\varphi(b) = D\varphi(b,s) = \sum_{n\geq 1} I_{n-1}(\hat{f}_n(\cdot, s)) = \sum_{n\geq 1} D\varphi_n(b,s)$$

where $\varphi_n = I_n(f_n)$ *and* $\hat{f}_n$ *is the symmetric extension of* f_n *to* $[0,1]^n$.

It is easy to check that $\|I_{n-1}(\hat{f}_n)\|^2 = n\|f_n\|^2$ (the first norm being the L^2 norm with respect to $W\times$Lebesgue on $\mathcal{C} \times [0,1]$), and hence $\|D\varphi\|^2 = \sum_{n=1}^{\infty} n\|\varphi_n\|^2$.

For a Wiener integral $\varphi(b) = \int_0^1 f(t)\mathrm{d}b_t$ (i.e. $\varphi \in Z_1$) it is straightforward to see that $D\varphi(b,s) = f(s) =$ "$\partial\varphi/\partial(\mathrm{d}b_s)$", which accords with the intuitive description of D in the Introduction. It is not so clear in general, however, that D is a derivative. This becomes clearer from the following nonstandard approach.

For $\Phi \in {}^*L^2(\Omega)$ we have the (internal) classical derivative, the gradient $\nabla\Phi$, given by

$$\nabla_t\Phi(x) = \nabla\Phi(x,t) = \frac{\partial\Phi}{\partial x_t}(x)$$

for *differentiable Φ, and extended to general Φ by means of the *Hermite polynomial expansion of Φ. (See the Appendix for details.) In all that follows, when we write $\nabla\Phi$ or $\nabla_t\Phi$ we are assuming implicitly that $\Phi \in \mathrm{dom}(\nabla)$ or $\mathrm{dom}(\nabla_t)$ respectively.

We can now characterise $\mathbb{D}^{2,1}$ using ∇:

Theorem 3.2 *Let* $\varphi \in L^2(W)$. *Then* $\varphi \in \mathbb{D}^{2,1}$ *if and only if* φ *has an* SL^2 *lifting* Φ *with* $\|\nabla\Phi\| < \infty$.

This follows almost immediately once we have established the following lemma.

Lemma 3.3 *Let* $\varphi = I_n(f) \in Z_n$ *and let* F *be an* SL^2 *lifting of* f, *and let* $\Phi = I_n(F)$. *Then* $\nabla\Phi$ *is an* SL^2 *lifting of* $D\varphi$.

Proof Let $\hat{F}$ be the symmetric extension of F to T^n, with $\hat{F}(t_1,\ldots,t_n)=0$ whenever $t_i=t_j$ for some $i\neq j$. Then $\hat{F}$ is an SL^2 lifting of $\hat{f}$ and it is straightforward to check that $\nabla\Phi(x,s)=I_{n-1}(\hat{F})$. ∎

The next result is quite routine:

Theorem 3.4 *Suppose that $\varphi\in\mathbb{D}^{2,1}$ and let Φ be any SL^2 lifting of φ. Then*

(a) *$\nabla\Phi^{(K)}$ is an SL^2 lifting of $D\varphi$ for sufficiently small infinite K.*

(b) $\|D\varphi\|\leq {}^\circ\|\nabla\Phi\|$

(c) *$\nabla\Phi$ is an SL^2 lifting of $D\varphi$ if and only if $\|\nabla\Phi\|\approx\|D\varphi\|$.*

The following definition is useful:

Definition 3.5 *Let $\Phi\in{}^*L^2(\Omega)$; then Φ is $S\mathbb{D}^{2,1}$ if $\|\nabla\Phi\|$ is finite and $\|\nabla(\Phi-\Phi^{(K)})\|\approx 0$ for all infinite K.*

We have:

Theorem 3.6 *Let $\varphi\in L^2(W)$. Then*

(i) *$\varphi\in\mathbb{D}^{2,1}$ if and only if φ has an $S\mathbb{D}^{2,1}$ lifting Φ;*

(ii) *if $\varphi\in\mathbb{D}^{2,1}$ with SL^2 lifting Φ, then $\nabla\Phi$ is an SL^2 lifting of $D\varphi$ if and only if Φ is $S\mathbb{D}^{2,1}$.*

The next result has several applications.

Theorem 3.7 *Suppose that $\varphi\in L^2(W)$ has SL^2 lifting Φ, and $\nabla\Phi$ is an SL^2 lifting of $\psi\in L^2(W\times[0,1])$. Then $\varphi\in\mathbb{D}_{2,1}$ and $\psi=D\varphi$ (so Φ is an $S\mathbb{D}^{2,1}$ lifting of φ).*

We can illustrate the use of this result with the following example.

Example 3.8 *Let $f\in L^2[0,1]$; then $\varphi=\exp(I_1(f))\in\mathbb{D}^{2,1}$ and $D\varphi(b,s)=\varphi(b)f_s$.*

To see this, take F that is an SL^2 lifting of f, so that $\Phi=\exp(I_1(F))=\exp(\sum_{t\in T}F_tx_t)$ is an SL^2 lifting of φ. Then $\nabla\Phi(x,t)=\Phi(x)F_t$ and this is an SL^2 lifting of $\varphi(b)f_s$.

Theorem 3.7 gives an easy proof of the following chain rule for D:

Theorem 3.9 *Suppose that $\varphi\in\mathbb{D}^{2,1}$ and $f\in C^1$ with bounded derivative f'. Then $f\circ\varphi\in\mathbb{D}^{2,1}$ and $D(f\circ\varphi)=f'(\varphi)D\varphi$.*

Proof Let Φ be an $S\mathbb{D}^{2,1}$ lifting of φ. Then $f(\Phi)$ is an SL^2 lifting of $f(\varphi)$, and by the transfer of classical calculus, $\nabla f(\Phi) = f'(\Phi)\nabla\Phi$, which is an SL^2 lifting of $f'(\varphi)D\varphi$. ∎

The chain rule for functions f of several variables follows in the same way.

Another easy application of Theorem 3.7 is

Theorem 3.10 *D is a closed operator.*

Proof Let $\varphi_n \to \varphi$ in $L^2(W)$, and suppose that $\varphi_n \in \mathbb{D}^{2,1}$ and $D\varphi_n \to \psi$ in $L^2(W)$. Take $S\mathbb{D}^{2,1}$ liftings Φ_n of φ_n. Then for sufficiently small infinite K, Φ_K is an SL^2 lifting of φ *and* $\nabla\Phi_K$ is an SL^2 lifting of ψ. By Theorem 3.7, $\varphi \in \mathbb{D}^{2,1}$ and $D\varphi = \psi$. ∎

From this it easy to show that D is a *local* operator; i.e. if $\varphi \in \mathbb{D}^{2,1}$ then $\{\varphi = 0\} \subseteq \{D\varphi = 0\}$ a.s.

Ocone [13] used the operator D to give a pleasant description of the integrand in the integral representation of $\varphi \in \mathbb{D}^{2,1}$, and we can give an intuitive proof using our methods.

Theorem 3.11 *Let $\varphi \in \mathbb{D}^{2,1}$. Then the integrand g in the stochastic integral representation*

$$\varphi(b) = \varphi_0 + \int_0^1 g(b,t)\,db_t \tag{3.1}$$

is given by $g(b,t) = E(D\varphi(b,t)|\mathcal{F}_t)$.

Proof (sketch) Let $\varphi = \sum_{n=0}^{\infty} \varphi_n = \varphi_0 + \sum_{n=1}^{\infty} I_n(f_n)$ be the chaos expansion of φ; taking appropriate liftings we have $\varphi = {}^\circ\Phi$ where

$$\Phi = \varphi_0 + {}^\circ\sum_{n=1}^{K} I_n(F_n)$$

and Φ is an $S\mathbb{D}^{2,1}$ lifting of φ. We can write Φ as

$$\Phi = \varphi_0 + \sum_{t\in T} G(x,t)x_t$$

where $G(x,t) = \sum_{n=1}^{K} I_{n-1}(F_n(\cdot,t))$ (with $F_n(\mathbf{s},t) = 0$ for $(\mathbf{s},t) \notin \Delta_n^T$). Clearly G is a nonanticipating SL^2 lifting of the adapted integrand $g \in L^2(W \times [0,1])$. Now

$$\nabla_s\Phi(x) = \sum_{t<1} \nabla_s(G(x,t)x_t) = G(x,s) + \sum_{s<t<1} \nabla_s G(x,t)x_t \tag{3.2}$$

(since G is nonanticipating, $\nabla_s G(x,t) = 0$ if $s \geq t$). Using the internal isometry property, this means that

$$\|\nabla\Phi\|^2 = \|G\|^2 + \|\nabla G\|^2$$

so that (using Theorems 3.2 and 3.4) we see that $g(\cdot,t) \in \mathbb{D}^{2,1}$ for a.a. t and $\nabla G(x,s,t)$ is an SL^2 lifting of $Dg(b,s,t)$. From (3.2) it follows that

$$D_s\varphi(b) = g(b,s) + \int_s^1 D_s g(b,t)\mathrm{d}b_t$$

and taking conditional expectations gives the result. ∎

The following is a strengthening of Theorem 3.7, whose proof we omit:

Theorem 3.12 *Suppose $\varphi \in L^1(W)$ has SL^1 lifting $\Phi \in {}^*L^2(\Omega)$, and $\nabla\Phi \in {}^*L^2(\Omega\times T)$ is an SL^1 lifting of $\psi \in L^2(W \times [0,1])$. Then $\varphi \in \mathbb{D}^{2,1}$ and $\psi = D\varphi$.*

Using this we can prove the product rule for D.

Theorem 3.13 *Suppose that $\varphi, \psi \in \mathbb{D}^{2,1}$ and $\varphi D\psi + \psi D\varphi \in L^2(W \times [0,1])$. Then $\varphi\psi \in \mathbb{D}^{2,1}$ and $D(\varphi\psi) = \varphi D\psi + \psi D\varphi$.*

Proof Take $S\mathbb{D}^{2,1}$ monomial liftings Φ and Ψ of φ and ψ. Classical calculus gives $\nabla(\Phi\Psi) = \Phi D\Psi + \Psi D\Phi \in {}^*L^2 \cap SL^1$. So $\varphi\psi \in L^1(W)$ and $\varphi D\psi + \psi D\varphi \in L^2(W \times [0,1])$ have SL^1 liftings $\Phi\Psi$ and $\nabla(\Phi\Psi)$ and Theorem 3.12 gives the result. ∎

It is routine to extend the above theory to iterations of D, so that for suitable functionals $\varphi \in L^2(W)$ we have

$$D^2\varphi(b,s,t) = D(D\varphi(\cdot,s))(b,t).$$

We define $\mathbb{D}^{2,2} = \mathrm{dom}(D^2)$ and it can be easily checked that $\varphi \in \mathbb{D}^{2,2}$ if and only if $\sum n(n-1)\|\varphi_n\|^2 < \infty$ (and this expression gives $\|D^2\varphi\|^2$). Similar remarks hold for higher derivatives D^m. Clearly we have $\varphi \in Z^{(n)}$ if and only if $\varphi \in \mathbb{D}^{2,n}$ and $D^n\varphi = 0$; and if $\varphi \in Z^{(n)}$ then $D^{n-1}\varphi = D^{n-1}\varphi_n = \hat{f}_n$ where $\varphi_n = I_n(f_n)$.

The iteration of the local property of D has the following useful consequence.

Theorem 3.14 *Let $0 \neq \varphi \in Z^{(n)}$. Then $\varphi \neq 0$ a.s.*

Proof Without loss of generality we may assume that $\varphi_n = I_n(f) \neq 0$ (otherwise $\varphi \in Z^{(n-1)}$). Let $A = \{\varphi = 0\}$. Then by the localisation property, $D^n\varphi = 0$ on $A \times [0,1]^n$. But $D^n\varphi(b,t_1,\ldots,t_n) = \hat{f}(t_1,\ldots,t_n)$, so if $P(A) > 0$ then $\hat{f} = 0$ a.s., and the same is true for f, contrary to assumption. Hence $P(A) = 0$. ∎

Directional Derivatives We conclude this section with a brief mention of directional versions of the operator D. Let $h \in L^2[0,1]$, so that $\bar{h} = \int h$ belongs to the Cameron-Martin subspace of $\mathcal{C}$. There are several ways to formalise the idea of "differentiation in $L^2(W)$ in the direction $\bar{h}$ (or h)". One way - which we present using the above framework - is as follows.

Let H be an SL^2 lifting of h. A densely defined operator D_h with domain $\mathbb{D}^{2,h}$ is defined by:

Definition 3.15 *Let $\varphi \in L^2(W)$. Then $\varphi \in \mathbb{D}^{2,h}$ if and only if φ has an SL^2 lifting Φ with $\|\langle \nabla\Phi, H\rangle\| < \infty$, where $\langle G, H\rangle = \sum_{t\in T} G(t)H(t)\Delta t$.*

If $\varphi \in Z_n$, $D_h\varphi = {}^\circ\nabla_H\Phi$ for any SL^2 lifting Φ, where

$$\nabla_H\Psi(x) = \langle \nabla\Psi(x), H\rangle = \sum_{t\in T} \frac{\partial\Psi(x)}{\partial x_t} H(t).$$

For $\varphi \in \mathbb{D}^{2,h}$, $D_h\varphi = \lim_{n\to\infty} D_h\varphi^{(n)} = {}^\circ\nabla_H\Phi^{(K)}$ for all sufficiently small infinite K.

From this it is clear that $\mathbb{D}^{2,1} \subset \mathbb{D}^{2,h}$ and for $\varphi \in \mathbb{D}^{2,1}$, we have $D_h\varphi = \langle D\varphi, h\rangle$, with $\langle\cdot,\cdot\rangle$ here denoting the inner product in $L^2[0,1]$.

There are basic results for D_h that are parallel to Theorems 3.4, 3.7 and 3.12, and D_h is closed, and has a chain rule and product rule like D.

4 The Malliavin Covariance

Suppose that $\varphi \in \mathbb{D}^{2,1}$. Then the *Malliavin covariance* $\sigma(\varphi) \in L^1(W)$ is given by

Definition 4.1

$$\sigma(\varphi)(b) = \langle D\varphi, D\varphi\rangle = \int_0^1 D\varphi(b,t)^2\,dt \tag{4.1}$$

More generally, for $\varphi = (\varphi_1, \ldots, \varphi_n)$ each $\varphi_i \in \mathbb{D}^{2,1}$, $\sigma(\varphi) = (\sigma_{i,j}(\varphi))$ is a matrix valued r.v. given by $\sigma_{i,j}(\varphi) = \langle D\varphi_i, D\varphi_j\rangle$.

One of the chief applications of the Malliavin calculus has been its use in establishing the existence of densities of Brownian functionals, and the following is a simple example of this in our framework (c.f. Prop. 2.2.1 of [18] and [15] Theorem 3.1).

Theorem 4.2 *Suppose that $\varphi \in L^2(W)$ with*

(i) $\varphi \in \mathbb{D}^{2,2}$ *(i.e. $\varphi \in \mathbb{D}^{2,1}$ and $D\varphi \in \mathbb{D}^{2,1}$)*

(ii) $\sigma(\varphi) \in \mathbb{D}^{2,1}$

(iii) $\sigma(\varphi) \neq 0$ *a.s.*

Then the measure μ induced by φ on $\mathbb{R}$ is absolutely continuous with respect to Lebesgue measure.

Proof For any real $\varepsilon > 0$ let P_ε be the measure on Ω defined by

$$\frac{\mathrm{d}P_\varepsilon}{\mathrm{d}P} = \frac{\sigma}{\sigma+\varepsilon}. \tag{4.2}$$

where we set $\sigma = \sigma(\varphi)$. Note that for any $C \subseteq \Omega$ we have $\lim_{\varepsilon\to 0} P_\varepsilon(C) = P(C)$, and so (considering in particular $C = \varphi^{-1}(N)$ where N is Lebesgue null) it is sufficient to show that $\mu_\varepsilon \ll$ Lebesgue, where $\mu_\varepsilon(\cdot) = P_\varepsilon(\varphi^{-1}(\cdot))$. For this it is well known that we need only show that there is a constant c such that

$$\int_{\mathbb{R}} \psi'(\xi)\mathrm{d}\mu_\varepsilon(\xi) \leq c\|\psi\|_\infty \tag{4.3}$$

for any $\psi \in \mathrm{C}_b^1$.

Take a $S\mathbb{D}^{2,1}$ lifting Φ of φ such that $\nabla^2\Phi$ is SL^2, and a $S\mathbb{D}^{2,1}$ lifting S of σ. Since $\sigma \geq 0$ a.s. we can take S with $S \geq -\frac{1}{2}\varepsilon$ surely. Then (since $\langle\nabla\Phi,\nabla\Phi\rangle$ is an SL^1 lifting of σ) we have

$$\begin{aligned}
\int \psi'\mathrm{d}\mu_\varepsilon &\approx \mathrm{E}_\Omega\left(\psi'(\Phi(x))\frac{\langle\nabla\Phi,\nabla\Phi\rangle}{S+\varepsilon}\right) \\
&= \mathrm{E}\Big(\langle\nabla\psi(\Phi),(S+\varepsilon)^{-1}\nabla\Phi\rangle\Big) \\
&= \mathrm{E}(\psi(\Phi)\delta((S+\varepsilon)^{-1}\nabla\Phi)) \\
&\leq \|\psi\|_\infty\mathrm{E}(|\delta((S+\varepsilon)^{-1}\nabla\Phi)|
\end{aligned}$$

where δ is the internal classical divergence operator (the dual of ∇) in ${}^*\mathbb{R}^T$ given by

$$\delta U(x) = \sum_{t\in T} U(x,t)x_t - \sum_{t\in T}\frac{\partial U(x,t)}{\partial x_t}\Delta t$$

for $U \in {}^*L^2(\Omega\times T)$ (see the Appendix). The transfer of classical calculus to ${}^*\mathbb{R}^T$ gives

$$\begin{aligned}
\delta((S+\varepsilon)^{-1}\nabla\Phi) &= (S+\varepsilon)^{-1}\delta\nabla\Phi - \langle\nabla((S+\varepsilon)^{-1}),\nabla\Phi\rangle \\
&= (S+\varepsilon)^{-1}\delta\nabla\Phi - (S+\varepsilon)^{-2}\langle\nabla S,\nabla\Phi\rangle
\end{aligned}$$

which is SL^1 (it is easy to see that $\|\delta\nabla\Phi\|^2 = \|\nabla\Phi\|^2 + \|\nabla^2\Phi\|^2$ - see the Appendix). So $c = {}^\circ\mathrm{E}(|\delta((S+\varepsilon)^{-1}\nabla\Phi)|)$ is the required constant in (4.3). ∎

We can apply this to show that functionals in the finite chaos' have densities (c.f. [15] Theorem 5.1)

Theorem 4.3 *Let $0 \neq \varphi \in Z^{(n)}$. Then the measure induced by φ is absolutely continuous with respect to Lebesgue measure.*

Proof Checking the conditions of Theorem 4.2, we have $\varphi \in \mathbb{D}^{2,2}$ clearly. The Lemma below shows that $\sigma(\varphi) \in Z^{(2n-2)}$ and so $\sigma(\varphi) \in \mathbb{D}^{2,1}$; and by Theorem 3.14 we have the condition 4.2(iii), as required. ∎

Lemma 4.4 *If $\varphi \in Z^{(n)}$ then $\sigma(\varphi) \in Z^{(2n-2)}$.*

Proof Let $\varphi_m = I_m(f_m)$ for $m \leq n$ and take an SL^2 liftings F_m of f_m. So $\nabla\Phi = \sum_{m\leq n} I_{m-1}(\hat{F}_m)$ is an SL^2 lifting of $D\varphi$ and $S = \langle \nabla\Phi, \nabla\Phi \rangle$ is an SL^1 lifting of $\sigma(\varphi)$. It can be shown that there is a finite constant $c = c_n$ such that for all $k_1, k_2 < n$ and $G_i : \Delta^T_{k_i} \times T \to {}^*\mathbb{R}$

$$\mathrm{E}_\Omega \left(\langle I_{k_1}(G_1), I_{k_2}(G_2) \rangle^4 \right) \leq c \|G_1\|^4 \|G_2\|^4$$

Thus $\mathrm{E}(S^4) < \infty$ and so S is an SL^2 lifting of of $\sigma(\varphi)$. It is clear that the monomial terms in S have degree $\leq 2(n-1)$. ∎

5 The Skorohod Integral

The *Skorohod integral* δu (also called the *divergence* operator) is defined for certain integrands $u \in L^2(W \times [0,1])$ and is an extension of the Itô integral that allows u to be nonanticipating . It was first introduced in [16]. The usual standard definitions (see e.g. [10],[11],[14]) of δu obscure the fact that it is an integral, whereas the nonstandard approach we outline below makes this quite apparent.

First, for reference, we give the standard definition. A function $u \in L^2(W \times [0,1])$ has chaos expansion

$$u(b,t) = \sum u_n(b,t) = u_o(t) + \sum_{n=1}^{\infty} I_n(f_n(\cdot,t))$$

with $f_n \in L^2(\Delta_n \times [0,1])$. The *symmetrisation* of $f = f_n$ is the function $\tilde{f} \in L^2(\Delta_{n+1})$ given by

$$\tilde{f}(t_1,\ldots,t_{n+1}) = \sum_{i=1}^{n+1} f(t_1,\ldots,t_{i-1},t_{i+1},\ldots,t_i).$$

Definition 5.1 *The* Skorohod integral *δu is the $L^2(W)$ sum*

$$\delta u = \sum_{n=0}^{\infty} I_{n+1}(\tilde{f}_n)$$

provided this limit exists.

It is routine to see that δ is linear and $Z^{(n)} \subset \mathrm{dom}(\delta)$ for each n. Moreover, δu is explicitly given in terms of its chaos expansion, and from this we see that $\mathrm{E}(\delta u) = 0$.

For our nonstandard approach, consider first $u(b,t) = I_n(f(\cdot,t))$ which has SL^2 lifting $U(x,t) = I_n(F(\cdot,t))$, where F is an SL^2 lifting of f. Defining $\tilde{F}$ in the same way that $\tilde{f}$ was defined, we see that δu is SL^2 lifted by $I_{n+1}(\tilde{F})$.

Now we have the following crucial Lemma.

Lemma 5.2

$$I_{n+1}(\tilde{F})(x) = \sum_{t\in T} U(x,t)x_t - \sum_{t\in T} \frac{\partial U(x,t)}{\partial x_t} x_t^2$$

Proof Routine combinatorics. ∎

In view of this we make the following

Definition 5.3 *Let $U \in {}^*L^2(\Omega \times T)$. Then $\check{\delta} U \in {}^*L^2(\Omega)$ is defined by*

$$\check{\delta} U = \sum_{t\in T} U(x,t)x_t - \sum_{t\in T} \frac{\partial U(x,t)}{\partial x_t} x_t^2$$

Now it is straightforward to prove:

Theorem 5.4 *Let $u \in dom(\delta)$ have SL^2 lifting $U(x,t)$. Then for all sufficiently small infinite K, $\check{\delta} U^{(K)}$ is an SL^2 lifting of δu.*

Although $\check{\delta} U$ is a good representation for δu for many purposes, a nicer one is obtained by replacing the term x_t^2 by Δt. Of course in Anderson's construction of Brownian motion, $x_t^2 = \Delta B_t^2 = \Delta t$, but this is not true in our model (this is the price to be paid for having $x_t = \Delta B_t$ vary continuously through ${}^*\mathbb{R}$, to allow internal differentiation $\partial/\partial x_t$). Nevertheless, if we make the definition

Definition 5.5 *For $U \in {}^*L^2(\Omega \times T)$,*

$$\delta U = \sum_{t\in T} U(x,t)x_t - \sum_{t\in T} \frac{\partial U(x,t)}{\partial x_t} \Delta t$$

we have the following parallel to Theorem 3.4(a).

Theorem 5.6 *Let $u \in dom(\delta)$ with SL^2 lifting U. Then for sufficiently small infinite K*

$$E((\check{\delta} U^{(K)} - \delta U^{(K)})^2) \approx 0$$

and hence $\delta U^{(K)}$ is an SL^2 lifting of δu.

Remark. We have chosen to use the same symbol δ to denote the standard operator on $L^2(W \times [0,1])$ and the internal operator above on ${}^*L^2(\Omega \times T)$; this should not cause confusion since we will use u and U to denote members of the respective domains.

If U is nonanticipating (i.e. $U(x,t)$ depends only on x_s for $s < t$) then $\partial U(x,t)/\partial x_t = 0$. So the usual nonstandard theory of the Itô integral now gives immediately:

Corollary 5.7 *If $u \in L^2(W \times [0,1])$ is adapted, then $u \in dom(\delta)$ and $\delta u = \int_0^1 u\,db_t$ (the Itô integral).*

The following results parallel Theorems 3.2, 3.4 for D.

Theorem 5.8 *Let $u \in L^2(W \times [0,1])$. Then $u \in dom(\delta)$ if and only if u has an SL^2 lifting U with $\|\delta U\| < \infty$.*

If $u \in dom(\delta)$ and U is any SL^2 lifting of u, then

(a) $\|\delta u\| \leq {}^\circ\|\delta U\|$

(b) *δU is a lifting of δu if and only if $\|\delta u\| \approx \|\delta U\|$.*

The following parallels Theorem 3.7 and can be applied to show that δ is closed.

Theorem 5.9 *Suppose that $u \in L^2(W \times [0,1])$ has SL^2 lifting U and δU is an SL^2 lifting of $\psi \in L^2(W)$. Then $u \in dom(\delta)$ and $\delta u = \psi$.*

In the literature on the Malliavian calculus, there is much mention of various integration by parts formulae. The most fundamental of these is the following, which shows that δ and D are dual to one another. It is a simple consequence of classical integration by parts at the nonstandard level.

Theorem 5.10 *Let $u \in dom(\delta)$ and $\varphi \in \mathbb{D}^{2,1}$. Then*

$$E(\varphi\delta u) = E(D\varphi.u)$$

i.e. in terms of the inner product in $L^2(W)$ and $L^2(W \times [0,1])$ respectively

$$\langle\varphi, \delta u\rangle = \langle D\varphi, u\rangle.$$

Proof Let Φ be an $S\mathbb{D}^{2,1}$ lifting of φ and U be an SL^2 lifting of u such that δU is an SL^2 lifting of δu. Then we have to show that

$$\mathrm{E}(\Phi\delta U) = \mathrm{E}(\nabla\Phi.U)$$

i.e.

$$\mathrm{E}\Big(\Phi(x)\big(\sum_{t\in T} U(x,t)x_t - \sum_{t\in T}\frac{\partial U(x,t)}{\partial x_t}\Delta t\big)\Big) = \mathrm{E}\Big(\sum_{t\in T}\frac{\partial \Phi(x)}{\partial x_t}.U(x,t)\Delta t\Big).$$

This follows immediately from the classical by parts formula for the Gaussian measure on ${}^*\mathbb{R}$ (see the Appendix) for each fixed $t \in T$:

$$\mathrm{E}\Big(\Psi(x_t)\big(\Theta(x_t)x_t - \frac{\partial\Theta(x_t)}{\partial x_t}\Delta t\big)\Big) = \mathrm{E}\Big(\frac{\partial\Psi(x_t)}{\partial x_t}\Theta(x_t)\Delta t\Big).\blacksquare$$

There is no standard counterpart of the isometry property for the Itô integral that applies to *all* $u \in \mathrm{dom}(\delta)$. However, if $U \in {}^*L^2(\Omega \times T)$ then elementary calculations give the following internal isometry:

Theorem 5.11 *Let $U \in {}^*L^2(\Omega \times T)$; then*

$$\begin{aligned} \|\delta U\|^2_\Omega &= \|U\|^2_{\Omega\times T} + E\Big[\sum_{s,t\in T}\nabla_s U(x,t)\nabla_t U(x,s)\Delta t^2\Big] \\ &= \|U\|^2 + \tfrac{1}{2}\|\nabla_s U(t) + \nabla_t U(s)\|^2 - \|\nabla U\|^2 \\ &\le \|U\|^2 + \|\nabla U\|^2 \end{aligned}$$

This gives an isometry for the subspace $\mathbb{L}^{2,1}$ defined by

Definition 5.12 *Let $u \in L^2(W \times [0,1])$. Then $u \in \mathbb{L}^{2,1}$ if $u(\cdot,t) \in \mathbb{D}^{2,1}$ for a.a. t and $D_s u(b,t) \in L^2(W \times [0,1]^2)$.*

Hence

Theorem 5.13 *If $u \in \mathbb{L}^{2,1}$ then $u \in dom(\delta)$ and*

$$\begin{aligned} \|\delta u\|^2 &= \|u\|^2 + E\Big[\int_0^1\int_0^1 D_s u(b,t)D_t u(b,s)\,ds\,dt\Big] \\ &= \|u\|^2 + \tfrac{1}{2}\|D_s u(t) + D_t u(s)\|^2 - \|Du\|^2 \\ &\le \|u\|^2 + \|Du\|^2 \end{aligned}$$

Proof Simply take a lifting U of u such that $\nabla_s U(x,t)$ is SL^2 and apply Theorem 5.11.■

There is a natural 'inner product' type formula for $\mathrm{E}(\delta u \delta v)$ for $u, v \in \mathbb{L}^{2,1}$, obtained from Theorem 5.13 by polarisation, which we leave to the reader to formulate.

Although it is an open question whether δ is local on $\mathrm{dom}(\delta)$, it is known that δ is local on $\mathbb{L}^{2,1}$, and we sketch the proof of this in our framework (cf. [10] Prop. 4.5).

Theorem 5.14 *Let $u \in \mathbb{L}^{2,1}$ and $A \subseteq \mathcal{C}$ be measurable. If $u = 0$ a.s. on $A \times [0,1]$ then $\delta u = 0$ on A a.s.*

Proof (Sketch) Take an SL^2 lifting of u such that $\nabla_s U(x,t)$ is an SL^2 lifting of $D_s u(b,t)$ (and hence δU is an SL^2 lifting of δu). Let $g(b) = \int_0^1 u(b,t)^2 dt$ and $G(x) = \sum_{t\in T} U(x,t)^2 \Delta t$. Pick an infinite M such that for a.a. x,

$$g(°x) = 0 \quad \Rightarrow \quad G(x) \le \frac{1}{M}$$

$$g(°x) \ne 0 \quad \Rightarrow \quad G(x) \ge \frac{2}{M}$$

and take an internal smooth nondecreasing function $\theta: {}^*\mathbb{R} \to {}^*\mathbb{R}$ with $\theta(y) = 0$ for $y \le \frac{1}{M}$ and $y \ge \frac{2}{M}$, and $0 \le \theta'(y) \le 2M$, so that $y\theta'(y) \le 4$ for all y. Let $\bar{U}(x,t) = \theta(G(x))U(x,t)$. Then $\bar{U}$ is an SL^2 lifting of u and from the definition of δ we have

$$\delta\bar{U}(x) = \theta(G(x))\delta U(x) - H(x)$$

where $H(x) = \sum_t \nabla_t \theta(G(x))U(x,t)\Delta t = \sum_t \theta'(G(x))\nabla_t G(x) U(x,t)\Delta t = 0$ a.s.

Now $\theta(G(x))\delta U(x)$ is an SL^2 lifting of $1_{\{g\ne 0\}}\delta u$ and so (using Theorem 5.9) it suffices to show that H is SL^2. We have

$$\begin{aligned} |H(x)| &= \theta'(G(x))|\sum_t \Big(\sum_s 2U(x,s)\nabla_t U(x,s)\Delta t\Big)U(x,t)\Delta t| \\ &\le 2\theta'(G(x))\sum_t U(x,t)^2\Delta t\Big(\sum_{s,t}\nabla_t U(x,s)^2\Delta t^2\Big)^{\frac{1}{2}} \\ &\le 8\|\nabla U\| \end{aligned}$$

so H is SL^2 (since ∇U is). ∎

6 The Malliavin Operator

In the early development of the Malliavin calculus the Malliavin operator L (defined below) played a prominent rôle (see [17] for example), but it is now understood that it is not the most fundamental operator in the calculus. The simplest definition is

Definition 6.1 *The operator $L : L^2(W) \to L^2(W)$ is given by*

$$dom(L) = \{\varphi \in \mathbb{D}^{2,1} : D\varphi \in dom(\delta)\}$$

$$L = \delta D$$

Below we give other characterisations of L beginning with a nonstandard one using the following internal operator on $L^2(\Omega)$.

Definition 6.2 *Let $\Phi \in {}^*L^2(\Omega)$; define $\mathcal{L}\Phi \in {}^*L^2(\Omega)$ by*

$$\begin{aligned} \mathcal{L}\Phi(x) &= \sum_t \nabla_t \Phi(x) x_t - \sum_t \nabla_t^2 \Phi \Delta t \\ &= \delta \nabla \Phi \end{aligned}$$

for $\Phi \in dom(\nabla_t^2)$ for all t.

It is immediate that for $\Phi = I_n(F)$ we have $\mathcal{L}\Phi = n\Phi$, and so from the results on D and δ in Sections 3 and 5 we have:

Theorem 6.3 *L is the number operator on $L^2(W)$; i.e. $\varphi \in dom(L)$ if and only if $\sum n^2 \|\varphi_n\|^2 < \infty$ and for such φ*

$$L\varphi = \sum n\varphi_n.$$

Hence $dom(L) = \mathbb{D}^{2,2} = dom(D^2)$.

The following results are immediate from the corresponding results for D and δ:

Theorem 6.4 *Let $\varphi \in L^2(W)$. Then $\varphi \in \mathbb{D}^{2,2}$ if and only if φ has an SL^2 lifting with $\|\mathcal{L}\Phi\| < \infty$.*

If $\varphi \in \mathbb{D}^{2,2}$ and Φ is any SL^2 lifting then

(a) $\|L\varphi\| \leq {}^\circ\|\mathcal{L}\Phi\|$.

(b) $\mathcal{L}(\Phi^{(K)}) = (\mathcal{L}\Phi)^{(K)}$ *is an SL^2 lifting of $L\varphi$ for all sufficiently small infinite K.*

(c) $\mathcal{L}\Phi$ *is an SL^2 lifting of $L\varphi$ if and only if $\|\mathcal{L}\Phi\| \approx \|L\varphi\|$.*

We also have

Theorem 6.5 *If Φ is an SL^2 lifting of $\varphi \in L^2(W)$ and $\mathcal{L}\Phi$ is an SL^2 lifting of $\psi \in L^2(W)$ then $\varphi \in \mathbb{D}^{2,2}$ and $L\varphi = \psi$.*

Many of the properties of L (and $\mathcal{L}$) follow easily from those of D (resp. ∇) and δ (resp. δ internal), including the following consequence of the integration by parts result 5.10:

Theorem 6.6 *Let φ, $\psi \in \mathbb{D}^{2,2}$. Then*

$$E(\varphi L\psi) = E_{\mathcal{C}\times[0,1]}(D\varphi . D\psi) = E(\psi L\varphi)$$

The paper [12] gives a large number of identities involving D, δ and $\mathcal{L}$ which can be obtained easily using the nonstandard framework.

The following is a deeper property of L and is needed to prove the product and chain rules. It is a strengthening of Theorem 6.5 (cf. Theorem 3.12 for D).

Theorem 6.7 *Suppose that φ, $\psi \in L^2(W)$ have SL^1 liftings $\Phi, \Psi \in {}^*L^2(\Omega)$ with $\Psi = \mathcal{L}\Phi$. Then $\varphi \in \mathbb{D}^{2,2}$ and $L\varphi = \psi$.*

The proof of this involves some deeper analysis of $\mathcal{L}$ as a generator of the Ornstein-Uhlenbeck process on ${}^*\mathbb{R}^T$ (which has as standard projection the fact that L is the generator of the Ornstein-Uhlenbeck semigroup on $L^2(W)$ - see [14] for details).

This result gives straightforward proofs of the following (by transfer of classical calculus on ${}^*\mathbb{R}^T$ to suitable SL^2 liftings):

Theorem 6.8 *Suppose that $\varphi, \psi, \varphi\psi \in L^2(W)$ and $\varphi, \psi \in \mathbb{D}^{2,2}$ and $\theta \in L^2(W)$ where*

$$\theta = \varphi L\psi + \psi L\varphi - 2\langle D\varphi, D\psi\rangle.$$

Then $\varphi\psi \in \mathbb{D}^{2,2}$ and $L(\varphi\psi) = \theta$.

Theorem 6.9 *Suppose that $f \in C_b^2$ and $\varphi \in \mathbb{D}^{2,2}$ and $\sigma(\varphi) = \langle D\varphi, D\varphi\rangle \in L^2(W)$. Then $f \circ \varphi \in \mathbb{D}^{2,2}$ and*

$$L(f \circ \varphi) = f'(\varphi)L\varphi - f''(\varphi)\sigma(\varphi)$$

Applications to Stochastic Differential Equations In applications to existence of densities of the solutions to SDE's it is necessary to find equations for the evolution of functionals such as $D\varphi_t$, $L\varphi_t$ and $\sigma(\varphi_t) = \langle D\varphi_t, D\varphi_t\rangle$ when φ_t is the solution to an SDE. This is achieved quite easily by combining our approach with Keisler's hyperfinite difference approach to SDE's [8]. This will be explained in detail in [7].

7 Appendix - Hermite Polynomials and $L^2(\mathbb{R}, \gamma)$

We give a brief review of the basics of classical calculus on $L^2(\gamma) = L^2(\mathbb{R}, \gamma)$ which is transferred to ${}^*\mathbb{R}^T$ in the previous sections. Here γ is the usual Gaussian measure on $\mathbb{R}$ so that

$$\gamma(A) = (2\pi)^{-\frac{1}{2}} \int_A \exp(-\tfrac{1}{2}\xi^2)d\xi.$$

The *Hermite polynomials* $h_n(\xi)$ $(n = 0, 1, 2 \ldots)$ are real polynomials with degree$(h_n) = n$ forming an orthonormal basis for $L^2(\gamma)$, obtained from 1, ξ, ξ^2, ξ^3,... by the

Gram-Schmidt procedure, and we have $h_0(\xi) = 1$, $h_1(\xi) = \xi$, $h_2(\xi) = \frac{1}{\sqrt{2}}(\xi^2 - 1)$,... . Alternatively they can be characterised by

$$\frac{\mathrm{d}^n}{\mathrm{d}\xi^n} \exp\left(-\frac{\xi^2}{2}\right) = (-1)^n \sqrt{n!}\, h_n(\xi) \exp\left(-\frac{\xi^2}{2}\right).$$

There are numerous identities concerning the family $(h_n)_{n \geq 0}$; here we need:

Theorem 7.1 (a) $E(h_n) = 0$ *for* $n \geq 1$

(b) $h_n' = \sqrt{n} h_{n-1}$

(c) $\xi h_n = h_n' + \sqrt{n+1} h_{n+1}$, *so* $E(\xi h_n) = E(h_n')$

For $\theta \in L^2(\gamma)$ we have $\theta = \sum_{n=0}^{\infty} a_n h_n = \sum_n \theta_n$, say, where $a_n = \langle \theta, h_n \rangle$ in $L^2(\gamma)$; so $\|\theta\|^2 = \sum_n \|\theta_n\|^2 = \sum_n \|a_n\|^2$. The derivative is extended to $L^2(\gamma)$ by

Definition 7.2

$$\theta' = \nabla\theta = \sum_n \theta_n' = \sum_n a_n h_n' = \sum a_n \sqrt{n} h_{n-1}$$

for $\theta \in L^2(\gamma)$ *with* $\sum_n n a_n^2 < \infty$.

The basic integration by parts formula extending Theorem 7.1(c) is

Theorem 7.3 *If* θ, $\psi \in dom(\nabla)$ *then* $\xi\theta \in L^2(\gamma)$ *and*

$$E(\xi\theta\psi) = E(\theta\psi' + \theta'\psi)$$

The *divergence operator* δ and the operator $\mathcal{L}$ are defined by

Definition 7.4

$$\delta\theta = \xi\theta - \theta'$$

$$\mathcal{L}\theta = \delta\nabla\theta = \xi\theta' - \theta''$$

with $dom(\mathcal{L}) = dom(\nabla^2)$.

It is easy to check that the following identities hold.

Theorem 7.5 *For all θ, $\psi \in L^2(\gamma)$ in the appropriate domains*

(a) $E(\theta.\psi') = E(\delta\theta.\psi)$ *(so δ is the dual of ∇)*

(b) $\delta\theta = \sum_n \sqrt{n} a_{n-1} h_n$ *where* $\theta = \sum_n a_n h_n$

(c) $\|\delta\theta\|^2 = \sum_n (n+1) a_n^2 = \|\theta\|^2 + \|\nabla\theta\|^2$

(d) $\mathcal{L} h_n = n h_n$ *and so* $\mathcal{L}\theta = \sum_n n\theta_n$

(e) $\|\mathcal{L}\theta\|^2 = \|\delta\nabla\theta\|^2 = \|\nabla\theta\|^2 + \|\nabla^2\theta\|^2$

(f) $\delta(\theta\psi) = \theta.\delta\psi - \psi.\nabla\theta$

For application to the Malliavin calculus by transfer, we need an n-dimensional scaled version of this. For the scaling take real $\varepsilon > 0$ (we will set $\varepsilon = \Delta t$ after transfer) and put $x = \sqrt{\varepsilon}\xi$, so x is $\mathcal{N}(0, \varepsilon)$ under γ and induces the measure γ_ε say on $\mathbb{R}$. The Hermite polynomials are now $H_n(x) = h_n(x/\sqrt{\varepsilon})$ and we have $H_n'(x) = \sqrt{n/\varepsilon} H_{n-1}(x)$ etc., giving the integration by parts formula

$$\mathrm{E}_{\gamma_\varepsilon}(x\theta\psi) = \varepsilon \mathrm{E}_{\gamma_\varepsilon}(\theta\psi' + \theta'\psi)$$

and, defining $\delta\theta = x\theta - \varepsilon\theta'$, we have the duality

$$\mathrm{E}_{\gamma_\varepsilon}(\delta\theta.\psi) = \mathrm{E}_{\gamma_\varepsilon}(\theta\psi'\varepsilon).$$

Putting $\mathcal{L}\theta = \delta\nabla\theta = x\theta' - \varepsilon\theta''$ as before gives $\mathcal{L}H_n = nH_n$.

To extend to n-dimensions, we have the measure γ_ε^n on $\mathbb{R}^n$ with orthonormal basis for $L^2(\gamma_\varepsilon^n)$ given by the family (H_α) for α a *multi-index.* i.e. $\alpha = (\alpha_1, \alpha_2, \ldots, \alpha_n)$ with each α_i a non-negative integer and $H_\alpha(x_1, x_2, \ldots, x_n) = \prod_{i=1}^n H_{\alpha_i}(x_i)$. This gives derivatives $\nabla_i\theta = \partial\theta/\partial x_i$ and divergences $\delta_i\theta = x_i\theta - \varepsilon\nabla_i\theta$ for $\theta \in L^2(\gamma_\varepsilon^n)$. The divergence δ that is dual to the gradient $\nabla = (\nabla_1, \nabla_2, \ldots, \nabla_n)$ is $\delta\theta = \sum_i \delta_i\theta_i$ for $\theta = (\theta_1, \ldots, \theta_n)$. Then

$$\mathrm{E}(\delta\theta.\psi) = \mathrm{E}\Big(\sum_i \theta_i \nabla_i \psi \varepsilon\Big)$$

Again putting $\mathcal{L} = \delta\nabla$, the calculus outlined above extends in a natural way to $L^2(\gamma_\varepsilon^n)$.

For the application to the Malliavin calculus we simply replace the index set $\{1, 2, \ldots, n\}$ by the set T.

References

[1] S.Albeverio, J.-E.Fenstad, R.Høegh-Krohn, and T.Lindstrøm, *Nonstandard Methods in Stochastic Analysis and Mathematical Physics*, Academic Press, New York 1986.

[2] R.M.Anderson, A non-standard representation for Brownian motion and Itô integration, *Israel J. Math.* **25**(1976), 15-46.

[3] N.J.Cutland, Nonstandard measure theory and its applications, *Bull. London Math. Soc.* **15** (1983), 529-589.

[4] N.J.Cutland, Infinitesimals in action, *J. Lond. Math. Soc.* **35**(1987), 202-216.

[5] N.J.Cutland (ed.), *Nonstandard Analysis and its Applications*, Cambridge University Press, Cambridge 1988.

[6] N.J.Cutland & S.-A.Ng, On homogeneous chaos, *Math. Proc. Camb. Phil. Soc.* **110**(1991), 353-363.

[7] N.J.Cutland & S.-A. Ng, An infinitesimal approach to the Malliavin calculus, monograph in preparation.

[8] H.J.Keisler, An infinitesimal approach to stochastic analysis, *Mem. Amer. Math. Soc.* **297**(1984).

[9] H.P.McKean, Geometry of differential space, *Annals of Prob.* **1**(1973), 197-206.

[10] D. Nualart & E. Pardoux, Stochastic calculus with anticipating integrands, *Probab. Theor. Rel. Fields* **78**(1988), 535-581.

[11] D. Nualart & M. Zakai, Generalized stochastic integrals and the Malliavin calculus, *Probab. Theor. Rel. Fields* **73**(1986), 255-280.

[12] D. Nualart & M. Zakai, A summary of some identities of the Malliavin calculus, *Lecture Notes in Mathematics* **1390**, Springer-Verlag 1989, 192-196.

[13] D. Ocone, Malliavin's calculus and stochastic integral representation of functionals of diffusion processes, *Stochastics* **12**(1984), 161-185.

[14] D. Ocone, A guide to the stochastic calculus of variations. In *Lecture Notes in Mathematics* **1316,** Springer-Verlag.

[15] I. Shigekawa, Derivatives of Wiener functionals and absolute continuity of induced measures, *J. Mat. Kyoto Univ.* **20**-2(1980), 263-289.

[16] A.V.Skorohod, On a generalisation of a stochastic integral, *Theor. Prob. Appl.* **20**(1975), 219-233.

[17] D.W.Stroock, The Malliavin calculus and its applications. In *Stochastic Integrals,* Lecture Notes in Mathematics **851**, Springer-Verlag 1987, 394-432.

[18] M. Zakai, The Malliavin calculus, *Acta Appl. Math.* **3**(1985), 175-207.

ERGODIC TRANSFORMATIONS IN AST

Martin Kalina
Dept. of Mathematics, Slovak Tech. University, Radlinského 11
813 68 Bratislava, Slovakia

Abstract

In this article we investigate the ergodicity of set transformations within the Alternative Set Theory. The main result is that each ergodic transformation is homomorhic to the cycle of a suitable length.

Teturo Kamae in his paper [K] has showen that each "classical" ergodic transformation can be viewed at as being a factorization of a cycle. Our intention is to find out wether this is the only way how to get ergodic transformations.

The paper is written in the language of the Alternative Set Theory [V]. Preliminarily, let us list the cructial notions used in this article. First, an **indiscernibility equivalence** on a set a is an equivalence, which is a π-class (i.e. it is an intersection of an, at most, countable system of sets) and each infinite set contains at least two elements that are in the same equivalence-class. Such elements are also called **indiscernible**. If $\triangleq$ is an indiscernibility equivalence on the set a, then for each $x \in a$ the class $\{y \in a; x \triangleq y\}$ is called the **monad** of x and denoted by $mon(x)$. For each subclass of a, $X \subseteq a$, the class $\{y \in a; (\exists x \in X)(y \triangleq x)\}$ is called the **figure** of X and denoted by $Fig(X)$. So, $Fig(X)$ is just the union of all the monads of x, such that $x \in X$.

Further, let a be any nonempty set, having α elements, $a = \{x_i\}_{i=1}^{\alpha}$, and $c : a \to a$ be a set transformation. We will say that c is a **cycle** if $c(x_i) = x_{i+1}$ for $1 \leq i < \alpha$ and $c(x_\alpha) = x_1$. More precisely, in that case we will call c an α–**cycle**. We have already mentioned a factorization of a cycle. More generally, we define the factorization of a set transformation, $t : a \to a$, by an indiscernibility equivalence on the set a, $\triangleq$, by the following - let $T : M \to M$ be a transformation, M be the system of all monads of the elements of a, then T is the **factorization** of t by $\triangleq$, if for each $x \in a$ (possibly except of a class of μ-measure zero, where μ is the Loeb measure) there is an $y \in a$ such that $t\, mon(x) \subseteq mon(y)$ and $T(mon(x)) = mon(y)$ hold. The transformation t will be called **factorizable by** $\triangleq$.

Finally, a class A is said to be **Borel**, if it is an element of the least σ-algebra, containing all sets. A relation R is said to be **Borel**, if R is a Borel class.

Now, we are going to introduce the following notation - a will be a fixed infinite set and α the number of its elements, μ will denote the measure, induced by the counting measure and normed by α , i.e. the Loeb measure ([L], see also [K-Z]), r will always denote a real number (recall that the system of real numbers is the factor space $\mathbb{Q}/\doteq$). And finally t will always denote a set transformation with $dom(t) = a$ and $rng(t) \subseteq a$.

Lemma 1: *Let t be any set-transformation such that $\mu(x) = \mu(t^{-1}x)$ for each set $x \subseteq a$. Then there is a measurable class $X \subseteq a$, for which $\mu(X) \neq 0, 1$ and $t^{-1}X \supseteq X$ hold.*

Proof: Take a set $\emptyset \neq x \subseteq a$ and a $\gamma \in \mathbb{N}$ and put

$$X_\gamma = \bigcap\{\bigcup\{t^{-n}x; i \leq n \leq \gamma - i\}; i \in \mathbf{FN}\}$$

It is just a matter of routine to show that x and γ can be chosen in such a way that X_γ has all the properties required of X. ∎

It follows that if we are interested in getting, in some sense, ergodic transformations, we have to shrink our σ-algebra. The most natural way how to do it within the framework of the Alternative Set Theory, is to take an indiscernibility equivalence and to restrict our attention to its measurable figures, only.

Let $\triangleq$ be a fixed indiscernibility equivalence on a. Denote $\mathcal{A}$ the σ-algebra of all its measurable figures (i.e. $X \in \mathcal{A}$ iff X is μ-measurable and $X = Fig(X) = \{y; (\exists x \in X)(y \triangleq x)\}$). We will say that the transformation t is **measure-preserving** with respect to $\triangleq$ if each $X \in \mathcal{A}$ yealds $Fig(t^{-1}X) \in \mathcal{A}$, i.e. if the figure of its pre-image is measurable, and, moreover, $\mu(X) = \mu(Fig(t^{-1}X))$. A class $X \subseteq a$ will be called **t-invariant** if $X \subseteq t^{-1}X$.

Lemma 2: *Let the transformation t be measure-preserving with respect to $\triangleq$. Then t is factorizable by $\triangleq$.*

Proof: Let t not be factorizable by $\triangleq$. Then there is a class $X \in \mathcal{A}$ with $\mu(X) \neq 0$ such that for each $x \in X$ the image $t\,mon(x)$ has a nonempty intersection with, at least, two different monads. So, we can find two disjoint classes, Y, Z, such that $t^{-1}Y \supseteq X$ and $t^{-1}Z \supseteq X$, but, since $\mu(X) \neq 0$, t is not measure-preserving with respect to $\triangleq$. ∎

We will call a measure-preserving transformation t **ergodic** (with respect to $\triangleq$) if all the t-invariant classes $X \in \mathcal{A}$ yield $\mu(X) = 0$ or $\mu(X) = 1$.

A straightforward consideration gives

Lemma 3: *If there exists a measure-preserving (with respect to $\triangleq$) transformation t then either there are n monads (n being finite), each of them of measure $1/n$, or all monads are of measure 0.*

Obviously the first case leads to the model of a finite space, which we are not interested in, so, in the whole paper we will consider only indiscernibility equivalences, all monads of which are of measure 0.

The definition of an ergodic transformation says how the t-invariant classes are if they are figures in $\stackrel{\Delta}{=}$. Something more about t-invariant classes is proved in the following

Theorem 1: *Let t be ergodic and let X be a t-invariant class with $Fig(X) \in \mathcal{A}$. Then $\mu(Fig(X)) = 0$ or $\mu(Fig(X)) = 1$.*

Proof: Obviously $Fig(X) \subseteq Fig(t^{-1}(X))$. Construct the following chain of classes $X_0 = FigX$, and for $i \in \mathbf{FN}$ $\quad X_{i+1} = Fig(t^{-1}(X_i))$. So we get $X_0 \subseteq X_1 \subseteq \ldots \subseteq \subseteq X_i \subseteq \ldots$ and the class $\bigcup\{X_i; i \in \mathbf{FN}\}$ is invariant. The transformation t is measure-preserving, hence $\mu(X_0) = \mu(X_i)$ for each $i \in \mathbf{FN}$, therefore $\mu(X_i \backslash X_0) = 0$. It follows $\mu(\bigcup\{X_i; i \in \mathbf{FN}\}) = \mu(X_0)$, and we get $\mu(X_0) = \mu(Fig(X)) = 0$ or $\mu(Fig(X)) = 1$. ∎

Theorem 2: *If the transformation t is ergodic then the class of all finite cycles is of measure 0.*

Proof: Suppose the assertion does not hold. Then there exists an $n \in \mathbf{FN}$ so that the set x_n of all n-cycles is of a positive measure. Because of Theorem 1, $\mu(Fig(x_n)) = 1$. Take an $x \subseteq x_n$ so that $\mu(Fig(x)) = \frac{1}{2n}$. Obviously, the set $\bigcup_{i=1}^{n} t^{-i}(x) = y$ is invariant and there holds $\frac{1}{2n} \leq \mu(Fig(y)) \leq \frac{1}{2}$, and this is a contradiction. ∎

Theorem 3: *If the transformation t is bijective and the class of all finite cycles is of measure 0, then there exists an indiscernibility equivalence $\stackrel{\circ}{=}$ which makes t ergodic.*

Proof: The class of all finite cycles is a σ-class (i.e. the union of an, at most, countable system of sets) and therefore it can be extended to a set of measure 0, which is invariant (see [K-Z]), hence, without loss of generality, it can be regarded to be empty.

Fix an irrational real number $r \in [0,1]$. The transformation t consists of β cycles, each of them of the length γ_i, $i < \beta$. Define a map $m : a \to \mathbb{Q}$ by the following- fix a point x_i from the i-th cycle and for all $\kappa < \gamma_i$ put $m(t^{-\kappa}(x_i)) = \frac{\kappa \cdot \delta_i}{\gamma_i}$, where $\delta_i \in \mathbb{N}$ is chosen in such a way that $\frac{\delta_i}{\gamma_i} \in r$. Now, it is enough to define the indiscernibility equivalence $\stackrel{\circ}{=}$ by

$$(\forall x, y \in a)(x \stackrel{\circ}{=} y) \Leftrightarrow (m(x) \doteq m(y)/mod 1),$$

and we are done. ∎

Theorem 4: *Let t be bijective and ergodic. Then there exists a set $z \subset a$ of measure 0 so that $t|_{(a \backslash z)}$ and c are isomorphic, where $c : a \to a$ is an α-cycle. The isomorphism can be done to be Borel.*

Proof: The class of all finite cycles is σ, hence there exists its superset z, which is still of measure 0 and t-invariant, and therefore there exists a Borel bijection between a and $a \setminus z$ (see [K-Z]). So, without loss of generality, we can assume $z = \emptyset$.

Let t consist of β infinite cycles, each of them of the length γ_i. The isomorphism T will be constructed by the following - take the k-th cycle, i.e. the points x_δ up to $x_{\delta+\gamma_k-1}$ (in a suitable ordering), where $\delta = \sum_{i=1}^{k-1} \gamma_i$, and put

$$\begin{aligned} T(x_\varepsilon) &= \varepsilon \qquad \text{for} \qquad \varepsilon \in \bigcap \{\{\vartheta; \delta \leq \vartheta \leq \gamma_k - n\}; n \in \mathbf{FN}\} \\ T(x_{\delta+\gamma_k-i}) &= \delta - i \qquad \text{for} \qquad i \in \mathbf{FN} \setminus \{0\}, \end{aligned}$$

and similarly the other cycles, and we are done. ∎

In the remainder of this paper $c : b \to b$ will denote a cycle, where b will be a set, having a suitable number of elements.

Theorem 5: *Let there exist Borel classes A, B, both having measure 0, such that $t|_{(a\setminus A)}$ is homomorphic to $c|_{(b\setminus B)}$, where the homomorphism H is a set function, restricted to the Borel class $a \setminus A$. And let*

$$(\forall x, y \in b \setminus B)(| H^{-1}(x) | / | H^{-1}(y) | \doteq 1)$$

hold[2]. *Then there exists an indiscernibility equivalence $\overset{\circ}{=}$, which makes the transformation t ergodic.*

The proof of this Theorem is omitted, since the construction of the searched indiscernibility equivalence is similar to that used in the proof of Theorem 3.

Theorem 6: *Let the transformation t be ergodic. Then there exist Borel classes A, B, both having measure 0, such that $t|_{(a\setminus A)}$ is homomorphic to $c|_{(b\setminus B)}$. Moreover, the homomorphism H can be done to be a set function, restricted to the Borel class $a \setminus A$, and*

$$(\forall x, y \in b \setminus B)(| H^{-1}(x) | / | H^{-1}(y) | \doteq 1)$$

holds.

Proof: The Transformation t is ergodic, i.e. the class of all finite cycles is of measure 0 (due to Theorem 2), hence there exists a sequence of sets $\{s_j\}_{j\in\mathbf{FN}\setminus\{0\}}$ (subsets of a) with the following properties

$$| s_j | / \alpha \doteq 1/j \tag{0.1}$$

$$s_j \bigcap t^{-k} s_j = \emptyset \qquad \text{for} 1 \leq k \leq j-1 \tag{0.2}$$

$$| \bigcup_{i=0}^{j-1} t^{-i} s_j | / \alpha \doteq 1 \tag{0.3}$$

and these properties imply the following one

$$| s_j | / | t^{-k} s_j | \doteq 1 \qquad \text{for} \qquad 0 \leq k \leq j-1. \tag{0.4}$$

Now, we can prolong the sequence $\{s_j\}$. Since the equivalence $\doteq$ is π, there exists a $\nu \in \mathbb{N} \setminus \mathbf{FN}$ such that for s_ν Properties (1-4) still hold.

[2] For any set z $| z |$ denotes the number of its elements.

Denote $\tilde{a} = \bigcup\{t^{-\gamma} s_\nu; 0 \leq \gamma \leq \nu\}$, Property (3) implies $\mu(\tilde{a}) = 1$, and define an $H_1 : \tilde{a} \to \mathbb{N}$ by $H_1^{-1}(\gamma) = t^{-\gamma} s_\nu$ for $\gamma < \nu$. Since there is no guarantee that $t^{-\nu} s_\nu = s_\nu$ holds, we have to restrict H_1 in the following manner - denote

$\tilde{A} = \bigcup_{i \in \mathbf{FN}}(t^{-i} s_\nu \bigcup t^{-(\nu-i)} s_\nu)$ and put $H = H_1|_{(\tilde{a} \setminus \tilde{A})}$. Now it is just a matter of routine to show that H has all the properties required. ∎

Theorems 5 and 6 say that each ergodic transformation is, up to a homomorphism, an infinite cycle with the "almost" uniform probability distribution.

However, there is still one problem open. Namely, Theorem 2, together with the condition $(\forall x \in a)(\mu(t^{-1}x) = 0)$, give us an obvious necessary condition on t to be ergodic. But it is unknown (at least to the author) wether this condition is also sufficient or wether there exists a set transformation, fulfilling this condition, which, under no indiscernibility equivalence, is ergodic.

References

[K] Teturo Kamae: *A simple proof of the ergodic theorem using nonstandard analysis*, Israel J. Stat. 42, 1982, 284–290

[K-Z] Kalina, M., Zlatos, P.: *Borel classes in AST. Measurability, cuts and equivalence,* Comment. Math. Univ. Carolinae 30, 1989, 357–372

[L] Loeb, P.A.: *Conversion from nonstandard to standard measure spaces and applications in probability theory,* Trans. AMS 211, 1975, 113–122

[V] Vopenka, P.: *Mathematics in the Alternative Set Theory* , Teubner Texte Leipzig 1979.

NONSTANDRAD CHARACTERIZATION FOR A GENERAL INVARIANCE PRINCIPLE

Dieter Landers
Mathematisches Institut der Universität Köln,
Weyertal 86 - 90 D-50931 Köln, Germany

Lothar Rogge
Fachbereich Mathematik, Universität - GH Duisburg,
D-47048 Duisburg, Germany

Abstract

In this paper it is shown that every invariance principle of probability theory is equivalent to a nonstandard construction of internal S–continuous processes, which all represent - up to an infinitesimal error - the limit process. This can be applied e.g. to obtain Anderson's nonstandard construction of a Brownian motion on a hyperfinite set.

1 Introduction and Notation

In 1969 Müller [8] gave the first nonstandard proof of Donsker's invariance principle. In 1976 Anderson [1] constructed a Brownian motion with nonstandard methods. He defined a hyperfinite random walk whose standard part was a Brownian motion on a Loeb-measure space. From this construction he deduced Donsker's invariance principle.

In 1986 Stoll [9] constructed Lévy's Brownian motion with nonstandard methods from which he derived a new invariance principle. In 1989 Mendieta [7] constructed a Brownian bridge with nonstandard methods and deduced an invariance principle for the empirical process.

The purpose of this paper is to show that in all these cases the invariance principle is not only implied but even equivalent to such a nonstandard construction. More generally we prove that this equivalence holds for each weak convergence result for suitable processes.

The general framework for invariance principles of probability theory is the following:

Let $(\Omega_n, \mathcal{A}_n, P_n)$ be probability spaces. Let D be a space of functions from a "time" set $T \subset \mathbb{R}^k$ into $\mathbb{R}^m$, endowed with a topology $\mathcal{T}$. The Borel-algebra $\mathcal{B}(\mathcal{D})$ is the

σ-algebra $\sigma(\mathcal{T})$, generated by $\mathcal{T}$. Furthermore let $Y_n : \Omega_n \to D$ be $\mathcal{A}_n, \mathcal{B}(\mathcal{D})$-measurable processes. Then invariance principles are assertions of the form

$$W_n := (P_n)_{Y_n} \to W \text{ weakly,}$$

where $W|\mathcal{B}(\mathcal{D})$ are special probability measures (= p-measures) and $Y_n, n \in \mathbb{N}$ are special processes. $(P_n)_{Y_n}$ denote the p-measures on $\mathcal{B}(\mathcal{D})$, defined by

$$(P_n)_{Y_n}(B) := P_n(Y_n^{-1}(B)), B \in \mathcal{B}(\mathcal{D}),$$

and $W_n \to W$ weakly means convergence with respect to the weak topology, i.e.

$$\liminf_{n \in \mathbb{N}} W_n(O) \geq W(O) \text{ for all } O \in \mathcal{T}.$$

In many invariance principles W is the Wiener measure or the distribution of the Brownian bridge and the function space D is the space $C[0,1]$ of continuous functions on $[0,1]$ or the space $D[0,1]$ of functions on $[0,1]$, which are right continuous and have left hand limits.

To cover more general situations we consider a rather large class of function spaces, which are suitable for all invariance principles, known to the authors. To this aim let $T \subset \mathbb{R}^k$ and $D \subset (\mathbb{R}^m)^T$ and denote by $\mathcal{T}_K$ the topology (on D) of uniform convergence on compact subsets of T. Let furthermore $C(T, \mathbb{R}^m)$ be the system of all continuous functions $f : T \to \mathbb{R}^m$.

1.1 Definition: A topological space $(D, \mathcal{T})$ is called a sip (= suitable for the invariance principle) -space, if

(i) $D \subset (\mathbb{R}^m)^T$ where $0 \in T \subset \mathbb{R}^k$ is locally compact and connected;

(ii) $\mathcal{T}$ is regular and fulfills the first axiom of countability;

(iii) $C := C(T, \mathbb{R}^m) \subset D$ and $C \in \mathcal{B}(\mathcal{D})$;

(iv) $f_n \xrightarrow[\mathcal{T}]{} f \Leftrightarrow f_n \xrightarrow[\mathcal{T}_K]{} f$ ($f_n \in D, f \in C$).

The notation $f_n \xrightarrow[\mathcal{T}]{} f$ means that the sequence $f_n, n \in \mathbb{N}$, converges to f with respect to the toplogy $\mathcal{T}$.

1.2 Examples:
The following spaces $(D, \mathcal{T})$ are sip-spaces

(i) $D := C(\mathbb{R}^k, \mathbb{R}^m)$ with $\mathcal{T} := \mathcal{T}_K$;

(ii) $D := C[0,1]$ or $D[0,1]$ or $B[0,1]$ with $\mathcal{T} := \mathcal{T}_K$;

(iii) $D := D[0,1]$ with the Skorohod topology $\mathcal{T}$.

Here $B[0,1]$ denotes the set of bounded functions $f : [0,1] \to \mathbb{R}$; for the Skorohod topology see e.g. [3], pp. 111.

In order to apply nonstandard methods, we work in this paper with a polysaturated model.

If $(D, \mathcal{T})$ is a topological space denote by $ns({}^*D)$ the set of all near-standard points of ${}^*D$ with respect to $\mathcal{T}$. If $g \in ns({}^*D)$ and $f \in D$ we write $st(g) = f$ if $g \underset{\mathcal{T}}{\approx} f$, i.e. if g is infinitesimally close to f with respect to the topology $\mathcal{T}$.

For near-standard points y of ${}^*\mathbb{R}^m$ denote the standard part of y by ${}^\circ y$. Let $D \subset (\mathbb{R}^m)^T$ with $T \subset \mathbb{R}^k$. If $g \in {}^*D$ is a function, such that $g(t)$ is a near-standard point of ${}^*\mathbb{R}^m$ for each $t \in T$, then ${}^\circ g$ denotes the function from T into $\mathbb{R}^m$, defined by ${}^\circ g(t) := {}^\circ(g(t))$ for $t \in T$. A function $g \in {}^*D$ is called S-continuous, if ${}^*T \ni s \approx t \in T \Rightarrow g(s) \approx g(t)$.

If $(\Omega_n, \mathcal{A}_n, P_n)$ are probability spaces and $Y_n : \Omega_n \to D$ are $\mathcal{A}_n, \mathcal{B}(\mathcal{D})$-measurable processes for $n \in \mathbb{N}$, we denote for $h \in {}^*\mathbb{N} - \mathbb{N}$ by

$$(\Omega_h, \mathcal{A}_h, P_h, Y_h)$$

the nonstandard extension of the sequence

$$n \to (\Omega_n, \mathcal{A}_n, P_n, Y_n), n \in \mathbb{N},$$

at the point h. Then $(\Omega_h, \mathcal{A}_h, P_h)$ is an internal probability space and $Y_h : \Omega_h \to {}^*D$ is an internal function with $Y_h^{-1}(G) \in \mathcal{A}_h$ for $G \in {}^*\mathcal{B}(\mathcal{D})$.

Let $(\Omega, \mathcal{A}, Q)$ be an internal probability space. Put for $M \subset \Omega$:

$$\begin{aligned} \underline{Q}(M) &:= \sup\{{}^\circ Q(A) : M \supset A \in \mathcal{A}\}, \\ \overline{Q}(M) &:= \inf\{{}^\circ Q(A) : M \subset A \in \mathcal{A}\}. \end{aligned}$$

Then $L(\mathcal{A}, Q) := \{M \subset \Omega : \underline{Q}(M) = \overline{Q}(M)\}$ is a σ-algebra over Ω and $Q^L := \overline{Q} = \underline{Q}$ on $L(\mathcal{A}, Q)$ is the Loeb-measure of Q.

2 The Main Results

The following theorem is the main result of this paper. It gives a nonstandard equivalence to weak convergence of stochastic processes. It can be applied to obtain invariance principles as well as nonstandard constructions for the limit process.

2.1 Theorem:

Let $(D, \mathcal{T})$ be a sip-space. For $n \in \mathbb{N}$ let $(\Omega_n, \mathcal{A}_n, P_n)$ be probability spaces and let

$$Y_n : \Omega_n \to D \text{ be } \mathcal{A}_n, \mathcal{B}(\mathcal{D})\text{-measurable with } Y_n(\omega)(0) = 0 \text{ for all } \omega \in \Omega_n.$$

Let furthermore $W|\mathcal{B}(\mathcal{D})$ be a p-measure with $W(C) = 1$. Then (i) *is equivalent to* (ii)*:*

(i) $(P_n)_{Y_n} \to W$ *weakly.*

(ii) *For each* $h \in {}^*\mathbb{N} - \mathbb{N}$ *there holds:*

a) $Y_h(\omega)$ *is* S*-continuous for* P_h^L*-a.a.* $\omega \in \Omega_h$;

b) $P_h^L\{\omega \in \Omega_h : {}^\circ Y_h(\omega) \in B\} = W(B)$ *for all* $B \in \mathcal{B}(\mathcal{D})$.

Let us remark that the measures $(P_n)_{Y_n}|\mathcal{B}(\mathcal{D})$ are not necessarily Radon-measures. This excludes the application of classical techniques (as e.g. the application of Prohorov's theorem) for proving the weak convergence $(P_n)_{Y_n} \to W$ in Theorem 2.1.

At first we apply Theorem 2.1 to the case that $Y_n, n \in \mathbb{N}$, are normalized sum processes. For this application let $D = D[0,1]$ be endowed with the topology $\mathcal{T} = \mathcal{T}_\mathcal{K}$ and $\mathbf{W}|\mathcal{B}(\mathcal{D})$ be the Wiener measure. Then $(D[0,1], \mathcal{T}_\mathcal{K})$ is a sip-space (see Example 1.2 (ii)) and $\mathbf{W}(C) = 1$. For $x \in {}^*\mathbb{R}$ let $[x] = \max\{n \in {}^*\mathbb{Z} : n \leq x\}$.

2.2 Corollary: *Let* $(\Omega_n, \mathcal{A}_n, P_n), n \in \mathbb{N}$, *be p-spaces and* $\xi_{in} : \Omega_n \to \mathbb{R}$ *for* $i = 1, \ldots, n$ *be random variables with mean zero and* $0 < \tau_n :=$ *var* $(\sum_{i=1}^{n} \xi_{in}) < \infty$. *Let*

$$Y_n(\omega)(t) := \frac{1}{\sqrt{\tau_n}} \sum_{i=1}^{[nt]} \xi_{in}(\omega), \omega \in \Omega_n, t \in [0,1].$$

Then $Y_n : \Omega_n \to D[0,1]$ *are* $\mathcal{A}_n, \mathcal{B}(\mathcal{D})$*-measurable and* (i) *is equivalent to* (ii):

(i) $(P_n)_{Y_n} \to \mathbf{W}$ *weakly.*

(ii) *For each* $h \in {}^*\mathbb{N} - \mathbb{N}$ *there holds:*

a) $\frac{1}{\sqrt{\tau_h}} \sum_{i=1}^{[ht]} \xi_{ih}(\omega)$ *is* S*-continuous for* P_h^L*-a.a.* $\omega \in \Omega_h$;

b) $\mathbf{B}(t,\omega) := {}^\circ(\frac{1}{\sqrt{\tau_h}} \sum_{i=1}^{[ht]} \xi_{ih}(\omega)), t \in [0,1], \omega \in \Omega_h$, *is a Brownian motion on* $(\Omega_h, L(\mathcal{A}_h, P_h), P_h^L)$ *with* P_h^L*-a.a. continuous paths.*

Proof: It is well known that Y_n are $\mathcal{A}_n, \mathcal{B}(D)$-measurable. By transfer there holds

$$Y_h(\omega)(t) = \frac{1}{\sqrt{\tau_h}} \sum_{i=1}^{[ht]} \xi_{ih}(\omega), \omega \in \Omega_h, t \in {}^*[0,1].$$

Hence Corollary 2.2 follows from Theorem 2.1 using $\mathbf{W}(C) = 1$. ∎

Invariance principles for the sum process (i.e. assertions of the form (i) in Corollary 2.2) are known for many different classes of random variables ξ_{in}, $i = 1, \ldots, n$, as e.g. under certain conditions for independent random variables, for mixing random variables or for martingale difference sequences. In each of these cases Corollary 2.2 ((i) $\Rightarrow$ (ii)) leads by means of hyperfinite sums to a nonstandard construction of a Brownian motion with continuous paths. Let e.g.

$$\Omega_n := \{-1,1\}^n, \mathcal{A}_n \text{ be the power set of } \Omega_n, P_n(\{\omega\}) := 1/2^n$$

and

$$\xi_{in}(\omega) := \omega_i \text{ for } \omega \in \Omega_n.$$

Then part (i) of Corollary 2.2 is a special case of Donsker's invariance principle and hence for each $h \in {}^*\mathbb{N} - \mathbb{N}$

$$\mathbf{B}(t,\omega) := {}^\circ\Big(\frac{1}{\sqrt{h}}\sum_{i=1}^{[ht]}\omega_i\Big), t \in [0,1], \omega \in \Omega_h$$

is a Brownian motion on the hyperfinite set Ω_h with P_h^L-a.a. continuous paths. Hence we obtain Anderson's construction [1] of a Brownian motion and this construction is, as Corollary 2.2 shows, equivalent to the invariance pinciple for the sum process.

Similarly as in Corollary 2.2 we can apply Theorem 2.1 to stochastic processes other than the sum process. Now we give an application to empirical processes and obtain results of Mendieta [7]. To this aim let $(\Omega, \mathcal{A}, P)$ be a probability space and $\xi_i : \Omega \to [0,1], i \in \mathbb{N}$, independent random variables which are uniformly distributed over $[0,1]$. Let

$$F_n(t,\omega) := \frac{1}{n}\#\{i \le n : \xi_i(\omega) \le t\}, t \in [0,1], \omega \in \Omega$$

be the empirical process and put

$$Y_n(\omega)(t) := \sqrt{n}(F_n(t,\omega) - t).$$

Let $D[0,1]$ be endowed with the Skorohod topology $\mathcal{T}$ and let $\mathbf{W}^1$ on $\mathcal{B}(\mathcal{D})$ be the distribution of the Brownian bridge. Then $(D[0,1], \mathcal{T})$ is a sip-space (see Example 1.2 (iii)) and $\mathbf{W}^1(C) = 1$. Furthermore $Y_n : \Omega \to D[0,1]$ are $\mathcal{A}, \mathcal{B}(D)$-measurable. Since $P_{Y_n} \to \mathbf{W}^1$ weakly (see e.g. [3], p. 141), we obtain from Theorem 2.1, that for each $h \in {}^*\mathbb{N} - \mathbb{N}$:

$$\mathbf{B}^1(t,\omega) := {}^\circ(\sqrt{h}(F_h(t,\omega) - t)), t \in [0,1], \omega \in {}^*\Omega$$

is a Brownian bridge on $({}^*\Omega, L({}^*\mathcal{A}, {}^*P), {}^*P^L)$ with ${}^*P^L$-a.a. continuous paths. By transfer there holds:

$$F_h(t,\omega) = \frac{1}{h}\#\{i \le h : \xi_i(\omega) \le t\}, t \in {}^*[0,1], \omega \in {}^*\Omega.$$

For this application we have used Theorem 2.1 ((i) $\Rightarrow$ (ii)) to obtain a nonstandard representation of a Brownian bridge. On the other hand Theorem 2.1 ((ii) $\Rightarrow$ (i)) can be used to prove the invariance principle for the empirical process from a nonstandard construction of a Brownian bridge.

To obtain a nonstandard bridge on a hyperfinite space, we can use Theorem 2.1 in the following way:

Consider once more $D[0,1]$ with the Skorohod topology. Let Ω_n be the set of all permutations of $\{1,\ldots,n\}$ and put

$$P_n(\{\omega\}) := \frac{1}{n!} \text{ for } \omega \in \Omega_n.$$

Let $x_{1n},\ldots,x_{nn}$ be real numbers with

$$\sum_{i=1}^{n} x_{in} = 0, \quad \sum_{i=1}^{n} x_{in}^2 = 1, \quad \max_{1\leq i\leq n} |x_{in}| \underset{n\to\infty}{\longrightarrow} 0.$$

Put

$$Y_n(\omega)(t) := \sum_{i=1}^{[nt]} x_{\omega(i)n};$$

then (see [3], p. 209)

$$(P_n)_{Y_n} \longrightarrow \mathbf{W}^1 \text{ weakly.}$$

Hence it follows from Theorem 2.1 ((i) $\Rightarrow$ (ii)) that for each $h \in {}^*\mathbb{N} - \mathbb{N}$

$$\mathbf{B}^1(t,\omega) := {}^\circ\Big(\sum_{i=1}^{[ht]} x_{\omega(i)h}\Big), t \in [0,1], \omega \in \Omega_h$$

is a Brownian bridge on the hyperfinite set Ω_h with P_h^L-a.a. continuous paths.

3 Proof of Theorem 2.1 and Auxiliary Lemmata

Using the auxiliary Lemmata 1-4, which will be given later, we prove at first our main theorem.

Proof of Theorem 2.1: We show at first:

$$(C, \mathcal{T} \cap C) \text{ is a Polish space,} \tag{3.1}$$

$$W|\mathcal{B}(D) \text{ is a Radon measure.} \tag{3.2}$$

To (1): The space $(C, \mathcal{T}_\mathcal{K} \cap C)$ is a Polish space according to Lemma 4. As $\mathcal{T} \cap C$ and $\mathcal{T}_\mathcal{K} \cap C$ fulfill the first axiom of countability, we obtain $\mathcal{T} \cap C = \mathcal{T}_\mathcal{K} \cap C$ (use (ii) and (iv) of Definition 1.1).

To (2): Since $C \in \mathcal{B}(\mathcal{D})$ we obtain:

$$\sigma(\mathcal{T} \cap C) = \mathcal{B}(D) \cap C \subset \mathcal{B}(\mathcal{D}).$$

As $W(C) = 1$ and measures on Polish spaces are Radon measures, (2) follows using (1).

(i) $\Rightarrow$ (ii): Let $h \in {}^*\mathbb{N} - \mathbb{N}$ be fixed. As $W_n := (P_n)_{Y_n} \to W$ weakly we get that W_h is infinitesimally close to W with respect to the weak topology. Hence (2) and Lemma 3 imply:

$$W_h^L(st^{-1}(B)) = W(B) \text{ for all } B \in \mathcal{B}(\mathcal{D}) \tag{3.3}$$
$$ns({}^*D) = st^{-1}(D) \in L({}^*\mathcal{B}(\mathcal{D}), W_h) \tag{3.4}$$

Since $W_n = (P_n)_{Y_n}$ for all $n \in \mathbb{N}$, transfer implies

$$W_h = (P_h)_{Y_h} \tag{3.5}$$

According to (4) and (5) we can apply Lemma 2 to $(\Omega, \mathcal{A}, Q) := (\Omega_h, \mathcal{A}_h, P_h)$ and $Y := Y_h$ and obtain for all $B \in \mathcal{B}(\mathcal{D})$:

$$\begin{aligned} P_h^L((st \circ Y_h)^{-1}(B)) &= ((P_h)_{Y_h})^L(st^{-1}(B)) \\ &\underset{(5)}{=} W_h^L(st^{-1}(B)) \underset{(3)}{=} W(B) \end{aligned} \tag{3.6}$$

Applying (6) to $B = C \in \mathcal{B}(\mathcal{D})$ we have

$$P_h^L\{\omega \in \Omega_h : st(Y_h(\omega)) \in C\} = W(C) = 1. \tag{3.7}$$

Hence we obtain according to Lemma 1 that there holds for P_h^L-a.a. $\omega \in \Omega_h$:

$$Y_h(\omega) \text{ is } S\text{-continuous and } st(Y_h(\omega)) = {}^\circ Y_h(\omega). \tag{3.8}$$

From (8) and (6) we obtain (ii).

(ii) $\Rightarrow$ (i) We have to show that W_h is infinitesimally close to W for each $h \in {}^*\mathbb{N} - \mathbb{N}$. Let $h \in {}^*\mathbb{N} - \mathbb{N}$ be fixed; we have to show

$$W(O) \leq W_h^L({}^*O) \text{ for all } O \in \mathcal{T}.$$

As $Y_h(\omega)(0) = 0$ for all $\omega \in \Omega_h$ we obtain from assumption (ii)a and Lemma 1 that

$$st(Y_h(\omega)) = {}^\circ Y_h(\omega) \text{ for } P_h^L\text{-a.a. } \omega \in \Omega_h.$$

This implies for each $O \in \mathcal{T}$:

$$\begin{aligned} W(O) &\underset{(ii)_b}{=} P_h^L\{\omega \in \Omega_h : st(Y_h(\omega)) \in O\} \\ &\leq P_h^L\{\omega \in \Omega_h : Y_h(\omega) \in {}^*O\} \\ &= {}^\circ(P_h\{\omega \in \Omega_h : Y_h(\omega) \in {}^*O\} \\ &\underset{(5)}{=} {}^\circ(W_h({}^*O)) = W_h^L({}^*O). \qquad \blacksquare \end{aligned}$$

The following Lemma 1 is an adaptation of a well known result to our situation.

Lemma 1: *Let $(D, \mathcal{T})$ be a sip-space. Then the following two conditions are equivalent for each $g \in {}^*D$:*

(i) *g is S-continuous and $g(0)$ is finite;*

(ii) *$g \approx_{\mathcal{T}} f$ for some $f \in C$.*

If $g \approx_{\mathcal{T}} f \in C$ then $f = {}^\circ g$, i.e. $f(t) = {}^\circ g(t)$ for $t \in T$.

Proof: As $\mathcal{T}$ and $\mathcal{T}_K$ fulfill the first axiom of countability, the identity on D is $\mathcal{T}, \mathcal{T}_K$-continuous and $\mathcal{T}_K, \mathcal{T}$-continuous in each $f \in C$ (use (ii) and (iv) of Definition 1.1). This implies for $g \in {}^*D, f \in C$:

$$g \approx_{\mathcal{T}} f \quad \Leftrightarrow \quad g \approx_{\mathcal{T}_K} f \Longrightarrow f(t) = {}^\circ g(t) \text{ for } t \in T. \tag{3.9}$$

(i) $\Longrightarrow$ (ii): We prove at first:

$$g(t) \text{ finite for each } t \in T. \tag{3.10}$$

To this aim put

$$O_1 := \{t \in T : g(t) \text{ finite}\}, O_2 := \{t \in T : g(t) \text{ infinite}\}.$$

Since $0 \in O_1$ by assumption and since T is connected, for (10) it is sufficient to prove that

$$O_1, O_2 \text{ are open.}$$

Let $t \in O_1$. Then $||g(t)|| \leq r$ for some $r \in \mathbb{R}_+$ and S-continuity of g implies:

$$\{s \in {}^*T : s \approx t\} \subset \{s \in {}^*O_1 : ||g(s)|| \leq r + 1\}.$$

Hence O_1 contains a neighbourhood of t, whence O_1 is open. The proof that O_2 is open runs similarly.

According to (10) ${}^\circ g(t)$ exists for $t \in T$; put $f(t) := {}^\circ g(t)$ for $t \in T$. In the usual way it follows that $f \in C$ and $g(u) \approx {}^*f(u)$ for all compact points u, i.e. $g \approx_{\mathcal{T}_K} f \in C$. Hence (ii) follows by (9).

(ii) $\Rightarrow$ (i): According to (ii) and (9) we have $g \approx_{\mathcal{T}_K} f \in C$. As T is locally compact this implies

$$g(u) \approx {}^*f(u) \text{ for all } u \in ns({}^*T). \tag{3.11}$$

From (11) we obtain that $g(0)$ is finite and $f(t) = {}^\circ g(t)$ for all $t \in T$. S-continuity of g follows directly from (11), using $f \in C$. ∎

Lemma 2: *Let $(\Omega, \mathcal{A}, Q)$ be an internal probability space. Let $(D, \mathcal{T})$ be a regular topological space. Let $Y : \Omega \to {}^*D$ be an internal function with*

(α) *$Y^{-1}(F) \in \mathcal{A}$ for all $F \in {}^*\mathcal{B}(\mathcal{D})$,*

(β) *$ns({}^*D) \in L({}^*\mathcal{B}(\mathcal{D}), Q_Y)$.*

Then for all $B \in \mathcal{B}(\mathcal{D})$ holds

(i) $st^{-1}(B) \in L({}^*\mathcal{B}(\mathcal{D}), Q_Y)$ *and* $(st \circ Y)^{-1}(B) \in L(\mathcal{A}, Q)$,

(ii) $Q_Y^L(st^{-1}(B)) = Q^L((st \circ Y)^{-1}(B))$.

Proof: Let $B \in \mathcal{B}(D)$ be given. As $Q_Y|{}^*\mathcal{B}(\mathcal{D})$ is an internal p-content, Corollary 3 (iv) of [4] implies

$$st^{-1}(B) \in L({}^*\mathcal{B}(\mathcal{D}), Q_Y) \cap ns({}^*D) \underset{(\beta)}{\subset} L({}^*\mathcal{B}(\mathcal{D}), Q_Y).$$

Now we prove for each $G \in L({}^*\mathcal{B}(\mathcal{D}), Q_Y)$

$$\underline{Q}_Y(G) = (Q_Y)^L(G) = \overline{Q}_Y(G). \tag{3.12}$$

As $st^{-1}(B) \in L({}^*\mathcal{B}(\mathcal{D}), Q_Y)$ we obtain

$$\begin{aligned} (Q_Y)^L(st^{-1}(B)) &\underset{(12)}{=} \underline{Q}_Y(st^{-1}(B)) = \underline{Q}((st \circ Y)^{-1}(B)) \\ (Q_Y)^L(st^{-1}(B)) &\underset{(12)}{=} \overline{Q}_Y(st^{-1}(B)) = \overline{Q}((st \circ Y)^{-1}(B)), \end{aligned}$$

hence $(st \circ Y)^{-1}(B) \in L(\mathcal{A}, Q)$ and (ii) holds.

To (12) For $F \in {}^*\mathcal{B}(\mathcal{D})$ we have

$$(Q^L)_Y(F) \underset{(\alpha)}{=} {}^\circ Q(Y^{-1}(F)) = {}^\circ(Q_Y(F)) = (Q_Y)^L(F). \tag{3.13}$$

Let $\varepsilon \in \mathbb{R}_+$. Since $G \in L({}^*\mathcal{B}(\mathcal{D}), Q_Y)$ there exist $F_1, F_2 \subset {}^*D$ with

$$F_1 \subset G \subset F_2 \text{ and } F_1, F_2 \in {}^*\mathcal{B}(\mathcal{D}), \tag{3.14}$$

$$(Q_Y)^L(F_2) - \varepsilon/2 \le (Q_Y)^L(G) \le (Q_Y)^L(F_1) + \varepsilon/2. \tag{3.15}$$

Hence we obtain:

$$\begin{aligned} \underline{Q}_Y(G) &\le \overline{Q}_Y(G) \underset{(14)}{\le} \overline{Q}_Y(F_2) \underset{(\alpha)}{=} (Q^L)_Y(F_2) \\ &\underset{(13)}{=} (Q_Y)^L(F_2) \underset{(15)}{\le} (Q_Y)^L(G) + \varepsilon/2 \\ &\underset{(15)}{\le} (Q_Y)^L(F_1) + \varepsilon \underset{(13)}{=} (Q^L)_Y(F_1) + \varepsilon \\ &\underset{(\alpha)}{=} \underline{Q}_Y(F_1) + \varepsilon \underset{14}{\le} \underline{Q}_Y(G) + \varepsilon. \end{aligned}$$

With $\varepsilon \to 0$ this implies $\underline{Q}_Y(G) = \overline{Q}_Y(G) = (Q_Y)^L(G)$, i.e. (12). ∎

Lemma 3: *Let $(D, \mathcal{T})$ be a regular topological space. Let $W|\mathcal{B}(\mathcal{D})$ be a Radon measure and $Q|{}^*\mathcal{B}(\mathcal{D})$ be an internal p-content such that Q is infinitesimally close to W with respect to the weak topology. Then for all $B \in \mathcal{B}(\mathcal{D})$:*

$$st^{-1}(B) \in L({}^*\mathcal{B}(\mathcal{D}), Q) \text{ and } Q^L(st^{-1}(B)) = W(B).$$

Proof: For $st^{-1}(B) \in L({}^*\mathcal{B}(\mathcal{D}), Q)$ see [4], Corollary 3(iv). For $Q^L(st^{-1}(B)) = W(B)$ use a slight generalization of Lemma 2.6 in [5] or see Theorem 32.6 of [6]. ∎

Lemma 4: *Let $T \subset \mathbb{R}^k$ be a locally compact subspace. Then $C(T, \mathbb{R}^m)$, endowed with the topology of uniform convergence on compact sets, is a Polish space.*

Proof: A slight generalization of Theorem 31.6 in [2], p. 245

References

[1] R. M. Anderson, *A nonstandard representation for Brownian motion and Itô integration*, Israel Journal of Mathematics **25** (1976) 15 - 46.

[2] H. Bauer, *Ma*

- und Integrationstheorie (de Gruyter Lehrbuch, Berlin-New York 1990).

[3] P. Billingsley, *Convergence of Probability Measures* (John Wiley & Sons, New York-Toronto 1968).

[4] D. Landers and L. Rogge, *Universal Loeb-measurability of sets and of the standard part map with applications*, Transactions of the Amer. Math. Soc. **304** (1987) 229 - 243.

[5] D. Landers and L. Rogge, *Nonstandard methods for families of τ-smooth probability measures*, Proceedings of the Amer. Math. Soc. **103** (1988), 1151 - 1156.

[6] D. Landers and L. Rogge, *Nichtstandard-Analysis* (to appear in Springer Verlag, Berlin-New York 1993).

[7] G. R. Mendieta, *Two hyperfinite constructions of the Brownian bridge*, Stochastic Anal. Appl. **7** (1989) 75 - 88.

[8] D. W. Müller, *Nonstandard proofs of invariance principles in probability theory*, in Applications of Model Theory to Algebra, Analysis and Probability, ed. W. A. J. Luxemburg (Holt, Rinehart and Winston, 1969) pp. 186 - 194

[9] A. Stoll, *A nonstandard construction of the Lévy Brownian motion*, Prob. Th. Rel. Fields **71** (1986), 321 - 334.

ANDERSON'S BROWNIAN MOTION AND THE INFINITE DIMENSIONAL ORNSTEIN–UHLENBECK PROCESS

Tom Lindstrøm †
Department of Mathematics, University of Oslo,
Box 1053 N–0316 =slo, Norway

Abstract

Anderson's construction [2] of Brownian motion as the standard part of a random walk with infinitesimal increments is one of the success stories of nonstandard analysis. Almost every subsequenct development in nonstandard probability theory is inspired – directly or indirectly – by Anderson's work. The purpose of this paper is to point out how another extremely important process in stochastic analysis – the infinite dimensional Ornstein-Uhlenbeck process – can be derived very easily from Anderson's construction.

Introduction

Anderson's construction [2] of Brownian motion as the standard part of a random walk with infinitesimal increments is one of the success stories of nonstandard analysis. Almost every subsequenct development in nonstandard probability theory is inspired – directly or indirectly – by Anderson's work. The purpose of this paper is to point out how another extremely important process in stochastic analysis – the infinite dimensional Ornstein-Uhlenbeck process – can be derived very easily from Anderson's construction.

To explain the basic idea, let us first recall what Anderson did. Choose an infinitely large integer $N \in {}^*\mathbb{N}$ and let $\Delta t = 1/N$. Think of

$$T = \{0, \Delta t, 2\Delta t, \ldots, 1 - \Delta t\}$$

as a hyperfinite timeline. Let Ω be the set of all internal functions $\omega : T \to \{1, -1\}$ and denote the internal, uniform probability measure on Ω by P (i.e. $P(A) = |A|/|\Omega|$

†This research is supported by VISTA, a research cooperation between the Norwegian Academy of Science and Letters and Den Norske Stats Oljeselskap a.s. (STATOIL).

for all internal sets A). By $L(P)$ we shall mean the Loeb measure of P. Anderson's hyperfinite random walk $B : \Omega \times T \to {}^*\mathbf{R}$ is defined by

$$B(\omega, t) = \sum_{s<t} \omega(s)\sqrt{\Delta t}\,.$$

Anderson showed that for $L(P)$-a.a. ω, the internal function $B(\omega, \cdot)$ is S-continuous, and hence defines a continuous standard function $b(\omega, \cdot)$ ($= {}^\circ B(\omega, \cdot)$) from [0,1] to $\mathbf{R}$. Moreover, the standard process $b(\omega, t)$ is a Brownian motion on $(\Omega, L(P))$ (see, e.g. [1], [2], [4], [5], [6], or [10] for the details).

The infinite dimensional Ornstein-Uhlenbeck process u is a stochastic process taking values in the space $C([0,1])$ of continuous functions. Intuitively, it looks like a continuous, random modification of Brownian paths which keeps the Wiener measure invariant. Using Anderson's construction we can make this intuition rigorous in the following way.

Pick an initial element ω_0 in Ω (and do it in such a way that $B(\omega_0, \cdot)$ is S-continuous). At time 0, toss an unfair coin for each $s \in T$ to decide whether you want to reverse the sign of s-th component $\omega_0(s)$ or not; the probability of changing the sign should be $\Delta t/2$ and the probability of keeping it $(1 - \Delta t/2)$, The resulting path is $\omega_{\Delta t}$. At time Δt repeat the procedure; flip independent coins for each s and change the sign of each component with probability $\Delta t/2$ to obtain $\omega_{2\Delta t}$. Continuing in this way, we get a random sequence of elements $\omega_0, \omega_{\Delta t}, \dots, \omega_{n\Delta t}, \dots$ in Ω. Each of these elements corresponds to an Anderson path $B(\omega_{n\Delta t}, \cdot)$, and hence we have constructed a randomly moving sequence of such paths. Taking standard parts, we get a randomly moving process of Brownian paths. The purpose of the paper is to show that this random motion is an infinite dimensional Ornstein-Uhlenbeck process. Our tools will mainly be taken from nonstandard martingale theory, and all the results we shall need can be found in Chapter 4 of [1]. We shall use the notation and terminology of that reference, and we shall abuse it with the following shorthand: If $\omega \in \Omega$ and $B(\omega, \cdot)$ is S-continuous and induces a standard function $x(\cdot)$, ($= {}^\circ B(\omega, \cdot)$) in $C([0,1])$, we shall say that ω *is nearstandard with standard part* x.

This paper is organized as follows: In Section 1 we formalize the construction above and study the anatomy of the nonstandard process. In Section 2 we show that the process is S-continuous, and in Section 3 we give a (well-known) representation in terms of a time changed Brownian sheet. In Section 4 we show that the standard part of our process is an infinite dimensional Ornstein-Uhlenbeck process, and Section 5 contains a brief description of the hyperfinite Ornstein-Uhlenbeck operator and the corresponding semigroup. We end the paper with a few remarks on two other nonstandard approaches to the infinite dimensional Ornstein-Uhlenbeck process.

1 Construction and basic properties

To formalize the construction sketched above, we define the space Ξ of all internal function

$$\xi : T \to \{-1, 1\}^T$$

where $\{-1,1\}^T$ denotes the space of all *internal* functions from T to $\{-1,1\}$. We shall think of $\xi(t)$ as the sequence of coin tosses performed at time t, and $\xi(t)(s) = -1$ will represent the event of switching the s-th component at time t. Let Q be the internal probability measure on Ξ which makes all the events $\{\xi(t)(s)\}_{t,s \in T}$ independent, and which gives the event $\xi(t)(s) = -1$ probability $\Delta t/2$ (and, consequently, the event $\xi(t)(s) = 1$ probability $1 - \Delta t/2$). Given an initial value ω_0, we define a process

$$\Theta^{\omega_0} : \Xi \times T \to \Omega$$

by

$$\Theta^{\omega_0}(\xi, t + \Delta t)(s) = \xi(t)(s)\Theta^{\omega_0}(\xi, t)(s)$$

i.e., the state changes from t to $t + \Delta t$ by switching the s-th coordinate if and only if $\xi(t)(s) = -1$. When the initial value ω_0 is of little interest, we shall drop the superscript and write Θ for Θ^{ω_0}. We now define the *hyperfinite Ornstein-Uhlenbeck* process U $(= U^{\omega_0})$ by

$$U(\xi, t)(\cdot) = B(\Theta(\xi, t), \cdot)$$

In this section we shall study the basic properties of U as a nonstandard object. We begin with a simple observation.

1.1 Lemma. *For all $s, t \in T$ and $\omega_0 \in \Omega$*

$$Q\{\xi \in \Xi | \Theta^{\omega_0}(\xi, t)(s) = \omega_0(s)\} = [(1 - \Delta t)^{t/\Delta t} + 1]/2 \approx (e^{-t} + 1)/2$$

Proof: Let

$$p_t = Q\{\xi \in \Xi | \Theta^{\omega_0}(\xi, t)(s) = \omega_0(s)\}$$

Since the probability of switching the s-th coordinate between time t and time $t+\Delta t$ is $\Delta t/2$, we have

$$p_{t+\Delta t} = p_t(1 - \Delta t/2) + (1 - p_t)\Delta t/2$$

and hence

$$p_{t+\Delta t} = p_t(1 - \Delta t) + \Delta t/2$$

Solving this difference equation with initial condition $p_0 = 1$, we get

$$p_t = [(1 - \Delta t)^{t/\Delta t} + 1]/2 \approx (e^{-t} + 1)/2 \ \blacksquare$$

For each $s \in T$, let $\mathcal{F}_s$ be the internal algebra on Ξ generated by the random variables $\{\xi(t)(r) | t, r \in T, r < s\}$. Whenever we say that something is a *martingale* in the s-variable, we shall mean an internal martingale adapted to the filtration $\{\mathcal{F}_s\}$.

1.2 Lemma. *Fix t and ω_0. Then*

$$E_Q[U(\xi,t)(s+\Delta t) - U(\xi,t)(s)|\mathcal{F}_s] = \omega_0(s)\sqrt{\Delta t}\,(1-\Delta t)^{t/\Delta t}$$
$$= \omega_0(s)\sqrt{\Delta t}\,e^{-t} + o(\sqrt{\Delta t}\,)$$

where $o(\sqrt{\Delta t}\,)$ denotes a quantity which is infinitesimal compared to $\sqrt{\Delta t}$.

Proof: With p_t as in the preceding lemma, we have

$$E_Q[U(\xi,t)(s+\Delta t) - U(\xi,t)(s)|\mathcal{F}_s] = \omega_0(s)\sqrt{\Delta t}\cdot p_t - \omega_0(s)\sqrt{\Delta t}\cdot(1-p_t) =$$
$$\omega_0(s)\sqrt{\Delta t}\,(2p_t - 1) = \omega_0(s)\sqrt{\Delta t}\,(1-\Delta t)^{t/\Delta t} = \omega_0(s)\sqrt{\Delta t}\,e^{-t} + o(\sqrt{\Delta t}\,) \quad \blacksquare$$

1.3 Lemma. *Fix ω_0 and t. Then*

$$U(\xi,t)(s) = (1-\Delta t)^{t/\Delta t}(U(\xi,0)(s) + K^{(\omega_0,t)}(\xi,s))$$

where $K^{(\omega_0,t)}$ is an S-continuous martingale such that the standard part of $K^{(\omega_0,t)}/\sqrt{e^{2t}-1}$ is a Brownian motion. Moreover, for all $s, s' \in T$

$$[K^{(\omega_0,t)}](\xi,s) - [K^{(\omega_0,t)}](\xi,s') \le 4e^2|s-s'|$$

where $[K^{(\omega_0,t)}]$ denotes the quadratic variation of $K^{(\omega_0,t)}$.

Proof: The process defined by

$$K^{(\omega_0,t)}(\xi,s) := (1-\Delta t)^{-t/\Delta t}U(\xi,t)(s) - \sum_{r<s}\omega_0(r)\sqrt{\Delta t} =$$
$$= (1-\Delta t)^{-t/\Delta t}U(\xi,t)(s) - U(\xi,0)(s)$$

is a martingale by Lemma 1.2. Observe that since

$$|\Delta K^{(\omega_0,t)}(\xi,s)| \le 2\sqrt{\Delta t}\,(1-\Delta t)^{-t/\Delta t} \le 2e\sqrt{\Delta t}$$

we have

$$[K^{(\omega_0,t)}](\xi,s) - [K^{(\omega_0,t)}](\xi,s') \le 4e^2|s-s'|$$

where $[K^{(\omega_0,t)}]$ is the quadratic variation of $K^{(\omega_0,t)}$. Thus $[K^{(\omega_0,t)}]$ is S-continuous, and so is $K^{(\omega_0,t)}$ according to Theorem 4.4.16 in [1].

Observe next that

$$\begin{aligned} E_Q(\Delta K^{(\omega_0,t)}(s)^2|\mathcal{F}_s) &= [\omega_0(s)\sqrt{\Delta t}\,((1-\Delta t)^{-t/\Delta t}-1)]^2 p_t + \\ &+[\omega_0(s)\sqrt{\Delta t}\,((1-\Delta t)^{-t/\Delta t}+1)]^2(1-p_t) \\ &= \Delta t\Big\{[(1-\Delta t)^{-t/\Delta t}-1]^2(1+(1-\Delta t)^{t/\Delta t})/2 + \\ &+[(1-\Delta t)^{-t/\Delta t}+1]^2(1-(1-\Delta t)^{t/\Delta t})/2\Big\} \\ &= \Delta t((1-\Delta t)^{-2t/\Delta t}-1) = \Delta t(e^{2t}-1)+o(\Delta t) \end{aligned}$$

Hence the compensator process $\langle K^{(\omega_0,t)}\rangle$ satisfies

$$\langle K^{(\omega_0,t)}(s)\rangle := \sum E_Q(\Delta K^{(\omega_0,t)}(r)^2|\mathcal{F}_r) \approx s((1-\Delta t)^{-2t/\Delta t}-1) \approx s(e^{2t}-1)$$

and since $\langle K^{(\omega_0,t)}(s)/\sqrt{e^{2t}-1}\rangle \approx s$, we see from (the proof of) Theorem 4.4.18 in [1] that the standard part of $K^{(\omega_0,t)}/\sqrt{e^{2t}-1}$ is a Brownian motion. ∎

So far we have let s vary for fixed t. If we reverse the situation, we first obtain the following result.

1.4 Lemma. *Fix ω_0 and s. Then*

$$E_Q(U(\xi,t+\Delta t)(s)-U(\xi,t)(s)|U(\xi,t)(s)=x) = -x\Delta t$$

Proof: Let $\omega = \Theta^{\omega_0}(\xi,t)$. Since $B(\omega,s)=U(\xi,t)(s)=x$, we must have

$$|\{r\in T|r<s \text{ and } \omega(r)=1\} = \frac{s}{\Delta t}+\frac{x}{2\sqrt{\Delta t}}$$

$$|\{r\in T|r<s \text{ and } \omega(r)=-1\} = \frac{s}{\Delta t}-\frac{x}{2\sqrt{\Delta t}}$$

When we move from t to $t+\Delta t$, the expected number of switches among components belonging to the first set is $(s/\Delta t + x/2\sqrt{\Delta t})\cdot \Delta t/2 = s/2 + x\sqrt{\Delta t}/4$, and the expected number of switches among components belonging to the second set is $(s/\Delta t - x/2\sqrt{\Delta t})\cdot\Delta t/2 = s/2 - x\sqrt{\Delta t}/4$. Since each switch of the first kind changes U by $-2\sqrt{\Delta t}$ and each switch of the second kind changes it by $2\sqrt{\Delta t}$, the expected change is $-2\sqrt{\Delta t}\,(s/2+x\sqrt{\Delta t}/4)+2\sqrt{\Delta t}\,(s/2-x\sqrt{\Delta t}/4) = -x\Delta t$. ∎

1.5 Lemma. it Fix ω_0 and s, and assume that ω_0 is nearstandard. Then

$$U(\xi,t)(s) = (1-\Delta t)^{t/\Delta t}(U(\xi,0)(s)+\sum_{r=0}^{t-\Delta t}(1-\Delta t)^{-(r+\Delta t)/\Delta t}\Delta W(\xi,r))$$

where W is an S-continuous martingale whose standard part is $\sqrt{2s}$ times a Brownian motion. Moreover, if we let

$$N^{(\omega_0,s)}(\xi,t)) = \sum_{r<t}(1-\Delta t)^{-(r+\Delta t)/\Delta t}\Delta W(\xi,r),$$

then

$$[N^{(\omega_0,s)}(\xi,t)] - [N^{(\omega_0,s)}(\xi,t)] \leq 9e^2|t-t'|$$

for all $t, t' \in T$.

Proof: By Lemma 1.4, the process

$$W(\xi,t) = U(\xi,t)(s) - U(\xi,0)(s) + \sum_{r<t} U(\xi,r)(s)\Delta t$$

is a martingale (by which we mean an internal martingale with respect to the obvious filtration $\mathcal{G}_t$). Rewriting this definition as

$$U(\xi,t)(s) = U(\xi,0)(s) - \sum_{r<t} U(\xi,r)(s)\Delta t + W(\xi,t)$$

we get a difference equation for U. Solving it, we see that

$$\begin{aligned} U(\xi,t)(s) &= (1-\Delta t)^{t/\Delta t}U(\xi,0)(s) + \sum_{r<t}(1-\Delta t)^{(t-r-\Delta t)/\Delta t}\Delta W(\xi,r) \\ &= (1-\Delta t)^{t/\Delta t}(U(\xi,0)(s) + \sum_{r<t}(1-\Delta t)^{-(r+\Delta t)/\Delta t}\Delta W(\xi,r)) \end{aligned}$$

Returning to W, we first observe that since the maximal value of $U(\xi,r)(s)$ is $s/\sqrt{\Delta t} \leq 1/\sqrt{\Delta t}$, we always have

$$|\Delta W(\xi,t)| \leq |\Delta U(\xi,t)(s)| + |U(\xi,t)(s)\Delta t| \leq 2\sqrt{\Delta t} + \sqrt{\Delta t} = 3\sqrt{\Delta t}$$

and hence

$$[W(\xi,t)] - [W(\xi,t')] \leq 9|t-t'|$$

for all $t, t' \in T$. By Theorem 4.2.16 and Proposition 4.4.3 in [1], this means that W is S-continuous and S-square integrable. It follows that if $U(\xi,0)(s)$ is finite, then with probability one, $U(\xi,t)(s)$ remains finite for all t. From this we see that

$$\begin{aligned} E_Q(\Delta W(t)^2|\mathcal{G}_t) \leq E_Q(|\Delta U(t)(s) - U(t)(s)\Delta t|^2|\mathcal{G}_t) = \\ (4\Delta t + o(\Delta t))s/2 = 2s\Delta t + o(\Delta t) \end{aligned}$$

and hence

$$\langle W(t)\rangle \approx 2st$$

almost everywhere. By Theorem 4.4.18 in [1], the standard part of $W\sqrt{2s}$ is a Brownian motion.

Defining $N^{(\omega_0,s)}(\xi,t) = \sum_{r<t}(1-\Delta t)^{-(r+\Delta t)/\Delta t}\Delta W(\xi,r)$, we finally observe that

$$[N^{(\omega_0,s)}(\xi,t)] - [N^{(\omega_0,s)}(\xi,t')] \leq e^2([W(\xi,t)] - [W(\xi,t')]) \leq 9e^2|t-t'| \quad \blacksquare$$

Combining Lemmas 1.3 and 1.5, we get:

1.6 Proposition. *For any nearstandard* ω_0

$$U^{(\omega_0)}(\xi,t)(s) = (1-\Delta t)^{t/\Delta t}(U(\xi,0) + M^{(\omega_0)}(\xi,t,s))$$

where $M^{(\omega_0)}(\xi,t,s)$ *is an* S*-continuous martingale in each of the variables* s *and* t *when the other is kept fixed. For each* t*, the standard part of* $(\xi,s)\to M^{(\omega_0)}(\xi,t,s)$ *is of the form* $\sqrt{e^{2t}-1}\,w(\xi,s)$ *where* w *is a Brownian motion. For fixed* s*, the standard part of* $(\xi,t)\to M^{(\omega_0)}(\xi,t,s)$ *is of the form* $\sqrt{2s}\int e^r db(\xi,r)$ *where* b *is a Brownian motion.*

Proof: Just observe that in the notation of Lemma 1.3, we have $M^{(\omega_0)}(\xi,t,s) = K^{(\omega_0,t)}(\xi,s)$, and in the notation of Lemma 1.5, $M^{(\omega_0)}(\xi,t,s) = N^{(\omega_0,s)}(\xi,t)$. ∎

Remark. It may be helpful to rephrase the results of this section in standard terms. If we keep t fixed and vary s, then Lemma 1.3 tells us that the standard part $u(\xi,t)(s)$ of $U(\xi,t)(s)$ can be expressed as

$$u(\xi,t)(s) = e^{-t}(u(\xi,0)(s) + \sqrt{e^{2t}-1}\,w(s))$$

where $w(\cdot)$ is a Brownian motion (depending on t).

If we instead fix s and vary t, then the proof of Lemma 1.5 says that $u(\xi,\cdot)(s)$ is a solution of the Langevin equation

$$du(\xi,t)(s) = -u(\xi,t)(s)dt + \sqrt{2s}\,db(t)$$

where $b(\cdot)$ is a Brownian motion (depending on s), and hence $u(\xi,\cdot)(s)$ is a one-dimensional Ornstein-Uhlenbeck (velocity) process

$$u(\xi,t)(s) = e^{-t}(u(\xi,0)(s) + \sqrt{2s}\int_0^t e^r db(r))$$

(see, e.g., Nelson [9]).

Remark. There is an alternative way of expressing the results above in terms of time-changed Brownian motions (or – more precisely – in terms of a time-changed Brownian sheet), but for technical reasons it will be convenient to postpone that discussion till after we have taken a closer look at continuity properties.

2 Continuity

The results above tell us that for each s, the process $(\xi, t) \to U(\xi, t)(s)$ is S-continuous for $L(Q)$-a.a. ξ, and similarly for $(\xi, s) \to U(\xi, t)(s)$ when t is kept fixed. What they do *not* tell us, is how the exceptional sets are related, and thus it could happen that $U(\xi, t)(s)$ failed to be a.s. jointly continuous in s and t. This would certainly create problems for our intended interpretation of U as a $C([0,1])$-valued, continuous process, but in this section we shall show that U is, in fact, jointly S-continuous a.e. whenever ω_0 is nearstandard. Note that by Proposition 1.6 it suffices to prove that $M^{(\omega_0)}(\xi, t, s)$ is jointly continuous.

As our main tool we shall use the nonstandard version of Kolmogorov's Continuity Theorem (see [1], Proposition 4.8.5). According to this result, it suffices to find constants $p, K, \varepsilon \in \mathbf{R}_+$ such that

$$E_Q[|M(\xi,t,s) - M(\xi,t',s')|^p] \leq K\|(s,t) - (s',t')\|^{2+\varepsilon}$$

for all $s, s', t, t' \in T$. As a first step toward this inequality, we have the following observation.

2.1 Lemma. *Fix ω_0 and t. Then for any $p > 1$*

$$E_Q(|M(\xi,t,s) - M(\xi,t,s')|^p) \leq C_p|s - s'|^{p/2}$$

where C_p is a constant only depending on p (and not on ω_0 and t).

Proof: By the Burkholder-Davis-Gundy inequalities (see, e.g., page 126 of [1])

$$E_Q(|M(\xi,t,s) - M(\xi,t,s')|^p) \leq$$
$$\leq p\sqrt{12}\, E_Q([M](\xi,t,s) - [M](\xi,t,s'))^{p/2} \leq 2^p e^p p\sqrt{12}\,|s - s'|^{p/2}$$

where the last step uses that $[M](\xi,t,s) - [M](\xi,t,s') \leq 4e^2|s-s'|$ (recall Lemma 1.3 and that $M(\xi,t,s) = K^{(\omega_0,t)}(\xi,s)$). ∎

We then do the same with s and t interchanged.

2.2 Lemma. *Fix ω_0 and s. For each $p > 1$,*

$$E_Q(|M(\xi,t,s) - M(\xi,t',s)|^p) \leq K_p|t - t'|^{p/2}$$

where K_p is a constant depending only on p.

Proof: Just as above, but replacing Lemma 1.3 by Lemma 1.5. ∎

2.3 Proposition. *Assume that ω_0 is nearstandard. Then for almost all ξ, the process $(s,t) \to U(\xi,t)(s)$ is (jointly) S-continuous.*

Proof: As already observed, it suffices to show that $M(\xi, t, s)$ is jointly S-continuous. Choose $p > 4$, and note that by Lemmas 2.1 and 2.2

$$E_Q[|M(\xi,t,s) - M(\xi,t',s')|^p] \leq$$
$$\leq 2^p(E_Q[|M(\xi,t,s) - M(\xi,t',s)|^p] + E_Q[|M(\xi,t',s) - M(\xi,t',s')|^p]) \leq$$
$$\leq K\|(s,t) - (s',t')\|^{p/2}$$

for some constant K. The proposition follows from Kolmogorov's theorem. ∎

We can now define the standard part of U as the process $u : \Xi \times [0,1] \to C([0,1])$ given by

$$u(\xi,t)(s) = {}^{\circ}U(\xi,t')(s')$$

where $t' \approx t$ and $s' \approx s$. It follows immediately from Proposition 2.1 that u is continuous ($L(Q)$-a.e.).

3 The Brownian sheet representation

If m is the standard part of M, then clearly

$$u(\xi,t)(s) = e^{-t}(u(\xi,0)(s) + m(\xi,t,s))$$

If we fix t, then according to Proposition 1.6, the process $(\xi, s) \to m(\xi,t,s)$ is of the form $\sqrt{e^{2t}-1}\,w(\xi,s)$ for some Brownian motion w. Another way of expressing this relationship is to say that the process w' defined implicitly by $m(\xi,t,s) = w'(\xi, s(e^{2t}-1))$ is a Brownian motion. If we instead fix s, we know that $(\xi,t) \to m(\xi,t,s)$ is of the form $\sqrt{2s}\int e^r db(r)$, which means that it is a continuous martingale with quadratic variation

$$2s\int_0^t e^{2r}dr = s(e^{2t}-1)$$

Hence the process b' defined implicitly by $m(\xi,t,s) = b'(\xi, s(e^{2t}-1))$, is a Brownian motion.

With this observation in mind, it is natural to introduce a new random field $v(\xi,t,s)$ by

$$v(\xi, e^{2t}-1, s) = m(\xi,t,s)$$

or, put more explicitly,

$$v(\xi,t,s) = m(\xi, \ln(t+1)/2, s)$$

3.1 Proposition. *v is a Brownian sheet with $v(0,s) = v(t,0) = 0$.*

Proof: We have to show that v is a continuous, Gaussian field with covariance

$$E(v(t,s)v(t',s')) = \min(t,t') \cdot \min(s,s').$$

It is obvious that v is Gaussian, and the continuity was proved in the previous section. To compute the covariance, we may clearly assume that $t \le t'$. There are two cases to consider; $s \le s'$ and $s > s'$. Since they are easy and quite similar, we only treat the second one.

Observe that

$$\begin{aligned} E(v(t,s)v(t',s')) &= \\ = E([v(t,s') + (v(t,s) - v(t,s'))] \cdot [v(t,s') + (v(t',s') - v(t,s'))]) &= \\ = E(v(t,s')^2) \end{aligned}$$

since v is a martingale in each variable, and time evolution in the t- and the s-direction are independent. But then

$$\begin{aligned} E(v(t,s)v(t',s')) &= E(v(t,s')^2) = E(m(\ln(t+1)/2,s')^2) = \\ &= t \cdot s' = \min(t,t') \cdot \min(s,s') \qquad \blacksquare \end{aligned}$$

3.2 Corollary. *We have*

$$u(\xi,t)(s) = e^{-t}(u(\xi,0)(s) + v(\xi, e^{2t} - 1, s))$$

where v is a Brownian sheet with $v(0,s) = v(t,0) = 0$.

4 Infinite dimensional Ornstein-Uhlenbeck processes

Infinite dimensional Ornstein-Uhlenbeck processes were introduced by Malliavin [7] and plays a fundamental role in the Malliavin calculus. Put briefly, one might say that these processes and their infinitesimal generator – the Ornstein-Uhlenbeck operator – play the same part in infinite dimensional calculus as Brownian motion and the Laplace operator do in finite dimensions.

There are many ways of describing infinite dimensional Ornstein-Uhlenbeck processes. We could have taken the description in Corollary 3.2 as our definition (see Meyer [8]), but it is more conventional to use a characterization which says that an Ornstein-Uhlenbeck process u is a continuous, strong Markov process with values in $C([0,1])$ generating the semigroup

$$T_t f(x) = \int (e^{-t}x + \sqrt{1-e^{-2t}} \cdot y) dW(y) \tag{1}$$

where W is the Wiener measure on $C([0,1])$ (see, e.g., Watanabe [11]). We shall show that our process u satisfies these criteria.

4.1 Lemma. *Assume that ω_0 is nearstandard with standard part x. If $f : C([0,1]) \to \mathbf{R}$ is square integrable with respect to the Wiener measure W, then*

$$E_{L(Q)}[f(u^{(\omega_0)}(\xi,t)(\cdot)] = \int f(e^{-t}x(\cdot) + \sqrt{1-e^{2t}} \cdot y(\cdot))dW(y)$$

Proof: Observe that if f is bounded and continuous, then

$$\begin{aligned} E_{L(Q)}[f(u^{(\omega_0)}(\xi,t)(\cdot)] &= {}^{\circ}E_Q[{}^*f(U^{(\omega_0)}(\xi,t)(\cdot)] = \\ &= {}^{\circ}E_Q[{}^*f((1-\Delta t)^{t/\Delta t}(B(\omega_0,\cdot) + M(\xi,t,\cdot))] = \\ &= \int f(e^{-t}x(\cdot) + \sqrt{1-e^{-2t}}\, y(\cdot))dW(y) \end{aligned}$$

by nonstandard measure theory and Propositions 1.6 and 2.3. Using the Monotone Convergence Theorem, the result is easily extended to nonnegative, square integrable functions, and the general case follows by treating positive and negative parts separately. ∎

4.2 Theorem. *Assume ω_0 is nearstandard with standard part $x \in C([0,1])$. Then $u^{(\omega_0)}$ is an Ornstein-Uhlenbeck process starting at x.*

Proof: We know from Proposition 2.3 that u is continuous, and the lemma takes care of (1). It only remains to check that u is a strong Markov process. Observe that for any two initial conditions ω_0 and ω_0', $U^{\omega_0}(\xi,t)(s) - U^{\omega_0'}(\xi,t)(s) = U^{\omega_0}(\xi,0)(s) - U^{\omega_0'}(\xi,0)(s)$ for all t and s. Thus if the initial conditions are infinitely close, one process is just an infinitesimal translation of the other. From this the Markov property follows easily (for instance by an appeal to Theorem 5.4.17 in [1] or Theorem 6.8 in [5], but this is certainly an overkill). ∎

5 The hyperfinite Ornstein-Uhlenbeck operator

Having completed our serious work, we may amuse ourselves by taking a look at some nonstandard consequences. Define the *hyperfinite Ornstein-Uhlenbeck operator* L to be the infinitesimal generator

$$LF(\omega) = E_Q(F(U^{(\omega)}(\xi,0) - F(U^{(\omega)}(\xi,\Delta t)))/\Delta t$$

of U. Observe that L acts on the space $L^2(\Omega,Q)$ of all internal functions $F : \Omega \to \mathbf{R}$ with the inner product $\langle F,G\rangle = \Sigma F(\omega)G(\omega)Q(\omega)$. For each subset A of Ω, let $\mathcal{X}_A \in L^2(\Omega)$ be the function

$$\mathcal{X}_A(\omega) = \begin{cases} \prod \omega(s) & \text{if } A \neq \emptyset \\ 1 & \text{if } A = \emptyset \end{cases}$$

Since the set $\{\mathcal{X}_A\}$ is orthonormal and has the right cardinality, it must be a basis for $L^2(\Omega)$, and hence any element $F \in L^2(\Omega)$ can be written uniquely as a sum $F = \sum_{A\subset\Omega} F(A)\mathcal{X}_A$ where $F(A) \in {}^*\mathbf{R}$ (this is often called the *Walsh expansion* of F).

5.1 Lemma. *For all subsets A of Ω*

$$L\mathcal{X}_A = \lambda_{|A|}\mathcal{X}_A$$

where

$$\lambda_m = \sum_{\substack{k\leq m\\ k \text{ odd}}} \binom{m}{k} (\Delta t/2)^{k-1}(1 - \Delta t/2)^{m-k}$$

When A is finite, $\lambda_{|A|}$ is infinitely close to $|A|$, and when A is infinite, $\lambda_{|A|}$ is also infinite.

Proof:

$$L_{\mathcal{X}_A}(\omega) = E_Q[\mathcal{X}_A(U^{(\omega)}(\xi,0)) - \mathcal{X}_A(U^{(\omega)}(\xi,\Delta t))]/\Delta t =$$

$=$ (the probability that an odd number of components in A is switched) $\times$

$$\times\, 2\mathcal{X}_A(\omega)/\Delta t = \sum_{\substack{k\leq |A|\\ k \text{ odd}}} \binom{|A|}{k} (\Delta t/2)^{k}(1 - \Delta t/2)^{|A|-k}\cdot 2\mathcal{X}_A(\omega)/\Delta t$$

$$= \lambda_{|A|}\mathcal{X}_A(\omega)$$

When A is finite, the sum has only finitely many terms; the first is infinitely close to $|A|$, the others are infinitesimal. The argument for infinite A is similar. ∎

We have thus found an orthonormal basis of eigenvectors for L, and we see that the eigenvalues are (infinitely close to) the positive integers. The multiplicity of n is the number of subsets of Ω of cardinality n, that is $\binom{|\Omega|}{n}$. This is in agreement with the standard theory where the spectrum also consists of the positive integers (and where the eigenvectors can be expressed in terms of Hermite polynomials, see, e.g., [11]).

As an immediate corollary we get:

5.2 Corollary. *Assume that $F \in L^2(\Omega)$ has finite norm and Walsh expansion $\sum F(A)\mathcal{X}_A$. Then*

$$E_Q(F(U^{(\omega)}(\xi,t)) \approx \sum_{A\subset\Omega} F(A)e^{-t|A|}\mathcal{X}_A(\omega)$$

6 Two alternative constructions

Let us end the paper with a brief look at two other nonstandard constructions of the infinite dimensional Ornstein-Uhlenbeck process, which may be more convenient for some purposes. The first is a very slight variation on the theme we have already presented. As above we start with an element $\omega_0 \in \Omega$, and then pick at random an element $r_0 \in T$. Switching the sign of the r_0-th component $\omega_0(r_0)$, we obtain a new element $\omega_{2\Delta t}$ (we have to move in steps of size $2\Delta t$ to get the right scaling). Again we choose a random and uniformly distributed element $r_1 \in T$ and switch the r_1-th component $\omega_{2\Delta t}(r_1)$ to get $\omega_{4\Delta t}$. Continuing in this way, we get a random sequence of Anderson paths, and taking standard parts we get a randomly moving Brownian path. Not surprisingly, this motion is an infinitely dimensional Ornstein-Uhlenbeck process.

This construction has some advantages and some disadvantages compared to the original one. Conceptually, it is slightly simpler and more natural, but technically it is a little less convenient as we lose the Markov property in the s-direction (this will, e.g., make Lemma 1.2 somewhat harder to formulate and prove). On the other hand, the infinitesimal generator turns out to be

$$AF(\omega) = \sum |A| F_A \mathcal{X}_A(\omega)$$

which is certainly nicer than the approximative expression in Lemma 5.1. It also turns out that A is more naturally related to the discrete Malliavin operators (see [3]). Let us finally observe that since the alternative construction shows that the Ornstein-Uhlenbeck process is the standard part of a nearest neighbor random walk on the space of all Anderson paths (just as finite dimensional Brownian motion is the standard part of a finite dimensional, nearest neighbor random walk), it adds some extra force to the idea that the Ornstein-Uhlenbeck process is really the infinite dimensional counterpart of finite dimensional Brownian motion.

Let us now turn to the second alternative construction. We still keep the discrete timeline T, but this time the state space Ω will be the space of all internal maps $\omega : T \to \mathbf{R}$, and the internal probability measure P on Ω will be the one making all increments $\Delta\omega(t) = \omega(t + \Delta t) - \omega(t)$ independent with distribution $N(0, \sqrt{\Delta t}\,)$. Now let all the increments $\Delta\omega(t)$ perform independent, one-dimensional Ornstein-Uhlenbeck processes (scaled to keep the initial measure invariant). The standard part of this random motion will be an infinite dimensional Ornstein-Uhlenbeck process.

Again this alternative construction has some advantages and some disadvantages compared to the original one. It is less elementary, but it supports the intuitive idea that an infinite dimensional Ornstein-Uhlenbeck process is one where the infinitesimal increments perform independent, one-dimensional Ornstein-Uhlenbeck processes. It also lends itself more easily to the translation of finite dimensional calculus to an infinite dimensional setting. Finally, if one wants to study flows on Wiener space, the discrete setting is very restrictive and unnatural, while the continuous model offered by our alternative construction works quite nicely.

In my opinion, the three slightly different constructions above all deserve further study. Although none of them can claim to be the simplest approach to the infinite dimensional Ornstein-Uhlenbeck process (the one suggested by Corollary 3.2 is hard to beat), they all offer additional insight into the nature and structure of one of the most important processes in today's stochastic analysis.

7 References

[1] Albeverio, S., Fenstad, J. E., Høegh-Krohn, R., Lindstrøm, T.: *Nonstandard Methods in Stochastic Analysis and Mathematical Physics*, Academic Press, 1986.

[2] Anderson, R. M.: A Nonstandard Representation of Brownian Motion and Itô-Integration, *Israel J. Math.* **25** (1976), 15–46.

[3] Holden, H., Lindstrøm, T.,Øksendal, B., Ubøe, J.: Discrete Wick Calculus, to appear in T. Lindstrøm, B. Øksendal, S. Ustunel (eds.): *Stochastic Analysis and Applications*.

[4] Hurd, A. E. & Loeb, P. A.: *An Introduction to Nonstandard Real Analysis*, Academic Press, 1985.

[5] Keisler, H. J.: An Infinitesimal Approach to Stochastic Analysis, *Mem. Amer. Math. Soc.* **297** (1984).

[6] Lindstrøm, T.: An Invitation to Nonstandard Analysis, in N. J. Cutland (ed.): *Nonstandard Analysis and its Applications*, Cambridge University Press, 1988, 1–105.

[7] Malliavin, P.: Stochastic Calculus of Variation and Hypo-elliptic Operators, in K. Itô (ed.): *Proc. Inter. Symp. Stoch. Diff. Equations*, Kinokuniya-Wiley, 1978, 195–263.

[8] Meyer, P. A.: Notes sur les Processus d'Ornstein-Uhlenbeck, *Seminaire de Prob.*, **XVI**, Lect. Notes in Math. Vol. 920, Springer, 1982, 95–133.

[9] Nelson, E.: *Dynamical Theories of Brownian Motion*, Princeton University Press, 1967.

[10] Stroyan, K. D. & Bayod, J. M.: *Foundations of Infinitesimal Stochastic Analysis*, North-Holland, 1986.

[11] Watanabe, S.: *Lectures on Stochastic Differential Equations and Malliavin Calculus*, Tata Institute of Fundamental Research, Bombay, 1984.

Two applications of NSA in the theory of stochastic dynamical systems

A. Ponosov †
Fakultät für Mathematik, Ruhr–Universität
D–44780 Bochum, Germany

Abstract

Two examples of how NSA may be useful in stochastics are considered. In Section 1 a new infinite dimensional multiplicative ergodic theorem for generalized cocycles is obtained; in Section 2 a new method of finding stationary solutions (invariant measures) to nonlinear stochastic parabolic equations is proposed.

1 Multiplicative ergodic theorem for generalized cocycles

1.1 Introduction.

The importance of multiplicative ergodic theorems (MET) is now well–known due to nice applications not only in the theory of dynamical systems (see e.g. [9]), but also in algebra [10]. In this section we propose a new MET for so–called "singular stochastic dynamical systems" which naturally arise from infinite dimensional stochastic equations of various types.

Let (Ω, F, P) be a (standard) probability space. It is known that the fundamental solution $X_t(\omega), \omega \in \Omega$ to any linear finite dimensional stochastic differential equation with time–independent coefficients

$$dx_t = Ax_t dt + Bx_t dw_t \tag{1.1}$$

satisfies the so–called "cocycle property":

$$X_{t+s}(\omega) = X_t(\theta^s\omega)X_s(\omega) \quad a.s., \qquad t, s \geq 0 \tag{1.2}$$

where $\theta : \Omega \to \Omega$ is a measure–preserving brownian shift which may be assumed to be ergodic.

†DFG – Research Fellow

This property allows to define Lyapunov exponents which describe all possible exponential asymptotics of solutions to (1.1):

$$\lambda(x) \equiv \lim_{t\to+\infty} \frac{1}{t} \log \|X_t(\omega)x\| \quad a.s.$$

From Oseledec's multiplicative ergodic theorem it follows that as soon as θ is supposed to be ergodic the function $\lambda(x)$ assumes a finite number of values (not bigger than the order of X_t) which are called Lyapunov exponents corresponding to (1.1) (or to (1.2)) while the set $\{\lambda(x)\}$ is called the Lyapunov spectrum of (1.1) (or of (1.2)).

This nice picture totally changes if we consider infinite dimensional stochastic equations. Although there exists a number of infinite dimensional generalizations of Oseledec's MET (e.g. Ruelle's MET [14]) providing the existence of a Lyapunov spectrum, they only cover cocycles $X_t(\omega)$ consisting of bounded random linear operators. Such cocycles will be called *regular* in the sequel. But the thing is that in infinite dimensional spaces there also exist "singular cocycles" which are not covered by any classical MET just due to the fact that they do not consist of bounded random linear operators. The corresponding examples were found by Skorohod [15] and Mohammed [11] (see also [4]). Thus, the simple scalar delay equation

$$dx = ax(t-h)dt + bx(t-g)dw \qquad h, g \geq 0 \tag{1.3}$$

generates a regular cocycle iff $bg = 0$. Similar difficulties arise in the theory of stochastic partial differential equations (see e.g. [5]).

Our aim is the following: we will generalize the notion of the cocycle to be able to cover the singular case as well; for such generalized cocycles we will formulate a new MET which can be regarded as some kind of extension of Ruelle's MET for compact cocycles in Hilbert spaces.

1.2 Notation and assumptions

Let H be a separable Hilbert space; (Ω, f, P) be a probability space; $\theta^t : \Omega \to \Omega$ $(t \in T$ where $T = \mathbb{N} \cup \{0\}$ or $\mathbb{R}_+ \cup \{0\})$ be a semigroup of measure preserving ergodic maps; $f_0 \subset f$ be a σ–subalgebra such that $f_t \equiv \theta^{-t} f_0 \supset f_0$; $\mathcal{P}(\mathcal{A}, B)$ denotes the set of all random points (or more precisely, the set of all P–equivalent classes of random points) in a random subset $B = B(\omega)$ of H which are measurable w.r.t. a σ–algebra $\mathcal{A}$; the set $\mathcal{P}(\mathcal{A}, B)$ is always supposed to be endowed with the topology of convergence in probability. We also denote by $\Theta^t : \mathcal{P}(f, H) \to \mathcal{P}(f, H)$ the isometry given by

$$(\Theta^t x)(\omega) = x(\theta^t \omega).$$

Assume now $T_t : \mathcal{P}(f_0, H) \to \mathcal{P}(f_t, H)$ to be linear operators with the following properties:

A1. T_t are local: $x \mid A = 0$ a.s. implies $T_t x \mid A = 0$ a.s. and continuous in probability.

A2. T_1 is tight: T_1(bounded sets) $\subset$ (tight sets).

A3. The family T_t satisfies the following generalized cocycle property:

$$T_{t+s} = \Theta^s T_t \Theta^{-s} T_s \tag{1.4}$$

A4. $\sup_{\|x\|\leq 1} E \sup_{0\leq t\leq 1} \log^+ \|T_t x\|_H < \infty \quad x \in L^\infty(f_0, H)$.

Let us make some comments. Property A1 trivially holds for bounded random operators, i.e. for the case of T_t given by

$$(T_t x)(\omega) = X_t(\omega) x(\omega) \tag{1.5}$$

for some measurable family $X_t(\omega)$ of bounded operators. In this case the generalised cocycle property (1.4) is easily seen to coincide with the ordinary cocycle property (1.2). In finite dimensional spaces there are no linear and continuous (in probability) cocycles different from those given by (1.5). By the way, this roughly explains why singular systems cannot occur in the finite dimensional situation. It is important to observe however that it is not the case any more in infinite dimensional spaces.

On the other hand, it can be easily checked that an autonomous linear stochastic differential equation gives rise to a generalized cocycle if the equation possesses the property of existence and uniqueness of solutions continuously depending (in probability) on initial data. No regularity condition is required. For instance, the delay equation (1.3) gives rise to the generalized cocycle (1.4) for any $a, b \in \mathbb{R}; g, h \geq 0$ (and to the regular cocycle only if $bg = 0$). One may therefore say that it is the property of locality (i.e. A1) which enables us to cover the case of singular systems too.

Assumption A3 extends the notion of a compact cocycle. Assumption A4 corresponds to Oseledec's integrability condition.

1.3 Multiplicative ergodic theorem

Theorem: *Under assumptions A1–A4 there exist a decreasing sequence $\{\lambda_m\}$ which goes to $-\infty$ or contains it and a flag $\{E_m(\omega)\}, E_1(\omega) = H$ of random subspaces (Oseledec's subspaces) such that*

1) $T_t(\mathcal{P}(f_0, E_m)) \subset \Theta^{-t}(\mathcal{P}(f_0, E_m))$ (invariance of Oseledec's subspaces);

2) $\frac{1}{t} \log \| (T_t x)(\omega) \| \to \lambda_m$ a.s., $t \to +\infty$ if $x \in \mathcal{P}(f_0, E_m \backslash E_{m+1})$.

Some comments. One cannot make use here of Kingman's subadditive ergodic theorem here as Ruelle did, because the random values $\sup_{\|x\|\leq 1} \| T_t(\omega) x \|$ are infinite in the singular case. This is of course the serious obstacle which makes it impossible not only to use Ruelle's approach, but even to define Lyapunov exponents in the standard way as limits of norms of suitably chosen random operators. By this reason the only thing we can try to make use of is Oseledec's proof which is based on two ideas: a transformation of the cocycle to a triangular form and an extension of the shift. Of course, Oseledec's approach is very much related to the finite dimensional technique, for example operators in Hilbert spaces have no triangular representation, in general. But fortunately these ideas of Oseledec still work if we add some portion of NSA. Thus, it becomes possible to transform operators to the triangular form if we replace our Hilbert space H by a hyperfinite dimensional space F for which $H \subset F \subset {}^*H$.

The idea of extending the shift can also be adapted to our situation, although Oseledec's extensions cannot be used directly. Instead of them we take the so–called "Kamae's universal shift" or the hyperfinite cycle $\hat{\theta}$ (see [7] or [1] given by

$$\hat{\theta}(n) \equiv n+1 \quad (modN), \qquad 0 \le n \le N, \quad N \in {}^*\mathbb{N}_\infty$$

which is universal in the sense that any measure–preserving shift θ in a separable Borel space is a factor of $\hat{\theta}$. Kamae's shift proved to have other interesting properties. Let us briefly formulate one of them: in the case of finite dimensional cocycle the corresponding Lyapunov exponents can be computed as the shadows of the logarithms of the real nonnegative part of the spectrum (in a suitable internal hyperfinite dimensional space) of the operator $\hat{\Theta}^{-1}\bar{T}_1$ (where $\bar{T}_n$ is a properly chosen internal lifting of T_n and $(\hat{\Theta}\alpha)(\omega) \equiv \alpha(\hat{\theta}\omega)$). Notice that such a property does not hold for an arbitrary internal lifting of the shift θ.

1.4 Sketch of the proof

Basically, the proof consists of 3 parts briefly discussed below. For the sake of simplicity we will restrict ourselves to the discrete case, i.e. we will assume that $T = \mathbb{N} \cup \{0\}$.

i). First, we describe how to extend the given cocycle. For this purpose, let us fix a hyperfinite dimensional space F such that $H \subset F \subset {}^*H$, π being an internal orthogonal projection onto F. The shift $\theta \equiv \theta^1$ is replaced by Kamae's shift $\hat{\theta}$ defined on the probability space $\hat{\Omega} \equiv \{0, 1, ..., N-1\}$ equipped with the Loeb measure $\hat{P}$ generated by the counting measure. The corresponding completion $\hat{f}$ of the family of all internal sets will play the role of the extended σ–algebra. There also exist internal liftings $\hat{T}_n$ of the local operators T_n defined on the set $\mathcal{P}(\hat{f}_0, \hat{F})$ for some $\hat{f}_0 \subset \hat{f}$, the Hilbert space $\hat{F}$ being a factor–space of the set of all finite elements of F w.r.t. the subspace $\{x \in F : \|x\| \approx 0\}$. Moreover, since F is hyperfinite dimensional one may nonstandardly represent the cocycle $\hat{T}_n$ by an internal family of random hyperfinite dimensional matrices $C_n(\hat{\omega})$, $\hat{\omega} \in \hat{\Omega}$, namely as follows:

$$\hat{T}_n{}^o x(\hat{\omega}) \approx C_n(\hat{\omega}) x(\hat{\omega}) \quad \hat{P} - a.s.,$$

where $x : \hat{\Omega} \to F$ are internal and $\hat{f}_0$–measurable. This representation exists due to the nonstandard transfer principle. Of course, the matrices $C_n(\hat{\omega})$ may have infinitely large norms for particular ω, but globally, they still generate the continuous local operators $\hat{T}_n$. By making use of this representation we can find an internal unitary matrix $U \equiv U(\hat{\omega}, \kappa)$ depending on $\hat{\omega} \in \hat{\Omega}$ and also on an arbitrary internal unitary matrix κ in F with the property

$$\kappa^{-1} C(\hat{\omega}) U(\hat{\omega}, \kappa) = D(\hat{\omega}) \qquad (\hat{\omega} \in \hat{\Omega}) \tag{1.6}$$

where $D(\hat{\omega})$ is a lower triangular matrix with nonnegative diagonal elements (the idea of such a transformation in the finite dimensional case was proposed by Oseledec). As the next step, we replace the probability space $\hat{\Omega}$ by a new one $\{\hat{\omega}_0, ..., \hat{\omega}_{N-1}\}$ given by $\hat{\omega}_0 \equiv (\hat{\omega}', \kappa')$, $\quad \hat{\omega}_k \equiv \Phi \hat{\omega}_{k-1}$, $\quad \Phi(\hat{\omega}, \kappa) \equiv (\hat{\theta}\hat{\omega}, U(\hat{\omega}, \kappa))$ where $\hat{\omega}' \in \hat{\Omega}$ and an unitary matrix κ' in F are arbitrary, but fixed. Clearly, the projection $\Pi(\hat{\omega}, \kappa) = \hat{\omega}$ induces a bijection between $\hat{\Omega}$ and the new probability space equipped with the Loeb measure generated by the counting measure, Kamae's cycle on $\hat{\Omega}$ being transformed to the same cycle $\hat{\omega}_k \to \hat{\omega}_{k+1}(modN)$ on the new probability space. So bearing in mind this bijection we may replace $\hat{\Omega}$ by the probability space just constructed without changing the notation. This new space has at least one important advantage in comparison with the old one. Namely, thanks to the relation (1.6) we may now replace our cocycle by one having a triangular form.

ii). Following our observation in the finite dimensional case we define Lyapunov exponents of the cocycle T_n as the standard parts of the numbers $\log \lambda$ where $\lambda \in Sp(\hat{\Theta}^{-1}\bar{T}_1) \cap \{\lambda \geq 0\}$, the internal lifting $\bar{T}_n$ of $\hat{T}_n$ being given by

$$(\bar{T}_n x)(\hat{\omega}) \equiv C_n(\hat{\omega}) x(\hat{\omega}).$$

Then the invariant subspaces corresponding to the pieces $\{\mu : |\mu| < \lambda\}$ of the spectrum of $\hat{\Theta}^{-1}\bar{T}_1$ are checked to be of the form $\mathcal{P}(\hat{f}_0, \hat{E}_\lambda)$ for some measurable family $\hat{E}_\lambda(\hat{\omega})$ of subspaces of H. According to our definition the Lyapunov spectrum is not changed if we transform the cocycle to the triangular form and moreover the spectrum coincides with the union of Lyapunov exponents computed for all diagonal elements in the triangular representation. By making use of the tightness assumption A2 we can prove that the Lyapunov spectrum of a triangular cocycle is a discrete set, the function $\lambda \to \hat{E}_\lambda$ assumes a countable number of values $\mathcal{P}(\hat{f}_0, \hat{E}_m)$ and the corresponding Oseledec's subspaces $\hat{E}_m$ for the extended cocycle are of finite codimension.

iii). Finally, we come back to the very cocycle T_n and check that the subspaces $\mathcal{P}(f_0, H) \cap \mathcal{P}(\hat{f}_0, \hat{E}_m)$ are of the form $\mathcal{P}(f_0, E_m)$ for some random subspaces E_m of finite codimension in H. Now, it remains to verify the asymptotic property 2 of Theorem 1.3. To do this we first notice that our constructions depend on two infinitely large parameters: $dimF$ and N (the period of Kamae's cycle). Without loss of generality we can assume that $dimF = N$. Of course, the Lyapunov spectrum might a priori depend on N. In fact, by making use of the Oseledec's MET and of the already established fact that the invariant subspaces corresponding to each piece $\{\mu : |\mu| \geq \lambda, \quad \lambda \in \mathbb{R}_+\}$ of the spectrum of $\hat{\Theta}^{-1}\bar{T}_1$ consist of the random points concentrated in certain finite dimensional random subspaces of F we can prove that the Lyapunov spectrum does not depend on the extension of the cocycle.

From the construction of $\bar{T}_n$, it is easily follows that the the dependence $\bar{T}_1 \equiv \bar{T}_1(N)$ of N is internal and moreover, it also makes sense for all finite N for which the top Lyapunov exponent λ_{top} is easily seen to be expressed as follows:

$$\lambda_{top} \equiv \lambda_{top}(N) \approx \sqrt[n]{\hat{E}\|\bar{T}_n\|_F} \qquad \forall n \in {}^*\mathbb{N}_\infty$$

Applying now the permanence principle and omitting some technicalities we can check that the last relation is valid for some infinitely large N. This implies the asymptotical property in Theorem 1.3, at least for the top Lyapunov exponent. Now, replacing the space H by Oseledec's subspaces E_m defined as above we may reduce the case of an arbitrary Lyapunov exponent to the case of the top one.

2 Stationary solutions to nonlinear stochastic differential equations in Hilbert spaces

2.1 Introduction

In Section 2 we are going to describe a method of finding stationary solutions for nonlinear infinite dimensional stochastic differential equations indexequation!stochastic differential with a monotone principal part. Our main result (Theorem in 2.5) can be regarded as a generalization of one of Browder's results concerning deterministic elliptic equations.

The usual approach to constructing stationary solutions of infinite dimensional stochastic equations is based on two ideas: first of all, one tries to find stationary solutions to finite–dimensional approximations of the very equation and then to check that weak limit points of the set of distribution–measures corresponding to these solutions yield the stationary solutions (or maybe solutions–measures) to the equation in question.

Let us notice, however, that this approach usually requires the path–continuity of solutions on initial data and by this reason it seems to have only been applied to equations of a rather particular form [16], [3] (e.g. with an additive stochastic noise). In fact, this situation is quite similar to that considered in Section 1 because the property of the path–continuity in the linear case is nothing but the property of regularity. So we may suspect that the reason of forthcoming difficulties is determined by the necessity of replacing regular equations by singular ones.

Additional problems arise if we want to study equations containing monotone–type nonlinearities, because of a rather special way of proving convergence of approximations.

To overcome these and other drawbacks it is proposed here (as in Section 1) to involve some NSA in our game, namely, we intend to use the machinery of Loeb probability spaces. The motivation for our choice can be explained as follows. First, Loeb spaces enable us a nice opportunity to avoid both weak solutions and solutions–measures to the equation in question and to consider strong solutions only. In fact, this situation is quite similar to one in the theory of weak and strong solutions of stochastic differential equations: once we choose a Loeb probability space we do not have to worry about other probability spaces; Loeb spaces are rich enough to contain all stochastic processes which might be of interest (see e.g. [6]). Second, by making use of Loeb spaces we may forget about weak limits of measures. It is important for our purposes because nonlinearities of monotone type we are dealing with are hardly compatible with this kind of a limit. Finally, the third advantage of our approach is that it does not require any regularity properties of solutions and, in principle, even uniqueness of the solutions to the Cauchy problem is not needed.

2.2 Notation and assumptions

Let $V \subset H \subset V'$ be a triple, consisting of a separable Hilbert space H, a reflexive Banach space V and its conjugate V', continuously and densely imbedding each in other, the pairing $a \cdot b$ between V and V' coinciding with the inner product in H iff $b \in H$.

The following equation is the main object of the section 2:

$$dx = f(x)dt + g(x)dw \tag{2.1}$$

where w is a scalar (just for the sake of simplicity) Brownian motion, $f = a + p$, $g = b + q$, $a, p : V \to V'$, $b, q : V \to H$ are time–independent coefficients which are assumed to satisfy 5 conditions.

B1. Monotonicity of the principal part:

$$2(x - y) \cdot (a(x) - a(y)) + |b(x) - b(y)|^2 \leq 0,$$

$|\cdot|$ denoting $H-$ norm.

B2. Semicontinuity of the principal part:

$$x_1 \cdot a(x_2 + \lambda x_3) \text{ is continuous in } \lambda \in \mathbb{R} \quad (\forall x_i \in V).$$

B3. Compactness of the subordinate part:

$$p : V \to V', q : V \to V' \quad \text{are compact continuous operators.}$$

B4. Coerciveness:

$$2x \cdot f(x) + |g(x)|^2 \leq -\epsilon \|x\|_V^p + const \quad (\epsilon > 0, p \geq 2)$$

B5. Restriction On Growth:

$$\|p(x)\|_{V'} + \|a(x)\|_{V'} \leq const(1 + \|x\|_V^{p-1}).$$

Similar conditions were considered for deterministic case by Browder [2]. We have adapted them for the stochastic case in the spirit of works [12] and [8], where the Cauchy problem for monotone stochastic equations ($p = q = 0$ in our notation) was studied. Our main result states that under conditions B1–B5, there is at least one solution which is stationary and stationarily related to the increments of the given Wiener process. This assertion can be viewed as a generalization of Browder's results.

2.3 Nonstandard reformulation of the problem

As in the Section 1 we will fix a hyperfinite dimensional space F such that $V \subset F \subset^* V$.

By making use of hyperfinite dimensional projections onto F which are orthogonal with respect to the inner product in H and the pairing in $V \times V'$ respectively, we can find internal liftings $A, B, P, Q : {}^*V \to F$ of the coefiicients a, b, p, q satisfying hypotheses which are analogous to B1–B5.

The nonstandard version of (2.1) can be written down in the usual way:

$$\Delta X_\tau = FX_{\tau-}\Delta t + GX_{\tau-}\Delta W_\tau \tag{2.2}$$

where W_τ, $\tau \in T \equiv \{-m\Delta t, ..., 0, \Delta t, ..., m\Delta t\}$, $m\Delta t = 1$ is a suitable internal lifting of w_t defined on a certain hyperfinite probability space $(\hat{\Omega}, \mathcal{F}, \bar{P})$, equipped with an internal measure $\bar{P}$, the Loeb extension of which will be denoted by $\hat{P}$; Δt is a "sufficiently small" infinitesimal.

It is convenient to represent the condition of stationarity in terms of an internal lifting $\bar{\theta}^\tau$ of the brownian shift θ^t, namely, in the following manner:

$$X^\tau(\bar{\theta}^\sigma \hat{\omega}) = X^{\tau+\sigma}(\hat{\omega}) \quad (\hat{\omega} \in \hat{\Omega}) \quad \forall \tau, \sigma \geq 0; \quad \tau, \sigma \in T. \tag{2.3}$$

For our further purposes we want $\bar{\theta}^\tau$ to have two properties: 1) $\bar{\theta}^\tau(\mathcal{F}_0) \subset \mathcal{F}_0$ where $\mathcal{F}_0$ is an internal subalgebra of $\mathcal{F}$, playing the role of the σ–algebra of "initial events", and 2) the shift $\bar{\theta}^{\tau}$ should preserve the measure $\hat{P}$. This can be done if we choose a proper hyperfinite approximation of the Brownian motion w_t. It is interesting to remark that Anderson's random walk with the natural shift does not satisfy these two conditions. Also it is not possible to exploit here Kamae's cycle as a lifting of the Brownian shift. One could think that using the condition (2.3) is not a great trick, because one can easily write down its standard analog by means of the Brownian shift on the canonical Wiener space. Let us however notice that such a standard formulation would not be sufficient for our purposes. As a matter of fact, there is no a priori reason to assume a shift generated by a stationary solution we are looking for to coincide with the Brownian shift (and it is indeed not the case). On the other hand, due to nice features of Loeb probability spaces briefly described above we might expect that the NSA approach would allow use to avoid uncertainty in choosing shifts.

2.4 Solvability of the internal problem (2.2), (2.3)

To solve the equations (2.2) and (2.3) we replace (2.3) by the following condition:

$$\Theta X_0 = X_{\Delta t} \tag{2.4}$$

where $\Theta\alpha(\hat{\omega}) \equiv \alpha(\hat{\theta}^{\Delta t}\hat{\omega})$. We reduce therefore our problem to one of finding periodic (in law) solutions with infinitely small period Δt. Now, the problem (2.2), (2.3) can easily seen to be rewritten as the following internal equation:

$$\alpha = \Gamma\alpha, \tag{2.5}$$

with $\alpha = X_0, \Gamma\alpha = \Theta^{-1}(\alpha + A\alpha\Delta t + B\alpha\Delta W_{\Delta t})$.

In order to find a solution to the equation (2.5) we introduce a hyperfinite Banach space L to be a set of $\mathcal{F}_0$–measurable internal functions from Ω to F equipped with the norm $\sqrt{E|\cdot|^2}$. Clearly, the nonlinear internal operator Γ is *–continuous in L. Now we intend to apply a nonstandard version of the well–known finite dimensional fixed point theorem. Namely, if we manage to prove that for all $\lambda \geq 1$ the equation

$$\lambda\alpha = \Gamma\alpha \tag{2.6}$$

would have no solutions with sufficiently large but still finite norms, then we could be sure that the equation (2.5) has at least one internal solution with a finite norm. From the coerciveness it follows that

$$E\lambda^2\alpha^2 = E(\Theta\alpha)^2 + EF^2(\Theta\alpha)\Delta t^2 + EG^2(\Theta\alpha)\Delta W^2 +$$

$$+2E(\Theta\alpha F(\Theta\alpha)\Delta t) + 2E(\Theta\alpha G(\Theta\alpha)\Delta W) + 2EF(\Theta\alpha)G(\Theta\alpha)\Delta W \leq$$

$$\leq E\Theta\alpha^2 - (\epsilon E\|\Theta\alpha\|_V^p + const)\Delta t + o(\Delta t) \leq E\Theta\alpha^2(1 - \epsilon\Delta t) +$$

$$+ const\Delta t + o(\Delta t) \leq E\alpha^2(1 - \epsilon\Delta t) + const\Delta t + o(\Delta t) < E\alpha^2, \qquad (2.7)$$

if $E\alpha^2 > \epsilon^{-1}const + 1$. The inequalities (2.7) contradicts, however, the assumption $\lambda \geq 1$. This concludes the proof of solvability of the equation (2.5) and, hence, of the problem (2.2), (2.3). We only have to bear it in mind that the solution just obtained is not S–integrable, in general, so we cannot take its standard part in the strong topology of the space L. The only thing we can derive from (2.7) is an estimate $E^*\|\alpha\|_V^p < \infty$.

2.5 Standartisation of the internal solution

Let us introduce another Banach space

$$\mathcal{L} \equiv L_p(\Omega \times [-1, 1], L(\mathcal{F}_0 \otimes \mathcal{B}), L(P \times m), V),$$

where $L(\cdot)$ denotes the Loeb extension of the corresponding internal object, m is the counting measure on T, $\mathcal{B}$ stands for the family of internal subsets of T.

Proposition: *If X is an internal solution to the problem (2.2), (2.3) satisfying an equality $E^*\|x_0\|_V^p < \infty$, then the weak standard part of the process X in the space $\mathcal{L}$ is a stationary process which is also stationarily related to increments of the Wiener process w_t and which satisfies the equation (2.1).*

Of course, we will omit here the proof of the proposition (which takes about 5 pages). Let us only make some comments. The essential part of the argument is based on the monotonicity technique [2], [8], [12] arranged in the spirit of NSA. We cannot but remark that in some points our approach still differs very much from those presented in the above papers.

As an immediate corollary we get

Theorem: *Under the assumptions B1 – B5 there exists at least one (weak) stationary solution to the equation (2.1) defined on a Loeb probability space which is an extension of the initial probability space. This solution is also stationarily related to increments of the Brownian motion w_t.*

Remark: If we want to return to the probability space on which the equation (2.1) is originally defined, we have to take shadows of solutions in the topology of the weak convergence in measure. In this way, we get a "stationary measure" instead of the ordinary process. If now the equation (2.1) has the properties of existence and uniqueness of solutions continuously depending on initial data (i.e. if the equation

(2.1) gives rise to a "generalized flow of solutions" in the terminology of the previous section) then the "stationary measure" just obtained can be viewed as an invariant measure w.r.t. this generalized flow.

References

[1] S. Albeverio, J. Fenstad, R. Høegh-Krohn, T. Lindstrøm. *Nonstandard methods in stochastic analysis and mathematical physics.* Acad. Press. Orlando (1986).

[2] F. E. Browder. *Nonlinear elliptic boundary value problems.* Bull. AMS, Vol. 69, Nov. (1963), p. 862–874.

[3] A. B. Cruzeiro. *Solutions et mesures invariantes pour des équations d'évolution stochastiques du type Navier- Stokes.* Expositiones Mathematicae 7 (1989), p. 73–82.

[4] G. Da Prato, J. Zabczyk. *Stochastic equations in infinite dimensions.* Cambridge Univ. Press (1992).

[5] F. Flandoli. *Stochastic flows and Lyapunov exponents for abstract stochastic PDEs of parabolic type.* In: Proceedings of Conf. "Lyapunov exponents". Oberwolfach (1990). Eds. L. Arnold et al. Lect. Notes Math. 1486.

[6] D. N. Hoover, E. Perkins. *Nonstandard construction of the stochastic integral and applications to stochastic differential equations. I, II.* Trans. Amer. Math. Soc., 275 (1983).

[7] T. Kamae.*A simple proof of the ergodic theorem using nonstandard analysis.* Israel J. of Math. 42 (1982), p. 284–291.

[8] N. V. Krylov, B. L. Rozovskii. *Itô equations in Banach spaces.* J. of Soviet Math. Vol. 16, No. 4 (1981), p. 1233–1277.

[9] *Lyapunov exponents.* Eds. L. Arnold et al., Proceedings, Oberwolfach (1990), Lect. Notes in Math. 1486.

[10] G. A. Margulis *Discrete subgroups of semisimple Lie groups.* Springer (1991).

[11] S.E.A. Mohammed. *Stochastic functional differential equations.* Pitman, Boston (1984).

[12] E. Pardoux. *Sur des equations aux dérivées partielles stochastiques monotones.* C. R. Acad. Sci., 275, No. 2 (1972), A101 - A103.

[13] Ph. Protter. *Semimartingales and measure preserving flows.* Ann. Inst. Henry Poincare, 22, No. 2 (1986), p. 127–147.

[14] D. Ruelle.*Characteristic exponents and invariant manifolds in Hilbert space.* Ann. of Math. 115 (1982), p. 243–290.

[15] A.V. Skorohod. *Random linear operators.* Reidel (1984).

[16] M. I. Višik, A. I. Komech, A. V. Fursikov. *Some mathematical problems of statistical hydromechanics.* Russian Math. Surveys. V. 34, No. 5 (1979), p.149–234.

NONSTANDARD METHODS AND THE SPACE OF EXPERIMENTS

David A. Ross
Department of Mathematics,University of Hawaii,
Honolulu, Hawaii 96822, U.S.A.

Abstract

This note gives natural nonstandard proofs of two results that are traditionally considered difficult in the standard theory: the first one is the proof of compactness of the space of experiments in an appropriate topology. The other one is concerned with the connection between two different notions of eqivalence of experiments.

Keywords: nonstandard hull, experiment, statistical experiment

1 Introduction

Following Blackwell, a statistical experiment is a family of probability measures on some fixed measurable space. It is conventional to say that two experiments are equivalent if, for any decision made based on one experiment, one can make at least as good a decision based on the other. The collection of equivalence classes forms a set, which — properly topologized — is called the *space of experiments.*

This note gives natural nonstandard proofs of two results that are traditionally considered difficult in the standard theory:

(1) In §4, it is proved that the space of experiments is compact when given a certain weak topology. As usual for nonstandard compactness proofs, it suffices to find a standard part for every *experiment. In this case it turns out to be an artifact called the *nonstandard hull* of the *experiment, obtained by applying the Loeb construction to each of the *experiment's internal probability measures.

(2) In section 5, a different notion of equivalence for experiments is considered; two experiments are equivalent in this sense if each can be used to simulate the other. It is shown that this kind of equivalence implies the other. The proof here is also a kind of nonstandard hull argument: the operator which formally represents the simulation is the standard part of an internal operator.

The reader might notice that many standard results are stated without proof, and that some (true) results are omitted where their inclusion would be natural (for example, the converse to Theorem 5.1). This is done to maintain the brevity of this note; a companion article to this one will contain these and related results (see section 6.)

For standard proofs and general overview, the reader is referred to [L] and [S]. The reader is assumed to be familiar with the s tandard theory, for which [C] is a good reference.

2 Standard concepts

Fix once and for all a parameter set T.

An *experiment* (on T) is a family $E = (X, \mathcal{A}, P_t)_{t\in T}$ of probability measures on a common measurable space, indexed by T.

Denote by $\mathcal{E}(T)$ the class of experiments on T.

If $E \in \mathcal{E}(T)$ and $T' \subseteq T$, denote by $E(T')$ the restriction $(X, \mathcal{A}, P_t)_{t\in T'}$ of E to T'.

If $(X, \mathcal{A})$ is a measurable space, denote by $\mathcal{M}(X)$ the set of signed real measures on $(X, \mathcal{A})$, and by $\mathcal{M}_1(X)$ the set of probability measures on $(X, \mathcal{A})$.

If $(X, \mathcal{A})$ and $(Y, \mathcal{B})$ are measurable spaces, then a *transition kernel* from X to Y is a function $\rho : X \longrightarrow \mathcal{M}_1(Y)$ such that for any $B \in \mathcal{B}$, the function $x \longmapsto \rho_x B$ is measurable.

More generally, suppose $\mathcal{M}' \subseteq \mathcal{M}(X)$, and denote by $\mathcal{L}^b(Y)$ the space of bounded real-valued measurable functions on Y. A *generalized transition kernel* from $\mathcal{M}'$ to Y is a bilinear function $\rho : \mathcal{M}' \times \mathcal{L}^b(Y) \longrightarrow \mathbb{R}$ such that (1) if $f \geq 0$ and $\mu \geq 0$, then $\rho(\mu, f) \geq 0$; (2) $|\rho(\mu, f)| \leq \|f\|\|\mu\|$; and (3) $\rho(\mu, 1) = \mu(X)$. (In (2), use the supremum norm on $\mathcal{L}^b(Y)$ and the total variation norm on $\mathcal{M}(X)$.)

Observe that every transition kernel gives a generalized transition kernel in a natural way, namely $(\mu, f) \longmapsto \int_X \int_Y f(y) d\rho_x(y) d\mu(x)$. Moreover, a generalized transition kernel ρ induces an operator $M_\rho : \mathcal{M}' \longrightarrow \mathcal{M}(Y)$ by $M_\rho(\mu)(A) = \rho(\mu, \chi_A)$. (Here χ_A is the characteristic function of A.)

If $E = (X, \mathcal{A}, P_t)_T \in \mathcal{E}(T)$, then denote by $\mathcal{M}_{\ll}(E)$ the set of $\mu \in \mathcal{M}(X)$ such that whenever $\nu \in \mathcal{M}(X)$ and $\nu \perp P_t$ for all $t \in T$, $\nu \perp \mu$ as well. (Note that for finite T this is just the set of measures dominated by $\sum_T P_t$. The general concept is due to Le Cam, and called the *L-space* of E.)

A *decision problem* is a family $\mathcal{D} = (D, W_t)_{t\in T}$ where D is a measurable space (note the measure algebra is implicit) and each $W_t \in \mathcal{L}^b(D)$. (Often it is assumed that D is topological or even metric, and that each W_t is continuous; all the arguments below go through with little or no change under these alternative assumptions — indeed, in the standard theory it is proved that without loss of generality D is the space consisting of just two points.)

If $E = (X, \mathcal{A}, P_t)_T$ and $F = (Y, \mathcal{B}, Q_t)_T \in \mathcal{E}(T)$, $\epsilon \geq 0$, and $\mathcal{D}$ is as above, write $E \sqsupseteq^{\epsilon}_{\mathcal{D}} F$ provided for every generalized transition kernel $\rho : \mathcal{M}_{\ll}(F) \times \mathcal{L}^b(D) \longrightarrow \mathbb{R}$ there is a generalized transition kernel $\eta : \mathcal{M}_{\ll}(E) \times \mathcal{L}^b(D) \longrightarrow \mathbb{R}$ with $\eta(P_t, W_t) \leq \rho(Q_t, W_t) + \epsilon\|W_t\|$ for each $t \in T$. Write $E \sqsupseteq^{\epsilon} F$ if $E \sqsupseteq^{\epsilon}_{\mathcal{D}} F$ for every $\mathcal{D}$; note attention may be restricted to those $\mathcal{D}$ where $\|W_t\| = 1$ for each t. If $E \sqsupseteq^{\epsilon} F \sqsupseteq^{\epsilon} E$ then $E \sim^{\epsilon} F$. The relations $\sqsupseteq^0_{\mathcal{D}}$, $\sqsupseteq^0$, and $\sim^0$ will be written $\sqsupseteq_{\mathcal{D}}$, $\sqsupseteq$, and $\sim$, respectively.

Proposition 2.1: *$\mathcal{E}(T)/\sim$ is a set.*

Suppose $E = (X, \mathcal{A}, P_t)_T \in \mathcal{E}(T)$, and $\alpha \subseteq T$ is finite. Put $P_\alpha = \sum_{t\in\alpha} P_t$, and for $t \in \alpha$ let $\dfrac{dP_t}{dP_\alpha}$ be the Radon-Nikodym derivative of P_t with respect to P_α. Without loss of generality $\left(\dfrac{dP_t}{dP_\alpha}\right)_\alpha$ is a measurable function from X to $\mathcal{S}^\alpha = \{\vec{x} \in [0,1]^\alpha | \sum_\alpha x_t = 1\}$. For any $\mu \in \mathcal{M}(X)$, write $\tilde{\mu}$ for the image measure of μ on $\mathcal{S}^\alpha$ under the mapping $\left(\dfrac{dP_t}{dP_\alpha}\right)_\alpha$. Note (in particular) that $\sum_\alpha \widetilde{P}_t = \widetilde{P}_\alpha$, and that $\mu \longmapsto \widetilde{\mu}$ is a linear isometry from $\mathcal{M}_{\ll}(E(\alpha))$ onto $\mathcal{M}_{\ll}\left(\widetilde{E}(\alpha)\right)$, where $\widetilde{E}(\alpha)$ is the experiment $(\mathcal{S}^\alpha, \text{Borel}, \widetilde{P}_t)_\alpha$.

The following is standard and not difficult.

Lemma 2.2: *If $E = (X, \mathcal{A}, P_t)_T$ and $F = (Y, \mathcal{B}, Q_t)_T \in \mathcal{E}(T)$, and $\alpha \subseteq T$ is finite, then $E(\alpha) \sim F(\alpha)$ iff $\widetilde{P}_\alpha = \widetilde{Q}_\alpha$.*

Thus, for any finite α the map $E(T) \longmapsto \widetilde{E}(\alpha)$ induces a map from $\mathcal{E}(T)/\sim$ to $\mathcal{M}(\mathcal{S}^\alpha)$; the smallest topology on $\mathcal{E}(T)/\sim$ making all these maps continuous, where $\mathcal{M}(\mathcal{S}^\alpha)$ is given the weak topology, is called the *weak topology* on $\mathcal{E}(T)/\sim$. (In practice it will usually be convenient to abuse notation and refer to the weak topology on $\mathcal{E}(T)$.)

3 Nonstandard concepts

Assume that the model is saturated at least in the cardinality of $\mathcal{E}(T)$.

It will be convenient to adopt the following notation. If $(X, \mathcal{A}, P)$ is an internal *probability space, denote the corresponding (standard) Loeb space by $(X, \widehat{\mathcal{A}}, \widehat{P})$. If $(X, \mathcal{A}, P,)$ is a *standard* probability space, write $(\widehat{X}, \widehat{\mathcal{A}}, \widehat{P})$ instead of $(\widehat{{}^*X}, \widehat{{}^*\mathcal{A}}, \widehat{{}^*P})$.

If $E = (X, \mathcal{A}, P_t)_{t\in{}^*T} \in {}^*\mathcal{E}(T)$, let $\widehat{E} = (X, \widehat{\mathcal{A}}, \widehat{P}_t)_{t\in{}^*T}$; this is the *nonstandard hull* of E. Note that $\widehat{E} \in \mathcal{E}({}^*T)$, and so $\widehat{E}(T) \in \mathcal{E}(T)$. If $E = (X, \mathcal{A}, P_t)_{t\in T} \in \mathcal{E}(T)$, write $\widehat{E}$ for $\widehat{{}^*E}$.

4 Weak compactness

Theorem 4.1: *Let* $E = (X, \mathcal{A}, P_t)_{{}^*T} \in {}^*\mathcal{E}(T)$*; then* $\widehat{E}(T)/\sim$ *is a weak standard part of* $E/^*\sim$.

Proof: A subbasic open neighborhood of $\widehat{E}/\sim$ has the form $\mathcal{U} = \{F = (Y, \mathcal{B}, Q_t)_T \in \mathcal{E}(T) \mid \left| \int \varphi d\widetilde{\widehat{P}_\alpha} - \int \varphi d\widetilde{Q}_\alpha \right| < \epsilon\}/\sim$, where $\alpha \subseteq T$ is finite, $\epsilon > 0$, and $\varphi : \mathcal{S}^\alpha \longrightarrow \mathbb{R}$ is continuous. It is easily verified that $\frac{dP_t}{dP_\alpha}$ lifts $\frac{d\widehat{P}_t}{d\widehat{P}_\alpha}$ for each $t \in \alpha$, so that ${}^*\varphi \circ \left(\frac{dP_t}{dP_\alpha}\right)_\alpha$ lifts $\varphi \circ \left(\frac{d\widehat{P}_t}{d\widehat{P}_\alpha}\right)_\alpha$. Morever, $\varphi(\mathcal{S}^\alpha)$ is compact, so this lifting is S-bounded. It follows:

$$\begin{aligned} \int {}^*\varphi d\widetilde{P}_\alpha &= \int {}^*\varphi \circ \left(\frac{dP_t}{dP_\alpha}\right)_\alpha dP_\alpha \\ &\approx \int \varphi \circ \left(\frac{d\widehat{P}_t}{d\widehat{P}_\alpha}\right)_\alpha d\widehat{P}_\alpha \\ &= \int \varphi d\widehat{P}_\alpha \end{aligned}$$

(where the two equalities are from a simple change of variables). Thus, $E/^*\sim \in {}^*\mathcal{U}$. This proves the theorem. ∎

Corollary 4.2: $\mathcal{E}(T)/\sim$ *is weakly compact.*

Proof: Every element of ${}^*\mathcal{E}(T)/\sim$ is nearstandard by 4.1. ∎

5 Comparison of Experiments

Theorem 5.1: *Let* $E = (X, \mathcal{A}, P_t)_T$*,* $F = (Y, \mathcal{B}, Q_t)_T \in \mathcal{E}(T)$*, and suppose* $E \sim F$*. Then there is a generalized transition kernel* $\eta : \mathcal{M}_\ll(E) \times \mathcal{L}^b(Y) \longrightarrow \mathbb{R}$ *with* $M_\eta(P_t) = Q_t$*,* $t \in T$.

Proof: Fix $\alpha \subseteq {}^*T$ hyperfinite with $t \in \alpha$ for every $t \in T$. Fix any $\nu \in^* \mathcal{M}_\ll(E(\alpha)) \cap {}^*\mathcal{M}_1(X)$, e.g. any ${}^*P_t$ with $t \in \alpha$. Note ${}^*\widetilde{P}_\alpha = {}^*\widetilde{Q}_\alpha$ on ${}^*\mathcal{S}^\alpha$. Define a map $\Theta : {}^*\mathcal{M}_\ll(E) \longrightarrow \mathcal{M}_\ll\left(\widetilde{E}(\alpha)\right)$ by $\Theta(\mu) = \widetilde{\mu}_\ll + \mu_\perp(X)\widetilde{\nu}$, where $(\mu_\ll, \mu_\perp)$ is the Lebesgue decomposition of μ with respect to ${}^*P_\alpha = {}^*Q_\alpha$, and $\mu_\ll \ll {}^*P_\alpha$. Define $\rho : {}^*\mathcal{M}_\ll(E) \times {}^*\mathcal{L}^b(Y) \longrightarrow {}^*\mathbb{R}$ by

$$\rho(\mu, f) = {}^*\int f(y) \left[\frac{d\Theta(\mu)}{d\widetilde{Q}_\alpha} \circ {}^*\left(\frac{dQ_t}{dQ_\alpha}\right)_\alpha\right] d^*Q_\alpha(y)$$

Note if $\mu \in \mathcal{M}_\ll(E)$ and $f \in \mathcal{L}^b(Y)$ then

$$\begin{aligned}
|\rho({}^*\mu,{}^*f)| &\le \|f\| \left| {}^*\!\int \frac{d\Theta(\mu)}{d\widetilde{Q}_\alpha} \circ {}^*\!\left(\frac{dQ_t}{dQ\alpha}\right)_\alpha d^*Q_\alpha \right| = \|f\|\,|\Theta({}^*\mu)({}^*\mathcal{S}^\alpha)| \\
&= \|f\|\,|{}^*\mu_\ll(X) + {}^*\mu_\perp(X)\nu(X)| \\
&= \|f\|\ |{}^*\mu(X)| \\
&\le \|f\|\|\mu\|
\end{aligned}$$

it follows that $\eta = {}^\circ\rho$ is a well-defined function from $\mathcal{M}_\ll(E) \times \mathcal{L}^b(Y)$ to $\mathbb{R}$ with $|\eta(\mu, f)| \le \|\mu\|\|f\|$. Furthermore,

$$\begin{aligned}
\eta(\mu, 1) &\approx \rho({}^*\mu, {}^*1) \\
&= \Theta({}^*\mu)({}^*\mathcal{S}^\alpha) \\
&= {}^*\mu_\ll(X) + {}^*\mu_\perp(X)\nu(X) \\
&= \mu(X),
\end{aligned}$$

and

$$\begin{aligned}
\eta(P_t, f) &\approx {}^*\!\int ({}^*f) \left[\frac{d\widetilde{P}_t}{d\widetilde{Q}_\alpha} \circ {}^*\!\left(\frac{dQ_t}{dQ_\alpha}\right)_\alpha\right] d^*Q_\alpha \\
&= {}^*\!\int ({}^*f) \left[\frac{d\widetilde{P}_t}{d\widetilde{P}_\alpha} \circ {}^*\!\left(\frac{dQ_t}{dQ_\alpha}\right)_\alpha\right] d^*Q_\alpha \\
&= {}^*\!\int ({}^*f)^*\!\left(\frac{dQ_t}{dQ_\alpha}\right) d^*Q_\alpha \\
&= {}^*\!\left(\int f dQ_t\right).
\end{aligned}$$

(where the penultimate equality follows from the observation that $\dfrac{d\widetilde{P}_t}{d\widetilde{P}_\alpha}$ is projection to the tth coordinate.)

Similarly, it is easy to verify that η is bilinear and that $\eta(\mu, f) \ge 0$ if $\mu, f \ge 0$; this proves the theorem. ∎

6 Discussion

As mentioned in the introduction, some of the results unproved in this note admit interesting nonstandard proofs; these and other results will appear in [R]. For example, $E \sim F$ if and only if $E(\alpha) \sim F(\alpha)$ for every finite $\alpha \in T$. This in turn implies Proposition 2.1, and also implies that $\mathcal{E}(T)/\sim$ is weakly Hausdorff (so that in Theorem 4.1, $\widehat{E}/\sim$ is *the* weak standard part of E).

There are also questions which arise in the nonstandard theory with no standard counterparts; for example:

(I) Is every standard experiment equivalent to its nonstandard hull? (Answer: Yes.)

(II) If two standard experiments are randomization equivalent (i.e., using generalized transition kernels), are their nonstandard hulls randomization equivalent using non-generalized transition kernels? (Answer: Yes.)

(III) If two nonstandard hulls of experiments are randomization equivalent, are they randomization equivalent using non-generalized transition kernels? (Answer: Unknown.)

References

[C] Cutland, Nigel (ed.): *Nonstandard Analysis and its Applications,* Cambridge University Press, Cambridge, England, 1988

[L] Le Cam, Lucien: *Asymptotics in Statistics,* Springer-Verlag, New York, 1990

[S] Strasser, Helmut: *Mathematical Theory of Statistics,* de Gruyter, Berlin, 1985

[R] Ross, David A.: *The Nonstandard Hull of an Experiment*, in preparation

Part III:

MATHEMATICAL PHYSICS

INFINITE RANGE FORCES AND STRONG L^1-ASYMPTOTICS FOR THE SPACE-HOMOGENEOUS BOLTZMANN EQUATION

L. Arkeryd
Dept. of Mathematics, Chalmers University of Goteborg,
Sven Gultius Gata 6, S–41296 Goteborg, Sweden

Abstract

The usefulness of NSA in kinetic theory both as a technical and conceptual tool is by now well documented. In the present paper we summarize the picture, concerning NSA methods to prove strong L^1-convergence with time to Maxwellian limits for standard solutions of the Boltzmann equation starting far from equilibrium. As a new result we prove the strong L^1 convergence for the space homogeneous BE *without* angular cutoff.

1 Introduction

This paper discusses the Boltzmann equation without exterior forces

$$(\partial_t + v\partial_x)f = Q(f, f)$$

with collision operator

$$Q(f, f) = \int_{\mathbb{R}^3 \times S^2} (f_1' f_2' - f_1 f_2) B(v_1, v_2, \theta) dn dv_2.$$

The discussion makes sense in any $\mathbb{R}^d$, $d \geq 2$, but we stick to the case $d = 3$ for convenience. Then the space and velocity arguments of f_1', f_2', f_1, f_2 are $(x, v_1'), (x, v_2')$, $(x, v_1), (x, v_2)$ with $v_1' = v_1 + n|v_1 - v_2|\cos\theta$, $v_2' = v_2 - n|v_1 - v_2|\cos\theta$, and n is a unit vector in S^2; in angular coordinates

$$\{(\theta, \zeta); 0 \leq \theta \leq \pi/2,\ 0 \leq \zeta \leq 2\pi\},\ dn = \sin\theta d\theta d\zeta.$$

The weight function B is

$$B(v_1, v_2, \theta) = b(\theta)|v_1 - v_2|^\beta, -3 < \beta \leq 1,$$

($\beta = (k-5)/(k-1)$ for inverse k:th power forces) and $0 < b(\theta)$, a measurable function on $(0, \pi/2)$.

Suppose in the nonstandard context that

$$^*e(f) := \int_{^*(\mathbb{R}^3 \times \mathbb{R}^3 \times S^2)} (f_1' f_2' - f_1 f_2) \log(f_1' f_2' / f_1 f_2)^* B^* dv_1 dv_2 dn \approx 0. \qquad (1)$$

If f is nonnegative and S-integrable on $^*R^3$, then Loeb a.e. on $ns^*(\mathbb{R}^3 \times \mathbb{R}^3 \times S^2)$

$$f_1 f_2 \approx f_1' f_2'. \qquad (2)$$

This relation provides useful information about f.

Proposition 1 *Let $q \in {}^*L^1_+(\mathbb{R}^3)$ be given with*

$$\int q(v)(1+v^2)^* dv, \quad \int q|v| \log q(v)^* dv$$

finite, and with q satisfying (2) for Loeb a.e. $(v_1, v_2, n) \in {}^*(\mathbb{R}^3 \times \mathbb{R}^3 \times S^2)$. Then either $q \approx 0$ for Loeb a.e. $v \in ns^*\mathbb{R}^3$ or $^\circ q(v) > 0$ for Loeb a.e. $v \in ns^*\mathbb{R}^3$.

For a proof see [3]. Having this result, the well-known proof that equality in (2) implies q Maxwellian, can easily be adapted to prove

Proposition 2 *Under the hypothesis of Proposition 1, there are $a, b \in \mathbb{R}_+$, and $c \in \mathbb{R}^3$, such that for Loeb a.e. $v \in ns^*\mathbb{R}^3$*

$$q(v) \approx a \exp(-b(v-c)^2).$$

For a detailed proof, see [2]. We next discuss the (standard) background of the assumption (1). Start from nonnegative functions $f, f_n (n \in \mathbb{N})$ in $L^1(\mathbb{R}^3)$ with

$$\mathrm{w}\lim_{n\to\infty} f_{n1} f_{n2} = f_1 f_2, \quad \mathrm{w}\lim_{n\to\infty} f_{n1}' f_{n2}' = f_1' f_2',$$

w lim meaning weak limit in $L^1(B_\lambda \times B_\lambda \times S^2)$ for $\lambda > 0$,

$$B_\lambda = \{v \in \mathbb{R}^3; |v| \leq \lambda\}.$$

Essentially by a convexity argument it holds for $(f_n)_1^\infty$ and f that

Proposition 3 [6] *If b is integrable then*

$$e(f) \leq \underline{\lim} e(f_n)$$

where e is defined in (1).

In particular if $\underline{\lim} e(f_n) = 0$, then it follows that

$$f_1' f_2' = f_1 f_2 \quad a.e. \quad \text{in} \quad \mathbb{R}^3 \times \mathbb{R}^3 \times S^2,$$

and so (e.g. employing Proposition 2) the standard function f is a Maxwellian,

$$f(v) = a \exp(-b(v-c)^2) \quad \text{a.e. in } \mathbb{R}^3.$$

This can be used to prove weak L^1 convergence to a Maxwellian when time tends to infinity for solutions to the Boltzmann equation in various initial(-boundary) value problems. In this paper the corresponding strong L^1 convergence by NSA techniques will be discussed. We shall consider in detail a new strong L^1 convergence result in the space homogeneous case *without* angular cutoff for inverse k:th power forces, $k > 3$, and later comment on space dependent (as well as space homogeneous) cases with angular cutoff.

2 Strong L^1 convergence to equilibrium

Consider the space homogeneous, nonlinear Boltzmann equation in the absence of exterior forces

$$D_t f(t,v) = Q(f,f)(t,v),\ t > 0,\ v \in \mathbb{R}^3$$

with Cauchy initial data

$$f(0,v) = f_0(v) \geq 0,\ v \in \mathbb{R}^3.$$

For inverse k:th power forces ($k > 1$), $b \in L^1_{loc}([0,\pi/2)]$ but

$$b(\theta) \sim |\pi/2 - \theta|^{(k+1)/(k-1)}$$

for θ near $\pi/2$. In this case the integral

$$\int_0^{\pi/2} b(\theta) d\theta$$

does not converge. As a consequence, the two terms in the collision operator Q do not converge separately. But we can use a suitable version of the weak form of the collision term;

$$\begin{aligned}\int_{\mathbb{R}^3} \varphi(t,v) f(t,v) dv &= \int_{\mathbb{R}^3} \varphi(0,v) f_0(v) dv + \int_0^t \int_{\mathbb{R}^3} \partial_t \varphi(s,v) f(s,v) dv ds + \\ &+ \int_0^t \int_{\mathbb{R}^3 \times \mathbb{R}^3 \times S^2} (\varphi(s,v_1') - \qquad\qquad (3) \\ &- \varphi(s,v_1)) f(s,v_1) f(s,v_2) B(v_2,v_2,\theta) ds dv_1 dv_2 dn,\end{aligned}$$

where

$$\varphi \in C^{1,\infty} := \{g \in C^1([0,\infty) \times \mathbb{R}^3);\ |g|_1 = \sup|g| + \sup|\partial_t g| + \sup|\partial_v g| < \infty\}.$$

For $k > 3$, $f \in L^\infty(\mathbb{R}_+; L^1(\mathbb{R}^3))$, the collision integral converges. The presently known solutions of (3) for $k > 3$ are weak L^1 limits of solutions f_n with $b(\theta)$ truncated near $\theta = \pi/2$, say $b_n(\theta) = b(\theta) \wedge n$, when

$$\int f_0(v)(1 + v^2 + |\log f_0|)dv < \infty. \tag{4}$$

The integral (4) of $f_n(t)$ is then bounded, independently of n and t, together with

$$\int_0^\infty e(f_n(t))dt < C < \infty,\ n \in \mathbb{N}. \tag{5}$$

The weak L^1 compactness ensuing from (4) can be used to construct ([1]) solutions $f = \mathrm{w}\lim f_{n'}$, where $(f_{n'})$ is a converging subsequence of $(f_n)_1^\infty$. Moreover, f satisfies (4) with a uniform bound for $t \in \mathbb{R}_+$ and using Proposition 3, also (5). For such solutions f, strong L^1 convergence to Maxwellians holds, when time tends to infinity.

Theorem 4 *Given any sequence $(t_k)_{\mathbb{N}}$, $t_k \nearrow \infty$, there is a subsequence $(t_{k'})$, and a global Maxwellian*

$$M(v) = a\exp(-b(v-c)^2),$$

such that for $T > 0$, strongly in $L^1([0,T] \times \mathbb{R}^3)$,

$$f(\cdot + t_{k'}) \to M.$$

Proof It follows from the uniformly in t bound on (4) for $f(t)$ that, given any sequence (t_k) of times increasing to infinity, and from the ensuing weak L^1 compactness of $(f(.+t_k))_{k\in\mathbb{N}}$, that there is a subsequence $(t_{k'})$ such that $f(.+t_{k'}) \rightharpoonup g(.)$, weakly in $L^1(\mathbb{R}^3 \times [0,T])$ for $T > 0$. The proof will demonstrate that g equals a global Maxwellian, and discuss the strong L^1 convergence. The first part of the proof is to show that g equals a time dependent local Maxwellian

$$M(v,t) = a(t)\exp(-b(t)(v-c(t))^2).$$

For this, notice that there is a countable sequence $\phi_1, \phi_2, \dots$ of functions with bounded support in $\mathbb{R}^3 \times \mathbb{R}_+$, such that $g = M$ in L^1, if

$$\int g\phi_j dvdt = \int M\phi_j dvdt,\ j \in N.$$

Also let the sequence contain $\chi_{\nu\rho}$, $v\chi_{\nu\rho}$, $v^2\chi_{\nu\rho}$, $(\nu,\rho \in N)$ where $\chi_{\nu\rho}(v,t) = 1$ for $v^2 \le \nu^2$, $t \le \rho$, $\chi_{\nu\rho}(v,t) = 0$ otherwise. Set $M_k = \{\phi_1, \dots, \phi_k\}$.

The subsequence $(t_{k'})$ can be so chosen that $f(t+t_{k'})$ converges weakly in $L^1(\mathbb{R}^3)$ for rational $t > 0$. Given $\varphi \in C^{1,\infty}$, (3) implies that the family $\int_{\mathbb{R}^3} f(t+t_{k'})\varphi dv$ is uniformly equicontinuous in t, and so $f(t+t_{k'}) \rightharpoonup g(t)$ weakly in $L^1(\mathbb{R}^3)$ for all $t > 0$. Using (3) for $f(t+t_{k'})$ it follows that g satisfies (3) for $t > 0$, and $\int g(t)\varphi dv$ is uniformly continuous in t. Moreover, for some subsequence of $(t_{k'})$, which will from here on be denoted (t_k),

$$\int_0^k ds|\int_{\mathbb{R}^3} f(.+t_k)\phi - g(.)\phi dv| < 1/k,\ \phi \in M_k. \tag{6}$$

We also assume that (t_k) was so chosen that

$$\int_{t_k}^{t_k+k} ds \int ((f_1f_2)' - (f_1f_2))\log((f_1f_2)'/(f_1f_2))Bdv_1dv_2dn < 1/k, \tag{7}$$

which is possible by (5).

A large part of the proof from here on relies on NSA. By transfer, in the nonstandard context, (6) and (7) hold for all $k \in {}^*N$. Given $\kappa \in {}^*\mathbb{N}_\infty$, (6,7) implies in particular for $k = \kappa$ that

$$\int_0^\kappa ds|\int_{\mathbb{R}^3}(f(t_\kappa+.)^*\phi - {}^*g^*\phi)dv| \approx 0, \quad \phi \in M_\kappa, \tag{6'}$$

$$\int_{t_\kappa}^{t_\kappa+\kappa} ds \int ((f_1f_2)' - (f_1f_2))\log((f_1f_2)'/(f_1f_2)Bdv_1dv_2dn \approx 0. \tag{7'}$$

From (7') it follows that the integrand in the left hand side is infinitesimal; for Loeb a.e. $t \in ns^*\mathbb{R}_+$

$$f(v_1,t+t_\kappa)f(v_2,t+t_\kappa) \approx f(v_1',t+t_\kappa)f(v_2',t+t_\kappa) \tag{8}$$

for Loeb a.e. $(v_1,v_2,n) \in ns^*\mathbb{R}^3 \times \mathbb{R}^3 \times S^2$. Using Proposition 1-2, we may conclude from (8) that for Loeb a.e. $t \in ns^*\mathbb{R}_+$

$$f(v,t+t_\kappa) \approx \bar{M}(v,t) \quad \text{for Loeb a.e. } v \in ns^*\mathbb{R}^3,$$

with $\bar{M}$ a local Maxwellian,

$$\bar{M}(v,t) = \bar{a}(t)\exp(-\bar{b}(t)(v-\bar{c}(t))^2).$$

This result together with (6') gives for $\phi \in \cup_{k\in N}M_k(\subset M_\kappa)$ and $T \in \mathbb{R}_+$

$$\begin{aligned} 0 &= {}^o\int_{[0,T]} {}^*dt|\int_{*\mathbb{R}^3} f(v,t+t_\kappa)*\phi*dv - {}^*\int_{\mathbb{R}^3} g\phi dv| = \\ &= \int_{*[0,T]} Ldt|\int_{ns^*\mathbb{R}^3} {}^of(v,t+t_\kappa)^{o*}\phi Ldv - {}^{o*}\int_{\mathbb{R}^3} g\phi dv| = \\ &= \int Ldt|\int {}^o\bar{M}(v,t)^{o*}\phi Ldv - {}^{o*}\int g\phi dv|. \end{aligned} \tag{9}$$

Taking $\phi = \chi_{\nu\rho}1,\ \chi_{\nu\rho}v,\ \chi_{\nu\rho}v^2$ and letting ν tend to infinity, we get

$$\begin{aligned} \bar{a}(t) &= {}^{o*}a(t), \bar{b}(t) = {}^{o*}b(t), \\ \bar{c} &= {}^{o*}c(t) \quad \text{Loeb a.e. } t \in ns^*\mathbb{R}_+, \end{aligned}$$

where a, b, c are the Lebesgue measurable functions on $\mathbb{R}_+$. It follows that ${}^o\bar{M} = {}^{o*}M$ with $M = a\exp(-b(v-c)^2)$, i.e., for Loeb a.e. $t \in ns^*\mathbb{R}_+$, $\bar{M}$ is infinitesimally close to a standard local Maxwellian.

In particular, for $\phi \in \cup_{k\in N} M_k$

$$
\begin{aligned}
|\int_{\mathbb{R}^3\times[0,T]} (g\phi - M\phi)dvdt| &= |\int_{ns^*\mathbb{R}^3\times^*[0,T]} ({}^{o*}g^{o*}\phi - {}^{o*}M^{o*}\phi)Ldvdt| = \\
&= |\int_{^*[0,T]} Ldt \int_{ns^*R^3} Ldv({}^{o*}g^{o*}\phi - {}^o\bar{M}^{o*}\phi)| \leq \\
&\leq \int_{^*[0,T]} Ldt|\int_{ns^*\mathbb{R}^3} Ldv({}^{o*}g^{o*}\phi - {}^o\bar{M}^{o*}\phi| = 0,
\end{aligned}
$$

where the last equality is (9). It follows that $g = M$, i.e. g is a timedependent Maxwellian.

We have thus proved

$$f(v, t_\kappa + t) \approx \bar{M}(v,t) \approx M(v,t) = g(v,t)$$

for oeb a.e. $(v,t) \in ns^*\mathbb{R}^3 \times ns^*\mathbb{R}_+$. This implies for $T \in ns^*\mathbb{R}_+$, that

$$\int_{\mathbb{R}^3\times[0,T]} {}^*\chi_\nu|^*f(t+t_\kappa) - {}^*M|^*dvdt \approx 0.$$

From here it follows in the standard context that

$$\lim_{k\to\infty} \int_{\mathbb{R}^3\times[0,T]} |f(t+t_k) - M|dvdt = 0, \quad T \in \mathbb{R}_+.$$

Finally, for $f = M$ the collision term vanishes in (3), so M is independent of t. ∎

Remark

1) No standard proof of this standard result is presently known.

2) If besides (4), $|v|^s f_0 \in L^1(\mathbb{R}^3)$ for some $s > 2$, then in the case of hard and Maxwellian forces, i.e. $k \geq 5$, the s-moment is globally in time bounded in L^1, [7]. It follows that the energy is conserved, which makes the limit M unique, hence $f(t,\cdot) \to M(\cdot)$, strongly in $L^1(\mathbb{R}^3)$ as $t \to \infty$.

3) The above result holds in particular in the case of angular cutoff, where the first proof was the nonstandard one in [2]. In this case a standard proof [8] has later followed.

Using the averaging technique instead of the above equicontinuity arguments, an analogous NSA proof of strong L^1 convergence can be obtained in various space dependent cases with angular cutoff. Start, e.g., from the setup in the introduction of this paper for inverse k:th power forces with $k > 2$, angular cutoff, and the space variable x in a torus $\mathbb{R}^3/Z^3$ (thus avoiding boundary problems). Consider a DiPerna Lions solution f of the problem with initial value f_0 satisfying

$$\int f_0(1 + v^2 + |\log f_0|)dxdv < \infty. \tag{4'}$$

Theorem 5 [4] *Given any sequence (t_k), $t_k \nearrow \infty$, there is a subsequence $(t_{k'})$ and a global Maxwellian*

$$M(v) = a\exp(-b(v-c)^2)$$

such that for $T > 0$,

$$f(\cdot + t_{k'}) \to M$$

strongly in $L^1(\mathbb{R}^3/Z^3 \times \mathbb{R}^3 \times [0,T])$, and for $t > 0$,

$$f(\cdot, t + t_{k'}) \to M$$

strongly in $L^1(\mathbb{R}^3/Z^3 \times \mathbb{R}^3)$.

An analogous result holds under specular or reverse reflexion for a large class of boundaries.

References

[1] Arkeryd, L., *Intermolecular forces of infinite range and the Boltzmann equation, Arch. Rat. Mechs. Anal.* 77 (1981), 11-21.

[2] Arkeryd, L., *On the Boltzmann equation in unbounded space far from equilibrium, Comm. Math. Phys.* 105 (1986), 205-219.

[3] Arkeryd, L., *The nonlinear Boltzmann equation far from equilibrium,* in *Nonstandard Analysis and its applications*, Cambr. UP 1988.

[4] Arkeryd, L., *On the strong L^1 trend to equilibrium for the Boltzmann equation, Stud. Appl. Math.* 87 (1992), 283-288.

[5] DiPerna, R.J., Lions, P.L., *On the Cauchy problem for Boltzmann equations: global existence and weak stability,* Ann. Math. 130 (1989), 321-366.

[6] DiPerna, R. J., Lions, P. L., *Global solutions of Boltzmann's equation and the entropy inequality,* Arch. Rat. Mechs. Anal. 114 (1991), 47-55.

[7] Elmroth, T., *Global boundedness of moments of solutions of the Boltzmann equation for forces of infinite range,* Arch. Rat. Mechs. Anal. 82 (1983), 1-12.

[8] Gustafsson, T., *Global L^p properties for the spatially homogeneous Boltzmann eqution,* Arch. Rat. Mechs. Anal. 103 (1988), 1-38.

A Nonstandard Analysis Approach to the Theory of Quantum Meanfield Systems †

Manfred P. H. Wolff
Mathematisches Institut, Universität Tübingen
Auf der Morgenstelle 10, D–72076 Tübingen

Abstract

In this article we develop an approach to the theory of meanfield systems in which the relevant algebras of observables are quotients of subalgebras of a hyperfinite tensor product of the one–particle algebra under consideration. So this aproach corresponds exactly to the intuitive picture of meanfield systems as many–particle systems, in particular it underlines the statistical character of meanfield phenomena. As practical applications we bring the treatment of general meanfield dynamics in this paper whereas the fluctuation phenomena are dealt with in **[WH]** on the basis of this approach.

Introduction

About 20 years ago K. Hepp and E.H. Lieb laid the foundations for a mathematically rigorous treatment of the dynamics of meanfield systems, including already fluctuation phenomena (see **[HL]**). However the algebras of intensive observables and of fluctuation observables were not constructed explicitly in that paper. The problem concerning the intensive observables was attacked later by different research groups as we will quickly review now:

To this end, let $\mathcal{A}$ be a unital C^*–algebra, the one – particle algebra, and let $\mathcal{A}_n$ denote the n–fold injective tensorproduct $\mathcal{A}^{\otimes n}$ (cf **[TA]**, p. 207 and **[BR1]**, p. 144 where it is called the C^*–tensorproduct). Consider the canonical embeddings $j_{n+p,n} : \mathcal{A}_n \to \mathcal{A}_{n+p}$ ($a \to a \otimes \mathbf{1}_p$, where $\mathbf{1}_p = 1 \otimes \cdots \otimes 1$ p times). Then the C^*–inductive limit $\underrightarrow{\lim}(\mathcal{A}_n, j_{n+p,n}) =: \mathcal{A}_\infty$ is called the quasilocal algebra of the system. The group around A. Rieckers (see e.g. **[DR]**, **[RI]**, **[UN]**) as well as P. Bóna **[BO1]** and others extended this algebra, roughly speaking, by selecting first of all the folium F in the state space $\mathcal{S}(\mathcal{A}_\infty)$ generated by the permutation invariant states and then taking the quotient $\mathcal{A}''_\infty/J$ of the second dual $\mathcal{A}''_\infty$ by the annihilator

†This reseach is part of the project Operator Algebraic Quantum Optics supported by the Deutsche Forschungsgemeinschaft

ideal J of F. In this way the intensive observables appear as limits (in an appropriate weak topology) of means of observables at different sites of the lattice $\mathbb{N}$, say. This approach was generalized further by Morchio and Strocchi **[MS]**.

Another construction of an appropriate extension of $\mathcal{A}_\infty$ was given in **[RW]** and was used to develop a general theory of meanfield dynamics in **[DW1]**, **[DW2]**. In **[RW]** G. Raggio and R.F. Werner used a certain kind of discrete approximation. More precisely they introduced the space of approximate symmetric sequences. This approach was generalized to the space of approximate symmetric sequences with respect to so–called tagged nets. This generalization appeared necessary in order to consider local and intensive observables at the same time, an effort which is not necessary within the first - mentioned approach.

In this paper we want to propose a third approach which seems to us much easier than these other two — *taken knowledge of nonstandard analysis for granted.* Moreover our approach allows also to deal with fluctuation phenomena which is not possible in the second approach and which seems to be difficult in the first one. Finally our approach seems to be well–suited to discuss the connection between classical and quantum mechanics, as it is already indicated in **[Bo1]**, **[DW2]**, and worked out in **[BO2]**, cf. also **[DW2]**.

Even if $\mathcal{A}$ is commutative the results on the meanfield dynamics are of some interest: if the local dynamics is dissipative the corresponding dynamics on the algebra of all intensive observables is still deterministic (see 4.8), (2)). This effect was already noticed by Th. Unnerstall **[UN]** without any reference to the classical situation. The close connection between classical and quantum probability theory appears quite natural within our framework: if the one–particle algebra $\mathcal{A}$ is commutative, fluctuations show classical phenomena, if it is noncommutative and simple then fluctuations show also quantum mechanical phenomena (see **[WH]**).

The paper is organized as follows: the first section contains basic notions and results, the second one is devoted to the reconstruction of the algebra of all local and intensive observables. The third section contains basic definitions concerning meanfield dynamics, whereas in the fourth section we treat an important special case: the polynomial bounded generators. The final section is devoted to some generalizations. There we discuss also the connection between quantum mechanics and classical mechanics. The treatment of fluctuations is to be found in **[WH]**.

Acknowledgements: I would like to thank Prof. Albeverio, Bochum for his encouragement, Prof. A. Rieckers, Tübingen, from whom I learned meanfield theory, for many valuable suggestions, and Prof. W. Henson, Urbana, and Dipl.–Phys. Th. Gerisch for carefully reading the manuscript.

1 Basic notions and results

We adhere to the notions and notations of the introduction.

1.1) Let $\mathcal{V} = \mathcal{V}(X)$ be the superstructure over an arbitrary set X such that the one–particle algebra $\mathcal{A}$ as well as $\mathbb{C}$ are elements of some $V_r(X)$. (For notions from nonstandard analysis not mentioned here we refer to **[AFHL]**, **[SL]**, **[HULO]**, **[HM]**). Viewing the n–fold algebraic tensor product $\mathcal{A}^n$ of $\mathcal{A}$ as the subspace of all n–linear mappings on $\mathcal{A}'^{\mathbb{N}}$
($\mathcal{A}'$: the dual of $\mathcal{A}$) depending only on the first n coordinates we can easily see that all $\mathcal{A}^n$ as well as their completions $\mathcal{A}_n$ (with respect to the injective tensor norm) are elements of some $V_s(X)$ for a fixed $s > r$.

1.2) In the following let ${}^*\mathcal{V}$ be a polysaturated enlargement of $\mathcal{V}$ and let N be an arbitrary infinitely large positive integer. Then $\mathcal{A}^{\otimes N} =: \mathcal{A}_N$ is a well–defined element in ${}^*V_s(X)$. As usual we build $\mathcal{A}_{N,fin} = \{a \in \mathcal{A}_N : \|a\| \text{ is limited }\}$, $\mathcal{A}_{N,0} = \{a \in \mathcal{A}_N : \|a\| \approx 0\}$, and $\hat{\mathcal{A}}_N = \mathcal{A}_{N,fin}/\mathcal{A}_{N,0}$. Equipped with the norm $\|\hat{a}\| = {}^\circ\|a\|$ (" $\circ$ " : the standard part map in ${}^*\mathbb{C}_{fin}$) $\hat{\mathcal{A}}_N$ is a unital C^*–algebra.[3] We denote the quotient mapping $\mathcal{A}_{N,fin} \ni a \to \hat{a} \in \hat{\mathcal{A}}_N$ by Q.

1.3) As in the introduction, we consider the embeddings $j_{r,n} : \mathcal{A}_n \to \mathcal{A}_r \quad (r > n)$. In this way we may identify ${}^*\mathcal{A}_n$ for n standard with $j_{N,n}({}^*\mathcal{A}_n)$ whenever this is convenient, e.g. in order to define the algebra $\mathcal{A}_{(0)}$ of local observables as $\mathcal{A}_{(0)} = \bigcup_{n\in\mathbb{N}} j_{N,n}({}^\sigma\mathcal{A}_n)$, where as usual ${}^\sigma E = \{a \in {}^*E : a = {}^*b \text{ for some } b \in E\}$. Then $J_n := Qj_{N,n}|{}^\sigma\mathcal{A}_n$ embeds $\mathcal{A}_n$ into $\hat{\mathcal{A}}_N$ and satisfies $J_r = J_n \circ j_{n,r} \quad (r \leq n)$.

Then $Q|_{\mathcal{A}(0)}$ maps $\mathcal{A}_{(0)}$ isometrically onto $\bigcup_n J_n({}^\sigma\mathcal{A}_n)$, so we may identify this space with $\mathcal{A}_{(0)}$. Its closure is obviously isomorphic to $\mathcal{A}_\infty$. $\mathcal{A}_\infty$ is the infinite C^*–algebraic tensorproduct in the sense of **[BR1]**, p. 144.

2 The algebra of intensive observables

2.1) Let $\mathcal{S}(N)$ be the group of (internal) permutations of $\{1,\ldots,N\}$. Then for each $\pi \in \mathcal{S}(N)$ by $U_\pi(a_1 \otimes \cdots \otimes a_N) = a_{\pi^{-1}(1)} \otimes \cdot \otimes a_{\pi^{-1}(N)}$ there is uniquely determined a C^*–automorphism on $\mathcal{A}_N$ (use the transfer principle). Moreover $\pi \to U_\pi$ is a representation of $\mathcal{S}(N)$. So $P = \frac{1}{N!}\sum_{\pi\in\mathcal{S}(N)} U_\pi$ is a conditional expectation of $\mathcal{A}_N$ onto the (internal) C^*–algebra of all fixed points of this representation.

Externally $\mathcal{S}(N)$ is very large. Nevertheless $\pi \to \hat{U}_\pi$ is a representation and $\hat{P}$ is a conditional expectation onto the algebra of all fixed points of this representation. (Here the operator $\hat{T}$ is defined by $\hat{T}(\hat{x}) = \widehat{(Tx)}$.)

2.2) Proposition: *P has the following properties.*
(a) *For* $x \in {}^*\mathcal{A}_{fin}$ *let* $x(k) = \mathbf{1}_{k-1} \otimes x \otimes 1_{N-k}$.
Then

$$P(j_{N,1}(x)) = P(x(1)) = \frac{1}{N}\sum_{k=1}^{N} x(k) =: \bar{x}.$$

$\bar{x}$ *is limited.*

[3] This is obvious though we could not find an explicit reference for it.

(b) $P(j_{N,n}(x_1 \otimes \cdots \otimes x_n)) \approx \bar{x}_1 \cdots \bar{x}_n$ *for all standard* n *and all* $x_1, \ldots, x_n \in {}^*\mathcal{A}_{fin}$
(c) $[\bar{x}, \bar{y}] \approx 0$ *for all* $x, y \in {}^*\mathcal{A}_{fin}$. *Here* $[a, b] = ab - ba$ *denotes the commutator of* a *and* b.
(d) $[\bar{x}, j_{N,n}(a)] \approx 0$ *for all* $a \in {}^*\mathcal{A}_{n,fin}$ *(where* n *is standard) and for all* $x \in {}^*\mathcal{A}_{fin}$.

Proof: (a) is obvious.
(b) For $\vec{k} = (k_1, \ldots k_n) \in \{1, \ldots, n\}^N$ set $x(\vec{k}) = x_1(k_1) \cdot \cdot x_n(k_n)$. Moreover for $j = 2, \ldots, n$ let $\sum(j)$ denote the sum over all such $\vec{k}$ for which exactly j components are equal. Then

$$\bar{x}_1 \cdot \cdot \bar{x}_n = \frac{1}{N^n} \sum(0) x(\vec{k}) + \frac{1}{N^n} \sum_{j=2}^{n}(j) x(\vec{k}).$$

Now for each $j \geq 2$ the sum $\sum(j)$ has exactly $r(j) := N \cdot \binom{n}{j} \cdot (n-j)! \binom{N-1}{n-j}$ summands of norm less than $b = max_{\{j\}}(\|x_j\| + 1))^n$ which is limited by hypotheses. Since $\frac{r(j)}{N^n} \approx 0$ for $j \geq 2$ we obtain

$$\bar{x}_1 \cdot\cdot \bar{x}_n \approx \frac{1}{N^n} \sum(0) x(\vec{k}) = \frac{1}{N(N-1) \cdot \cdot (N-n+1)} \cdot \sum(0) x(\vec{k})(1 - \frac{1}{N}) \cdot\cdot (1 - \frac{n-1}{N}).$$

But

$$P(j_{N,n}(x_1 \otimes \cdots \otimes x_n)) = \frac{1}{N!} \sum_{\pi}(x(\vec{k})) = \frac{(N-n)!}{N!} \sum(0) x(\vec{k});$$

clearly this element is limited. So
$P(j_{N,n}(x_1 \otimes \cdots \otimes x_n)) - \bar{x}_1 \cdots \bar{x}_n \approx (1 - \Pi_{j=1}^n (1 - \frac{j}{N})) \cdot P(j_{N,n}(x_1 \otimes \cdots \otimes x_n)) \approx 0.$

(c) $[\bar{x}, \bar{y}] = \frac{1}{N^2} \sum_{k=1}^N [x, y](k) \approx 0.$

(d) $[\bar{x}, j_{N,n}(a)] = \frac{1}{N} \sum_1^n [x(k), a \otimes \mathbf{1}_{N-n}] \approx 0.$ ∎

2.3) Corollary: *The closure* $\mathcal{N}$ *of* $Q(P(\mathcal{A}_{(0)}))$ *in* $\hat{\mathcal{A}}_N$ *is a unital commutative* C^**–algebra which commutes with all quasilocal observables* $a \in \mathcal{A}_\infty$.

2.4) Definition: $\mathcal{N}$ is called the algebra of all intensive observables.

We can view $\bar{x}$ to be the strong limit of N identical, mutually independent observables. By the symmetrization procedure all local effects disappear. Obviously $\mathcal{N}$ is generated by $\{Q\bar{x} : x \in \mathcal{A} \text{ (standard) }\}$.

Let $X = \mathcal{S}(\mathcal{A})$ be the state space of $\mathcal{A}$, $X = \{\rho \in \mathcal{A}' : \rho(1) = \|\rho\| = 1\}$. It is a ω^*–compact set. For $\rho \in X$ let $\rho' = \rho^{\otimes N}$ denote the N–fold tensor product of ρ. Then $P'\rho' = \rho'$.

By $\hat{\rho}(\hat{a}) = {}^\circ(\rho'(a))$ there is defined a state on $\hat{\mathcal{A}}_N$. By 2.2 (b) we obtain
$\rho'(\bar{x}\bar{y}) \approx \rho'(P(j_{N,2}(x \otimes y)) = \rho(x)\rho(y)$.

This implies that $\hat{\rho}|_{\mathcal{N}}$ is multiplicative which proves the first part of the following theorem, due to Størmer **[ST]**.

2.5) Theorem: $\mathcal{N}$ *is* C^**–isomorphic to the* C^**–algebra* $C(\mathcal{S}(\mathcal{A}))$ *of all continuous functions on the state space of* $\mathcal{A}$.

Proof: Denote by f_x the continuous affine function $\rho \to \rho(x)$ on $\mathcal{S}(\mathcal{A})$. For $r \in \mathbb{N}$ (standard) and $x_1, \dots x_r \in \mathcal{A}$ set $S_0Q(P(x_1 \otimes \cdot\cdot \otimes x_r \otimes 1_{N-r})) = f_{x_1} \cdot\cdot f_{x_r}$.

S_0 is well-defined and uniquely linearly extendable to a map $S : Q(P(\mathcal{A}_{(0)})) \to C(\mathcal{S}(\mathcal{A}))$ which is obviously multiplicative and contractive, satisfying $S(R^*) = S(R)^*$ for all polynomials R. Hence S prossesses a unique extension to an equally denoted C^*–homomorphism of $\mathcal{N}$ into $C(\mathcal{S}(\mathcal{A}))$. But $\{f_x : x \in \mathcal{A}_{sa}\}$ separates points on $\mathcal{S}(\mathcal{A}))$. By Stone – Weierstrass's theorem S is onto.

Finally $x \to x(1) \to \bar{x} = P(x(1)) \to Q(\bar{x})$ is norm–continuous from $\mathcal{A}$ into $\mathcal{N}$. Moreover it is linear. So every point ξ in the Gelfand space K, say, of $\mathcal{N}$ defines a $\rho \in \mathcal{S}(\mathcal{A})$ by $\rho(x) = Q(\bar{x})(\xi)$. This shows that $\rho \to \hat{\rho}$ is surjective hence S is injective. ∎

2.6) From now on we identify $Q(\bar{x})$ and f_x for all (standard) $x \in \mathcal{A}$. We want to show that the C^*–algebra generated by $\mathcal{A}_\infty$ and $\mathcal{N}$ in $\hat{\mathcal{A}}_N$ is isomorphic to $\mathcal{A}_\infty \otimes \mathcal{N}$.

To this end we denote by $P_K(K \in {}^*\mathbb{N})$ the conditional expectation corresponding to the action of the permutation group $\mathcal{S}(K)$ on $\mathcal{A}_K$.

2.6.1) Lemma 1: *Let $x \in {}^*\mathcal{A}_{fin}$ be arbitrary, and choose $n \in \mathbb{N}$ standard. Then $P_N(j_{N,1}(x)) \approx \mathbf{1}_n \otimes P_{N-n}(j_{N-n,1}(x))$.*

Proof: The lefthand side u is equal to $\frac{1}{N}\sum_{k=1}^N x(k)$, the righthand side v equals $\frac{1}{N-n}\sum_{k=n+1}^N x(k)$. Hence

$$u - v = \frac{1}{N}\sum_{k=1}^{n} x(k) + \frac{n}{N(N-n)}\sum_{k=n+1}^{N} x(k).$$

Since n is standard $u - v \approx 0$. ∎

2.6.2) Corollary: *Let $x_1, \dots, x_r \in {}^*\mathcal{A}_{fin}$ be arbitrary, where $r \in \mathbb{N}$ is standard. Moreover let n be a standard positive integer. For an arbitrary x in $\mathcal{A}$ set*

$$\bar{x} = \frac{1}{N}\sum_{k=1}^{N} x(k), \quad \tilde{x} = \frac{1}{N-n}\sum_{k=1}^{N-n} x(k).$$

Then $\bar{x}_1 \cdots \bar{x}_r \approx \mathbf{1}_n \otimes \tilde{x}_1 \cdots \tilde{x}_r$

Proof: Since r and n are standard $\bar{x}_1 \cdot\cdot \bar{x}_r \approx \Pi_{l=1}^r(\mathbf{1}_n \otimes \tilde{x}_l)$ by the lemma. ∎

In the following let $n \in N$ be standard. For $y = \bar{x}_1 \cdot\cdot \bar{x}_r$ in $\mathcal{A}_n$ we set $\tilde{y} = \tilde{x}_1 \cdot\cdot \tilde{x}_r$ in $\mathcal{A}_{N-n}$.

2.6.3) Lemma 2: *Let $a_1, \dots a_p \in {}^*\mathcal{A}_{n,fin}$; let moreover $y_l \in \mathcal{A}_N$ be of the form $y_l = \Pi_{j=1}^{q_l}\bar{x}_{j,l}$, where $x_{j,l} \in {}^*\mathcal{A}_{fin}$.*
Then $\sum_1^p a_l y_l \approx \sum_1^p a_l \otimes \tilde{y}_l$

Proof: Since p is standard the assertion follows directly from 2.6.2. ∎

2.7) Now we are able to prove the following theorem. Its importance lies in the fact that we obtain a structure *which is independent of our special enlargement and our* *particular* N, but which is nevertheless something gained by a "generalized" limit process.

2.7.1) Theorem: *The* C^**-algebra generated by* $\mathcal{A}_\infty$ *and* $\mathcal{N}$ *in* $\hat{\mathcal{A}}_N$ *is* C^**-isomorphic to* $\mathcal{A}_\infty \otimes \mathcal{N} = C(\mathcal{S}(\mathcal{A}), \mathcal{A}_\infty)$.

Proof: The last equality is well-known. In order to prove the theorem let us consider the set E of elements of the form $\sum_1^p a_l h_l$, where all a_l are in $\mathcal{A}_{(0)}$, hence in some $\mathcal{A}_n$ for an appropriate n, and where $h_l = \Pi_{j=1}^q f_{x_{j,l}}$ for some $x_{j,l} \in \mathcal{A}$.

E is a norm–dense linear subspace of the C^*–algebra generated by $\mathcal{A}_\infty$ and $\mathcal{N}$ in $\hat{\mathcal{A}}_N$. Likewise the set F of elements of the form $\sum_1^p a_l \otimes h_l$ is a dense linear subspace of $\mathcal{A}_\infty \otimes \mathcal{N}$.

Choose such an element $w = \sum_1^p a_l \otimes h_l$, where a_l, n, and h_l are as above. Theorem 2.5 holds for all N as long as they are infinitely large. Thus we may assume that $f_{x_{j,l}} = \tilde{Q}(\tilde{x}_{j,l})$, where $\tilde{x}_{j,l}$ is as in 2.6.2 and $\tilde{Q} : \mathcal{A}_{N-n,fin} \to \hat{\mathcal{A}}_{N-n}$ is the corresponding quotient mapping. Then $w = Q(\sum_1^p a_l \otimes \tilde{y}_l)$ where $\tilde{y}_l = \Pi_1^{q_l} \tilde{x}_{j,l}$.

But by 2.6.3 $w = Q(\sum_1^p a_l y_l)$, where $y_l = \Pi_1^{q_l} \bar{x}_{j,l}$. Since Q is a homomorphism $w = \sum a_l h_l$. It follows that the map $U : \sum a_l \otimes h_l \to \sum a_l h_l$ is an isomorphic homomorphism of involutive algebras. So it may be extended to a C^*–isomorphism from $\mathcal{A} \otimes \mathcal{N}$ onto the C^*–algebra generated by $\mathcal{A}_\infty$ and $\mathcal{N}$ in $\hat{\mathcal{A}}_N$. ∎

2.7.2) Corollary: *There exists a conditional expectation* $\mathbf{E}$ *from* $\mathcal{A} \otimes \mathcal{N}$ *onto* $1 \otimes \mathcal{N}$ *satisfying* $\mathbf{E}(x_1 \otimes \ldots \otimes x_r) = 1 \otimes f_{x_1} \cdot\cdot f_{x_r}$ *(where* $x_1 \otimes \cdots \otimes x_r$ *is identified with its canonical image in* $\mathcal{A}_\infty$*).*

Proof: Let B denote the external involutive algebra generated by $\mathcal{A}_{(0)}$ and $P(\mathcal{A}_{(0)})$ in $\mathcal{A}_N$ and set $E_0 = QP|_B$. Then the continuous extension of this map to all of $\mathcal{A} \otimes \mathcal{N}$ is the desired $\mathbf{E}$.

3 Dynamics in meanfield systems

Since the fundamental paper of K. Hepp and E.H. Lieb ([**HL**]), research on meanfield dynamics has made substantial progress; see in particular by [**BO1**], [**RI**], [**UN**], [**DW1**], and [**DW2**].

In the following we want to give an easier proof of one of the main results of [**DW2**] (see theorem 4.6.). To this end we give first of all the relevant definitions and auxiliary results.

3.1) Definition: $B \in \mathcal{L}(\mathcal{A}_N)$ is called a meanfield operator if B has the following properties:

(1) $\mathbf{P}(B) := \frac{1}{N!} \sum_{\pi \in \mathcal{S}(N)} U_\pi^{-1} B U_\pi \approx B \quad$ (*i.e.* $\|B - \mathbf{P}(B)\| \approx 0$)
(2) $\|B\|/N$ is limited but $\not\approx 0$.
(3) For every $t \geq 0 \quad \exp(tB) = T_t$ is a completely positive unital operator.

Now denote by $X \bullet Y$ the (external) involutive subalgebra generated by X and Y.

3.2) Definition: A meanfield operator B is called a meanfield generator if
(1) $T_t(\mathcal{A}_{(0)} \bullet P(\mathcal{A}_{(0)})) \subset \overline{\mathcal{A}_{(0)} \bullet P(\mathcal{A}_{(0)})}^S$ (the S–closure with respect to the internal norm on $\mathcal{A}_N$).
(2) (T_t) is strongly S–continuous on $(\mathcal{A}_{(0)} \bullet P(\mathcal{A}_{(0)}))$.

3.3) Remarks: 1) $\mathbf{P}B \approx B$ implies $PB \approx BP$ and a fortiori $PT_t \approx T_tP$.
2) Let B be a meanfield generator. Then, by setting $\hat{T}_t\hat{u} = \widehat{(T_tu)}$ for all u in the algebraic tensor product $\mathcal{A}_{(0)} \circ \mathcal{N}$, there is uniquely defined a strongly continuous semigroup of completely positive unital operators $\hat{T}_t\,(t \geq 0)$ on $\mathcal{A}_\infty \otimes \mathcal{N}$. The generator of this semigroup is closely connected with B, as we shall see later on. This justifies our definition of a meanfield generator.

3.4) Examples: 1) Let $Q(\xi_1, \ldots, \xi_r) = \xi_1^{\alpha_1} \cdots \xi_r^{\alpha_r}$ be a monomial in r variables and let $x_1, \ldots, x_r \in \mathcal{A}_{sa}$ (standard) be arbitrary. Then $H = N \cdot P(x_1^{\otimes\alpha_1} \cdot \otimes x_r^{\otimes\alpha_r} \otimes \mathbf{1}_{N-|\alpha|})$ is selfadjoint ($|\alpha| = \sum \alpha_j$), and is denoted by $H^Q(x_1, \ldots, x_r)$. It is easy to check that by the commutator bracket $B^Q u = i[H, u]$ there is defined a meanfield operator B^Q. Now it is clear how to define B^Q for an arbitrary (standard) polynomial Q in r variables with real coefficients and for each standard r–tuple $(x_1, \ldots, x_r) \in \mathcal{A}_{sa}^r$. These operators were the starting point of the theory of meanfield systems. They are in fact generators in our sense as we will see later on.
2) Let $G \in \mathcal{L}(\mathcal{A}_g)$ (g and G standard) be such that $\exp(tG)$ is unital and completely positive for $t \geq 0$. Then $B = N\mathbf{P}(G \otimes I_{N-g})$ is a meanfield operator. This kind of operator corresponds to the polynomially bounded meanfield operators of **[DW1]**, **[DW2]**, so we shall use the same expression for our ones.
The example (1) fits into this framework. Indeed, set $h = x_1^{\otimes\alpha_1} \cdots \otimes x_r^{\otimes\alpha_r}$ and $G(u) = i[h, u]$ on $\mathcal{A}_g$ ($g = |\alpha|$). Then $B^Q = N\mathbf{P}(G \otimes I_{N-g})$.

In the next section we shall show that a meanfield operator as in example (2) is always a meanfield generator. The corresponding meanfield dynamics $(\hat{T}_t)$ (see 3.3) leaves $1 \otimes \mathcal{C}(\mathcal{S}(\mathcal{A}))$ invariant and there it is induced by a classical flow, which is best understood in the case where $\mathcal{A}$ is finite dimensional. The rest of this section contains the connection between internal one–parameter semigroups and their nonstandard hull which we need in the rest of the paper.

3.5) Proposition: *Let B be an internal, internally bounded linear operator on the internal Banach space E, and assume that there exists a standard constant M, such that $\|e^{tB}\| \leq M$ for all $t \geq 0$. Let $F \subset E_{fin}$ be the subspace of all x such that $0 \leq t \to e^{tB}x$ is S–continuous at 0. Then F is invariant under $T_t := e^{tB}$ $(t \geq 0)$, and by $\hat{T}_t\hat{x} = \widehat{(T_tx)}$ there is defined a strongly continuous one–parameter semigroup $(\hat{T}_t)_{t\geq 0}$ on $\hat{F} = F/(E_0 \cap F)$, which is norm bounded by M.*

Proof: Straigtforward (cf. also **[W]**, sect. 1 and prop. 4.1). ∎

We describe the generator of $(\hat{T}_t)$:

3.6) Corollary: *The generator A of $(\hat{T}_t)$ is given by $\mathcal{D}(A) = [(\lambda I - B)^{-1}]^\wedge(\hat{F})$ for some $\lambda > 0$ (standard), and $A\hat{x} = \widehat{(Bx)}$ for all $\hat{x} \in \mathcal{D}(A)$.*

Proof: (I) Let $\lambda > 0$ be standard. Then $(\lambda I - B)^{-1} =: R(\lambda) = \int_0^\infty e^{-\lambda t}T_t dt$

(by transfer) hence $\|R(\lambda)\| \leq \frac{M}{\lambda}$, that is $\hat{R}(\lambda)\hat{E} \subset \hat{E}$. Moreover for $y \in F$ and $x = R(\lambda)y$ we have
$T_t x = e^{\lambda t}(x - \int_0^t T_s y ds)$ for all t, which shows that $x \in F$, hence $\hat{R}_\lambda(\hat{F}) \subset \hat{F}$.
(II) Let y be in F. Then $T_s y \approx y$ implies $\frac{1}{t}\int_0^t T_s y ds \approx y$ for all $t \approx 0$, hence $\frac{T_t x - x}{t} \approx \lambda x - y$ for all these t. Since our model is saturated, we conclude that to every $\varepsilon > 0$ standard there exists $\delta > 0$ standard satisfying $\|(T_t x - x)/t - (\lambda x - y)\| < \varepsilon$ for all $0 < t < \delta$. This implies $\hat{x} \in \mathcal{D}(A)$ and $A\hat{x} = \lambda\hat{x} - \hat{y} = \widehat{(Bx)}$.
(III) By II $(A - \lambda)\hat{R}(\lambda) = -I$, and the corollary follows. ∎

3.7) Corollary 2: $F = \{x \in E_{fin} : Bx \in E_{fin}\}$

Proof: By the transfer principle we have $T_t x - x = \int_0^t T_s Bx ds$, hence for all $t \approx 0$ $\|T_t x - x\|$ $(\leq t \cdot M \cdot \|Bx\|) \approx 0$ holds iff $\|Bx\|$ is limited. ∎

4 Polynomially bounded meanfield operators

In the following let $g \in \mathbb{N}$ be a fixed standard integer and let $G \in \mathcal{L}(\mathcal{A}g)$ be a standard, bounded linear operator on $\mathcal{A}_g$ such that for every $t \geq 0$ $\exp(tG)$ is unital and completely positive. This implies in particular $G\mathbf{1} = 0$.

First of all we want to establish another formula for $B = N\mathbf{P}(G \otimes I_{N-g})$. To this end let $L \subset \{1, \ldots, N\}$ be a fixed set of size g, and denote by Φ_L the unique permutation mapping L onto $\{1, \ldots g\} =: K$ and $N\backslash L$ onto $N\backslash K$ in such a manner that it is order preserving on L and order preserving on $N\backslash L$. Let π be another permutation mapping L onto K. Then there exists a unique permutation ρ on K and a unique permutation σ on $N\backslash K$ such that $\pi = (\rho \oplus \sigma) \circ \Phi_L$ where $\rho \oplus \sigma$ is the permutation acting as ρ on K and as σ on $N\backslash K$. Let i denote the identity (on the set under consideration). Then $\rho \oplus \sigma = (\rho \oplus i)(i \oplus \sigma) = (i \oplus \sigma)(\rho \oplus i)$. so for $\underline{G} = G \otimes I_{N-g}$ we have

$$\begin{aligned} B &= \frac{1}{(N-1)!}\sum_\pi U_\pi^{-1}\underline{G}U_\pi \\ &= \frac{g!}{(N-1)!}\sum_{|L|=g} U_{\Phi_L}^{-1}\sum_\sigma U_{i\oplus\sigma}^{-1}\left(\frac{1}{g!}\sum_\rho U_{\rho\oplus i}^{-1}\underline{G}U_{\rho\oplus i}\right)U_{i\oplus\sigma}U_{\phi_L}. \end{aligned}$$

Setting

$$\bar{G} = \frac{1}{g!}\sum_{\rho\in S(g)} V_\rho^{-1}GV_\rho$$

where V_ρ is the automorphism on $\mathcal{A}_g$ induced by the permutation ρ we obtain

$$B = \frac{g!(N-g)!}{(N-1)!}\sum_{|L|=g} U_{\Phi_L}^{-1}\underline{\bar{G}}U_{\Phi_L}.$$

Denoting $U_{\Phi_L}^{-1}\underline{\bar{G}}U_{\phi_L}$ by G_L we can summarize our considerations as follows:

4.1) Lemma: $B := N\mathbf{P}(G \otimes I_{N-g}) = \frac{N}{\binom{N}{g}} \sum_{|L|=g} G_L$.

It follows in particular that we can assume $G = \bar{G}$ without loss of generality whenever it is convenient. We put together some easy consequences of this lemma:

4.2) Lemma: *Let $K = \{1, \dots, r\}$ for some standard $r \geq g$, and set for $j = 0, 1$ $B_j = \frac{N}{\binom{N}{g}} \sum_{\substack{|L|=g \\ |L\cap K|=j}} G_L$. Then the following assertions are true:*

(1) $\|B - (B_0 + B_1)\| \approx 0$

(2) $\|B_1\| \lessapprox r \cdot g \cdot \|G\|$

(3) $B_0 = 0$ *on* $^*\mathcal{A}_r$ *(identified with* $j_{N,r}(^*\mathcal{A}_r)$*)*

(4) $\|B_1 P|_{^*\mathcal{A}_r}\| \approx 0$

Proof: From 4.1 we obtain $B = \frac{N}{\binom{N}{g}} \sum_{j=0}^{g} \sum_{\substack{|L|=g \\ |L\cap K|=j}} G_L$. So we first have to show $\|\frac{N}{\binom{N}{g}} \sum_{\substack{|L|=g \\ |L\cap K|=j}} G_L\| \approx 0$ for $j \geq 2$. Now for $0 \leq j \leq g$ there are $S(j,g,r) = \binom{r}{j}\binom{n-r}{g-j}$ subsets L of size g satisfying $|L\cap K| = j$. Moreover $\|G_L(a)\| \leq \|G\|\|a\|$ independently of L as follows easily from the definition of G_L. Finally

$$\begin{aligned}
\frac{S(j,g,r)\cdot N}{\binom{N}{g}} &= \frac{\binom{r}{j}(N-r)!g!(N-g)!}{(g-j)!(N-r-g+j)!(N-1)!} \\
&= \frac{\binom{r}{j}g!}{(g-j)!} \cdot \frac{(N-r)\cdot\cdot(N-r-g+j+1)}{(N-1)\cdot\cdot(N-g+1)} \\
&=: t(j,g,r).
\end{aligned}$$

If $j \geq 2$ then $t(j,g,r) \leq \frac{r!g!}{(N-r+1)} \approx 0$, which proves (1).
If $j = 1$ then $t(j,g,r) \approx rg$ which implies 2.
If $a \in {}^*\mathcal{A}_r$ and $\tilde{a} = a \otimes \mathbf{1}_{N-r}$ then for $L \cap K = \emptyset$ $U_{\phi_L}(\tilde{a}) = \mathbf{1}_r \otimes b$ for some $b \in \mathcal{A}_{N-r}$.
But $G\mathbf{1} = 0$, hence $G_L = 0$, so (3) follows.
It remains to prove (4). Claim: If $b = P(c \otimes \mathbf{1}_{N-r})$ is limited then $b \approx \frac{1}{\binom{N}{r}} \sum_{\phi, \phi(K)\cap K=\emptyset} U_\phi(c \otimes \mathbf{1}_{N-r}) =: \tilde{b}$. For considerations similar to those of lemma 4.1 show

$$P(c \otimes \mathbf{1}_{N-r}) = \frac{1}{\binom{N}{r}} \sum_{|L|=r} U_{\Phi_L}(c \otimes \mathbf{1}_{N-r})$$

$$= \frac{1}{\binom{N}{r}} \sum_{k=1}^{r} \sum_{\substack{|L|=r \\ |L\cap K|=k}} U_{\phi_L}(c \otimes \mathbf{1}_{N-r})$$

Now $\frac{S(k,r,r)}{\binom{N}{r}} \approx 0$ for $k \geq \mathbf{1}$, so the claim is proved. But since B_1 is limited we have
$B_1 P(c \otimes \mathbf{1}_{N-r}) \approx B_1(\tilde{b}) = 0$ since $G1 = 0$. ∎

So far we have all the ingredients needed to prove that B behaves almost as a derivation – at least for a restricted set of products:

4.3) Proposition: (a) *Let* $a_1, \ldots a_r \in {}^*\mathcal{A}_{fin}$ *be arbitrary* (r : *standard*). *Then*

$$B(\Pi_{j=1}^{r} a_j(j)) = \sum_{j=1}^{r} \Pi_{k\neq j} a_k(k) \cdot B(a_j(j))$$

(b) *Let* $r \geq g$ *be standard and let* $a, b \in \mathcal{A}_r$ *standard. Set* $\tilde{a} = a \otimes \mathbf{1}_{N-r}$, $c = P(b \otimes \mathbf{1}_{N-r})$. *Then*
(1) $B_0(\tilde{a}c) = \tilde{a}B_0(c)$
(2) $B_1(\tilde{a}c) = B_1(\tilde{a})c$
(c) *Let* r, s *be standard, and let* $a \in \mathcal{A}_r$, $b \in \mathcal{A}_s$ *be standard and* $c = P(b \otimes 1_{N-s})$.
Then $B(\tilde{a}c) \approx B(\tilde{a})c + \tilde{a}B(c)$.
(d) *Let* a, b *as in (c) and set* $d = P(a \otimes \mathbf{1}_{N-r})$, $c = P(b \otimes \mathbf{1}_{N-s})$.
Then $B(cd) \approx B(c)d + cB(d)$.

Proof: (a) Using 4.2 we obtain for $a = \Pi_1^r a_j(j)$
$$B(a) \approx B_1(a) = \frac{N}{\binom{N}{g}} \sum_{k=1}^{r} \sum_{\substack{L\cap K=\{k\} \\ |L|=g}} U_{\Phi_L^{-1}} G U_{\Phi_L}(a).$$
Now if $|L| = g$ and $L \cap K = \{k\}$ $(1 \leq k \leq r)$ then
$U_{\Phi_L}^{-1} G U_{\Phi_L}(a) = \Pi_{l\neq k} a_l(l) U_{\Phi_L}^{-1}(G(a_k \otimes \mathbf{1}_{g-1}))$.
Inserting this into the formula above we obtain the result.
(b) (1): Since $U_{\Phi_L}(c) = c$ we obtain $B_0(\tilde{a}c) = \frac{N}{\binom{N}{g}} \sum_{\substack{|L|=g \\ |L\cap K|=\emptyset}} U_{\Phi_L}^{-1}(G((U_{\Phi_L}(\tilde{a}))c))$.

By the formula for P in the proof of 4.2 (4) we get
$$c = P(b \otimes \mathbf{1}_{N-r}) = \frac{1}{\binom{N}{r}} \sum_{|L'|=r} U_{\Phi_{L'}}(b \otimes \mathbf{1}_{N-r}).$$

If $L \cap K = \emptyset = L' \cap K$ then $U_{\Phi_L}(\tilde{a}) U_{\Phi_{L'}}(b \otimes \mathbf{1}_{N-r}) = \mathbf{1}_r \otimes d$ for some $d \in \mathcal{A}_{N-r}$ hence G vanishes on such an element. Moreover $U_{\Phi_L}^{-1}(\tilde{a}) = \mathbf{1}_r \otimes d_L$ for some $d_L \in \mathcal{A}_{N-r}$ whenever $L \cap K = \emptyset$. This alltogether gives

$$\begin{aligned} B_0(\tilde{a}c) &= \tilde{a} \frac{N}{\binom{N}{g}} \sum_{L\cap K=\emptyset} U_{\Phi_L}^{-1}(G(\frac{1}{\binom{N}{r}} \sum_{|L'|=r} U_{\Phi_{L'}}(b \otimes \mathbf{1}_{N-r}))) \\ &= \tilde{a}B_0(c). \end{aligned}$$

(2): Since B_1 is norm–limited we can use the approximation of $P(\tilde{b})$ in the proof of 4.2 obtaining $B_1(\tilde{a}c) \approx \frac{N}{\binom{N}{g}\binom{N}{r}} \sum_{|L\cap K|=1} \sum_{L'\cap K=\emptyset} U_{\Phi_L}^{-1}(G((U_{\Phi_L}(\tilde{a}))U_{\Phi_{L'}}(\tilde{b})))$.

A careful but straightforward calculation similar to that one for B_0 above will give the formula.
(c) Let $t = max(r, s, g)$. Then one may view a, b as elements of $\mathcal{A}_t$ and G as an operator on $\mathcal{A}_t$ (replacing G by $G \otimes I_{t-g}$ if necessary). But then 4.2 and (b) will give

$$\begin{aligned} B(\tilde{a}c) &\approx B_0(\tilde{a}c) + B_1(\tilde{a}c) \approx \tilde{a}B_0(c) + B_1(\tilde{a})c \\ &\approx \tilde{a}B(c) + B(\tilde{a})c, \end{aligned}$$

where B_0, B_1 are built with respect to $K = \{1, \ldots t\}$.
(d) Apply the conditional expectation P to (c) and notice $PB = BP$. ∎

We need one further lemma before we can prove our first main theorem.

4.4) Lemma: *For $r, s \in \mathbb{N}$ standard denote the S–closure of the external $\mathbb{C}$–linear hull of $\{a \otimes \mathbf{1}_{N-r}P(c \otimes \mathbf{1}_{N-s}) : a \in \mathcal{A}_r,\ c \in \mathcal{A}_s,\ \text{standard } r\}$ by $\mathcal{E}_{r,s}$. Then B maps $\mathcal{A}_r$ (identified with $j_{N,r}(\mathcal{A}_r)$) into $\mathcal{E}_{r,g-1}$ and $\|B|_{\mathcal{A}_r}\| \stackrel{<}{\approx} \|G\| g \cdot r$.*

Proof: We build B_0, B_1 as in 4.2 and obtain $B|_{\mathcal{A}_r} = B_1|_{\mathcal{A}_r}$ by 4.2 (3), hence the norm estimate follows from 4.2 (2). Since B_1 is norm–limited we need only to prove $B_1(\Pi_{j=1}^r a_j(j)) \in \mathcal{E}_{r,g-1}$ for all $a_1, \ldots a_r \in \mathcal{A}$ standard. By 4.3 (1) we can reduce this further to $B_1(a(j)) \in \mathcal{E}_{r,g-1}$ for all $1 \leq g \leq r$ and $a \in \mathcal{A}$. But

$$B_1(a(j)) = \frac{N}{\binom{N}{g}} \sum_{k=1}^{r} \sum_{\substack{L\cap K=\{k\} \\ |L|=g}} U_{\Phi_L}^{-1} \underline{G} U_{\Phi_L}(a(j)),$$

where $\underline{G} = {}^*G \otimes I_{N-g}$.
Since $\Phi_L^{-1}(l) \geq r+1$ for $1 < l \leq r$ and $\Phi_L^{-1}(1) = k$ we obtain

$$B_1(a(j)) = \frac{N}{\binom{N}{g}} \sum_{\substack{|L|=g \\ |L\cap K|=\{j\}}} U_{\Phi_L}^{-1} \underline{G}(a(1)).$$

Now define for $1 \leq k \leq N$ $V_k : {}^*\mathcal{A} \to \mathcal{A}_N$ by $V_k(a) = a(k)(= 1_{k-1} \otimes a \otimes \mathbf{1}_{N-k})$ and for $L' = L_j \backslash j$ $V_{L'} = \otimes_{k\in L'} V_k : {}^*\mathcal{A}_{g-1} \to \mathcal{A}_N$. Then $U_{\Phi_L}^{-1}(\underline{G}(a(1))) = (V_j \otimes V_{L'})(G(a(1))$. Using 2.6.1 we obtain

$$B(a(j)) \approx (V_j \otimes P_{N-1})\underline{G}(a(1)) \in \overline{\mathcal{A}_g \bullet P(\mathcal{A}_{(g-1)})}^S \quad \text{(S–closure).} \ ∎$$

4.5) Corollary: *Let $\mathcal{E}_{r,\infty}$ be the S–closure ot the external $\mathbb{C}$–linear hull of $\{a \otimes \mathbf{1}_{n-r} \cdot P(\mathcal{A}_{(0)}) : a \in \mathcal{A}_r \text{ standard}\}$ or in other words: $\mathcal{E}_{r,\infty} = Q^{-1}(\mathcal{A}_r \otimes \mathcal{N})$. Then B_1 maps $\mathcal{E}_{r,\infty}$ into itself.*

Proof: B_1 is limited, hence S–continuous, and the assertion follows from 4.4. ∎

Now we are able to prove the following final theorem on polynomially bounded meanfield operators which is essentially due to N. Duffield and R. Werner **[DW2]**:

4.6) Theorem: *Let $G \in \mathcal{L}(\mathcal{A}_g)$ be a continuous linear (standard) operator on $\mathcal{A}_g$ such that for all $t \geq 0$ $\exp(tG)$ is a unital completely positive operator on $\mathcal{A}_g$. Then $B = N\mathbf{P}(G)$ is a meanfield generator.*
More precisely set $T_t = \exp(tB)$ for $t \geq 0$. Then for every standard t, $\hat{T}_t$ leaves invariant $\mathcal{D} := \mathcal{A}_\infty \otimes \mathcal{N}$ and the restriction of the semigroup $(\hat{T}_t)$ to $\mathcal{D}$ is strongly continuous. Moreover the following assertions are true:
(1) $\mathbf{E}\hat{T}_t = \hat{T}_t\mathbf{E}$ for all $t \geq 0$ and the restriction of $(\hat{T}_t)$ to $1 \otimes \mathcal{N} = \mathbf{E}(\mathcal{D})$ is a strongly continuous semigroup of C^*-homomorphisms, thus it is induced by a flow Φ_t on $\mathcal{S}(\mathcal{A})$ (notice $\mathcal{N} = C(\mathcal{S}(\mathcal{A}))$).
(2) For every $r \geq g$ $(\hat{T}_t)$ leaves invariant $\mathcal{A}_r \otimes \mathcal{N}$.

4.7) Corollary: *Let A denote the generator of $(\hat{T}_t)_{|\mathcal{D}}$. Then the following assertions hold:*
(1) The linear hull $\mathcal{D}_0$ of the set $\mathcal{A}_{(0)} \cdot \mathbf{E}(\mathcal{A}_{(0)}) = \{af : a \in \mathcal{A}_{(0)}, f \in \mathbf{E}(\mathcal{A}_0)\}$ is contained in the domain $\mathcal{D}(A)$ of A.
(2) For all $a_1, \dots a_r \in \mathcal{A}$

$$A(a_1 \otimes \cdot \cdot \otimes a_r) = \sum_{k=1}^{r} \otimes_{j \neq k} a_j \cdot A(a_k \otimes 1)$$

(modulo the obviously necessary identifications).
(3) For all $r \in \mathbb{N}$, $a \in \mathcal{A}_r$ and $f = \mathbf{E}(b) \in \mathbf{E}(\mathcal{A}_{(0)})$ $\quad A(af) = A(a)f + aA(f)$.

Proof of 4.7: This follows from 4.3, 4.6 and 3.6. ∎

Proof of 4.6: Let $F \subset \mathcal{A}_{N,fin}$ be the space on which (T_t) is strongly S–continuous (cf. 3.5). Then by 3.7 and 4.2, 4.3, 4.4 the external $\mathbb{C}$–linear hull of $\{a \otimes \mathbf{1}_{N-r} \cdot b :$ r standard $, a \in \mathcal{A}_r$ standard $, b \in P(\mathcal{A}_0)\}$ lies in the domain of the generator A of $(\hat{T}_t)$ on $\hat{F}(= Q(F))$. Moreover 4.4 and the equation $PB = BP$ implies that $\|B|_{P(\mathcal{A}_r)}\| \leq rg\|G\|$.
(I) Claim: $\mathcal{N} \subset \hat{F}$ and $\hat{T}_t(\mathcal{N}) \subset \mathcal{N}$
Proof: By the last inequality and 3.6 we obtain $Q(P(\mathcal{A}_r)) := \mathcal{D}_r \subset \mathcal{D}(A)$ and $\|A_{|\mathcal{D}_r}\| \leq rg\|G\|$. Since A is closed $\overline{\mathcal{D}_r} \subset \mathcal{D}(A)$ and $\|A_{|\overline{\mathcal{D}_r}}\| \leq rg\|G\|$.
By induction on n we obtain for $a \in \overline{\mathcal{D}_r}$ $\|A^n a\| \leq g^n \|G\|^n \Pi_{k=0}^{n-1}(r + kg)\|a\|$. Let $r' = min\{k \geq r/g : k \in \mathbb{N}\}$ Then

$$\begin{aligned} \Pi_{k=0}^{n-1}(r+kg) &\leq g^n \Pi_{k=0}^{n-1}(r'+k) = g^n \cdot n! \binom{r'+n}{r'} \\ &\leq g^n \cdot n! 2^{r'} \cdot 2^n. \end{aligned}$$

This shows that for $0 \leq t < (\|G\| \cdot g^2 \cdot 2 \cdot 2^r)^{-1}$ the series $\sum\limits_{k=0}^{\infty} \frac{t^k A^k a}{k!}$ converges, i.e. a is an analytic element in $\mathcal{D}(A)$, and moreover the series is equal to $\hat{T}_t(a)$, and lies in $\mathcal{N}$. Since $\bigcup_{r=1}^{\infty} \overline{\mathcal{D}_r} =: \mathcal{D}_\infty$ is dense in $\mathcal{N}$, the assertion follows.

(II) Fix an $r \in \mathbb{N}$, and consider B_0 and B_1 as in 4.2. B_1 is limited. Set $V_t = e^{tB_1}$. By 3.4 and 3.5 $\hat{V}_t$ is uniformly continuous on $\hat{\mathcal{A}}_N$. Moreover 4.5 tells us that $\hat{V}_t(\mathcal{A}_r \otimes \mathcal{N}) \subset \mathcal{A}_r \otimes \mathcal{N}$. Now consider also $S_t = \exp(tB_0)$ $B_0(\mathcal{A}_r) = 0$ by 4.2 (3) and $B_0|_{P(\mathcal{A}_0)} \approx B|_{P(\mathcal{A}_0)}$ by 4.2 (4).

This gives $\hat{S}_t|_{\mathcal{A}r\otimes\mathcal{N}} = I_{\mathcal{A}r} \otimes \hat{T}_t|_{\mathcal{N}}$, possessing the generator $A_0 = I_{\mathcal{A}r} \otimes A|_{\mathcal{N}\cap\mathcal{D}(A)}$. But then $A_2 = A_0 + A_1$ is a generator, where $A_1 = \hat{B}_1$ is bounded. The semigroup generated by A_2 is given by $W_t = \lim_{n\to\infty}(\hat{S}_{t/n} \cdot \hat{V}_{t/n})^n$ ([**BR1**], 3.2.31).

(III) Claim: $A_2 = A|_{\mathcal{D}(A)\,\cap\,\mathcal{A}_r\,\otimes\,\mathcal{N}}$.

Let $a \in \mathcal{A}_r$, $b \in P(\mathcal{A}_s)$ be arbitrary. Then $\widehat{(ab)} = \hat{a} \otimes \hat{b} \in \mathcal{D}(A)$ since $B(a \otimes \mathbf{1}_{N-r} \cdot b)$ is finite. Moreover 4.3 (b) and 4.2 (1) imply $B(a \otimes \mathbf{1}_{N-r} \cdot b) \approx B_1(a \otimes \mathbf{1}_{N-r}b) + a \otimes \mathbf{1}_{N-r}B_0(b)$. Thus $A(\hat{a}\otimes\hat{b}) = A_1(\hat{a}\otimes\hat{b}) + A_0(\hat{a}\otimes\hat{b})$. But $B_1(a\otimes\mathbf{1}_{N-r})b \approx B_1(a\otimes\mathbf{1}_{N-r}b)$, by 4.3 b (2). So the assertion follows easily.

(IV) Finally we have to prove that $\hat{T}_t|_{\mathcal{N}}$ consists of *–homorphisms. Since $\mathcal{N}$ is commutative, and $\hat{T}_t$ maps the selfadjoint part onto itself it is enough to show that A_0 is a derivation. But this follows from 4.3 c. ∎

4.8) Remarks: 1) As was pointed out already, theorem 4.6 and its corollary are essentially due to N. Duffield and R. Werner [**DW2**]. We have presented them here since our proof seems to be much easier. Moreover our approach reflects the physical intuition:
Without the use of the equivalence relation " $\approx$ ", i.e. without the use of the quotient $\hat{\mathcal{A}}_N$, we remain totally within the *microscopic picture.* Only the use of that equivalence relation yields the result. But this relation means "indiscernable by experiments of a macroscopic type". Loosely spoken: *classical effects result from quantum mechanical microscopic systems by observation methods suppressing the microscopic details or making them indiscernable.*
2) One of the concrete results coming out of 4.6 is the following observation due to Unnerstall [**UN**] : even if the given local action G is dissipative the induced dynamics on the "classical" part $\mathcal{N}$ is "deterministic", i.e. it comes from a flow, a result which supports again the macroscopic nature of the classical dynamical part.

5 Approximation of meanfield generators

Let $\mathcal{E}$ denote $Q^{-1}(\mathcal{A}_\infty \otimes \mathcal{N})$, and let B be a meanfield operator. An S–dense subspace $M \subset \mathcal{E}$ is called a core of B, if $\{(x, Bx) : x \in M\}$ is S–dense in $\{(x, Bx) : x \in \mathcal{D}(B)\}$ where $\mathcal{D}(B) = \{x \in \mathcal{E} : Bx \text{ is limited }\}$.

First of all we formulate the following "uniqueness theorem ":

5.1) Proposition: *Let $B \in \mathcal{L}(\mathcal{A}_N)$ be meanfield generator, and let $D \in \mathcal{L}(\mathcal{A}_N)$ be arbitrary such that for $\lambda > 1$ $R(\lambda, D)$ exists and is limited and moreover such that $Du \approx Bu$ for all u in some fixed core M of B. Then D is also a meanfield generator and both induce the same meanfield dynamics on $\mathcal{A}_\infty \otimes \mathcal{N}$.*

Proof: Since M is a core and 1 is in the resolvent set of B $(\lambda - B)(M)$ is S–dense in $\mathcal{E}$ for all standard $\lambda \geq \mathbf{1}$.
Now $R(\lambda, D) - R(\lambda, B) = R(\lambda, D)(B - D)R(\lambda, B)$. Hence $R(\lambda, D)x \approx R(\lambda, B)x$ for all $x \in (\lambda - B)M$. Since both operators are limited this approximate identity holds on $\mathcal{E}$, hence $\widehat{R(\lambda, D)} = \widehat{R(\lambda, B)}$ on $\mathcal{A}_\infty \otimes \mathcal{N}$ and the assertion follows. ∎

Now we want to calculate explicitly the action of a polynomially bounded operator B, induced by a hamiltonian. Let $R = R(\xi_1, \ldots, \xi_s)$ be a polynomial in s variables. Then the partial derivatives $D_j R = \frac{\partial R}{\partial \xi_j}$ are again polynomials. Now if $x_1, \ldots, x_s$ are elements of $\mathcal{A}_{sa}$ then we obtain $R(x_1, \ldots, x_s)$ by replacing ξ_j by x_j. We denote this polynomial by $R(\vec{x})$. If $R(\vec{x}) = x_1^{\alpha_1} \cdot\cdot x_s^{\alpha_s}$ then we define $\tilde{R}(\vec{x})$ by $x_1^{\otimes \alpha_1} \otimes \cdot\cdot x_s^{\otimes \alpha_s}$ as in section 2. For a general polynomial $R(\vec{x})$ we define $\tilde{R}(\vec{x})$ by linear extension. The element $P(\tilde{R}(\vec{x}))$ will be denoted by h_R. Clearly $Q(h_R) = R(f_{x_1}, \ldots, f_{x_s})$ (see section 2).
If R is a monomial, $R(\vec{x}) = x_1^{\alpha_1} \cdots x_s^{\alpha_s}$, then we define G_R on $\mathcal{A}_g$ $(g = \sum \alpha_j)$ by $G_R(a) = i[\tilde{R}(\vec{x}), a]$, and similarly for general polynomials. Then
$N\mathbf{P}(G_R) = B$ is given by $B(u) = Ni[h_R, u]$ (see 3.3) thus B is an internal inner derivation.

The next two formulas are easily proved by showing them at first for monomials:

5.2) Lemma:

(a) *For* $a \in \mathcal{A}$ $\quad B(a(1)) \approx i \sum_{j=1}^{s} [x_j, a](1) \cdot h_{\widetilde{D_j R}}$

(b) *For* $\bar{a} = P(a(1))$ $\quad B(\bar{a}) \approx i \sum_{j=1}^{s} \overline{[x_j, a]} \cdot h_{\widetilde{D_j R}}$

Proof: (a) Let $R(\xi_1, \ldots, \xi_s) = \xi_1 \cdots \xi_s$ and choose $x_1, \ldots x_g \in \mathcal{A}_{sa}$ not necessarily distinct.
Then $h_{\tilde{R}} = P(x_1(1) \cdots x_g(g)) = \frac{(N-g)!}{N!} \sum_{\vec{k}} (0) x_1(k_1) \cdot\cdot x_g(k_g)$ (see sect. 2). This gives

$$
\begin{aligned}
B(a(1)) &= i\frac{(N-g)!}{(N-1)!} \sum_{\vec{k}} (0)[x_1(k_1) \cdots x_g(k_g), a(1)] \\
&= i \cdot \sum_{j=1}^{g} [x_j, a](1) \otimes \frac{(N-g)!}{(N-1)!} \sum_{\vec{k}} (0) x_1(k_1) \cdot\cdot \widehat{x_j(k_j)} \cdot\cdot x_g(k_g).
\end{aligned}
$$

Here the roof denotes as usual that this factor does not occur and $\vec{k}'$ denotes all those indices where k_j does not occur. But this gives (cf. sect. 2)

$$
\begin{aligned}
B(a(1)) &= i \sum_{j=1}^{g} [x_j, a](1) \otimes P_{N-1}(x_1(1) \cdot\cdot \widehat{x_j(j)} \cdot\cdot x_g(g)) \\
&\approx i \sum_{j=1}^{g} [x_j, a](1) \cdot h_{\tilde{D_j R}}
\end{aligned}
$$

and (a) follows easily. Now we apply P to (a) and obtain (b). ∎

Because B is an internal inner derivation, it is completely determined by formulas (a) and (b) (cf. also 4.7). In particular we obtain the following corollary (see e.g. **[BO1]**, **[DW1]**).

5.3) Corollary 1: *The dynamics induced by $R(x_1,\ldots,x_g)$ on $\mathcal{N} = C(\mathcal{S}(\mathcal{A}))$ is completely described by the following Poisson bracket for polynomials $S(f_{y_1},\ldots,f_{y_s})$:*

$$\{R(f_{x_1},\ldots f_{x_g}),\, S(f_{y_1},\ldots,f_{y_s})\} = \sum_{\substack{k=1..g\\ l=1..s}} D_k R(f_{x_1},\ldots,f_{x_g}) D_l S(f_{y_1},\ldots f_{y_s}) f_{[x_k,\, y_l]}$$

By 4.6 and 4.7 the global dynamics on $\mathcal{A}_\infty \otimes \mathcal{N}$ leaves invariant all $\mathcal{A}_r \otimes \mathcal{N}$ whenever $r \geq g$. But since in the case under consideration B is a derivation this holds also for *all* r. Since we obtain B on $\mathcal{A}_r \bullet P(\mathcal{A}_{(0)})$ as a sum of operators acting on $\mathcal{A}_{\{l\}} \bullet P(\mathcal{A}_{(0)})$ we need only to consider this case. Lemma 5.2 yields that $B|_{\mathcal{A}_1 \bullet P(\mathcal{A}_{(0)})}$. can be decomposed into a bounded operator $B_1 : a(1)\cdot u \to i \sum_{j=1}^{s} [a_j, a](1) \cdot h_{D_j^{\sim} R} \cdot u$ and an unlimited one. Going into the quotient we obtain the following corollary ($\mathcal{A}_{\{k\}} = \{1_{k-1} \otimes x : x \in \mathcal{A}\} \subset \mathcal{A}_k = \mathcal{A}^{\otimes k}$) :

5.4 Corollary 2: *$(\hat{T}_t)$ leaves invariant all sub–algebras $\mathcal{A}_{\{k\}} \otimes \mathcal{N}$. It is uniquely determined by the action of its generator on $\mathcal{A}_1 \otimes \mathcal{N}$, which is given by $A|_{\mathcal{D}(A)\cap \mathcal{A}_1 \otimes \mathcal{N}} =: A_1 = A_2 + A_0$ where $A_0(a(1) \otimes f) = \sum [x_j, a] \otimes D_j^{\sim} R \cdot f$, i.e. A_0 is bounded, and $A_2 = I \otimes A_3$, where A_3 is uniqueley determined by the Poisson brackets for polynomials of coroll. 1.*

We come now to the final result of this section. For the sake of convenience we assume that $\mathcal{A}$ is finite–dimensional. Then we may select a base $x_1, \ldots x_s$ of $\mathcal{A}_{sa} \ominus \mathbb{R}\mathbf{1}$ over $\mathbb{R}$ which is orthonormal with respect to the trace (we consider $\mathcal{A}$ to be a C^*–subalgebra of some full matrix algebra). Then since $\mathcal{A}_{sa}$ is a Lie–algebra with respect to $[.,.]$, $\mathcal{S}(\mathcal{A})$ becomes a Poisson manifold in a natural manner (see **[AR]**, p. 457. All these cases of a Poisson structure occur in [BO1] and are already mentioned in [HL].). Moreover all polynomials $R(y_1, \ldots y_g)$ are expressable as polynomials in the usual sense, refering only to $\{x_1, \ldots, x_s\}$ (see also **[UN]**). The Poisson bracket is given for C^1–functions F, G by

$$\{F, G\} = \sum_{j,k=1}^{s} \frac{\partial F}{\partial x_j} \frac{\partial G}{\partial x_k} f_{[x_j,\, x_k]}$$

(see **[AR]**, p. 457). So every C^2–function F defines a flow Φ_F on V'.

In the following we suppose that F is a C^2–function such that $\mathcal{S}(\mathcal{A})$ is positively invariant set of Φ_F. We denote by (S_t) the semigroup of *–automorphisms on $C(S(\mathcal{A}))$ defined by $(S_t f)(\rho) = f(\Phi(t, \rho))$. Since F is of class C^2, Φ_F is of class C^1 with respect to $\rho \in \mathcal{S}(\mathcal{A})$, hence $\mathcal{C}^1(\mathcal{S}(\mathcal{A}))$ is a core of the generator of (S_t).

5.5) Theorem: *Let (R_n) be a sequence of real polynomials which converges to F uniformly with all its derivatives of order one. Let $(\hat{T}_{n,t})$ be the meanfield dynamics* *on $\mathcal{A}_\infty \otimes \mathcal{N}$ corresponding to R_n (see 5.3). Then there exists a strongly continuous semigroup $(\hat{T}_t)$ of *-homomorphisms on $\mathcal{A}_\infty \otimes \mathcal{N}$ such that $\hat{T}_{n,t}u \to \hat{T}_t u$ uniformly on every bounded interval. The restriction $\hat{T}_t|_{\mathcal{N}}$ is given by the flow induced by F on $\mathcal{S}(\mathcal{A})$.*

Roughly spoken: *every C^2-function F which induces a C^1-semiflow on $\mathcal{S}(\mathcal{A})$ via the Poisson bracket above determines a classical dynamics which is the restriction of a meanfield hamiltonian dynamics on $\mathcal{A}_\infty \otimes \mathcal{N}$.*

Proof: By 5.3, 5.4, 4.6 and 4.7 we have only to prove the convergence on $\mathcal{A}_1 \otimes \mathcal{N}$. We decompose the generators $A_n|_{\mathcal{D}(A_n) \cap \mathcal{A}_1 \otimes \mathcal{N}}$ into $A_{2,n} + A_{0,n}$ according to 5.4. Then $(A_{0,n})$ converges strongly to a bounded operator A_0, say, on $\mathcal{A}_1 \otimes \mathcal{N}$ since $D_j R_n \to D_j F$ uniformly by hypothesis. $A_{2,n} = I \otimes A_{3,n}$, where $A_{3,n}$ is given by the Poisson bracket $\{R_n, \cdot\}$. This bracket converges strongly to $\{F, \cdot\}$ on $C^1(\mathcal{S}(\mathcal{A}))$ by hypothesis. The latter is a derivation $G \to \delta(G) = \{F, G\}$ the closure of which is the generator of (S_t) (see above and note that C^1 is a core of δ). But then $(S_{n,t}) \to (S_t)$ strongly (uniformly on compact intervals), (where $(S_{n,t})$ corresponds to $\{R_n, \cdot\} = \delta_n = A_{3,n}$) by 5.7.11 in **[FAT]**.
So $A_{2,n} + A_{0,n} \to I \otimes \delta + A_0$ on $\mathcal{A}_1 \odot C^1(\mathcal{S}(\mathcal{A}))$ (where $\odot$ denotes the algebraic tensor product) which is a core of $I \otimes \delta + A_0$ since A_0 is bounded by the uniform boundednes theorem. 5.7.11 in **[FAT]** yields our theorem. ∎

The proof has shown how the generator of the limit looks like:

5.6) Corollary: *The generator A of the dynamics of 5.5 is given on $\mathcal{A}_1 \odot C^1(\mathcal{S}(\mathcal{A}))$ by* $A(a(1) \otimes G) = \sum_{j=1}^{s} [x_j, a](1) \otimes (D_j F) \cdot G + a(1) \otimes \sum_{j,k} D_j F D_k G f_{[x_j,\, x_k]}$.

5.7) Final remarks: 1) Obviously $\mathcal{A}_{(0)} \odot C^1(\mathcal{S}(\mathcal{A}))$ is a core of A. Since A is a derivation it is completely determined by its values on $\mathcal{A}_1 \odot C^1(\mathcal{S}(\mathcal{A}))$.
2) Let R be a polynomial such that $\mathcal{S}(\mathcal{A})$ is positively Φ_R - invariant. Then the classical dynamics on $\mathcal{C}(\mathcal{S}(\mathcal{A}))$ given by the Poisson bracket is nothing else than the "statistical limit" of the quantum dynamics given by the corresponding commutator bracket $[Nh_R, \cdot]$ on $\overline{P(\mathcal{A}_{(0)})}^S$. This establishes an idea of P. Bóna[4]
The more common idea that the transition from quantum mechanics to classical mechanics is given by letting $\hbar$ be standard ($\neq 0$) or infinitesimal (as is sketched in **[OO]**) seems to be more restrictive.
3) Within our framework it is possible to extend the correspondence of $\{R, \cdot\}$ and $[Nh_R, \cdot]$ from polynomials to arbitrary C^2-functions:
let F be approximated by the internal polynomial R, say, of degree K ($K \approx \infty$), i. e. $\| {}^*F - R\| + \sum \| {}^*D_j F - D_j R\| \approx 0$. Now choose $N = K!$ (or $N = K^K$). Then (cf. the paragraph preceeding 5.2) $N\mathbf{P}(G_R)$ is given by $Ni[h_R, \cdot] = B$ which gives $B(\bar{a}) \approx i \sum_1^N \overline{[x_j, a]} h_{\widetilde{D_j R}}$. So even in this case the correspondence between the two types of brackets, i.e. between classical and quantum mechanics is established. We do not want to go into the details.

[4]oral discussion during my stay at Bratislava in January 1993, cf. also **[BO2]**. I would like to thank the Komenius University at Bratislava for the kind hospitality.

References

[**AFHL**] S. Albeverio, J. E. Fenstad, R. Høegh–Krohn, T. Lindstrøm: Nonstandard methods in stochastic analysis and mathematical physics, Academic Press, Orlando etc. 1986.

[**AR**] V. I. Arnold: Mathematical methods of classical mechanics, second ed. Springer Verlag Berlin etc. 1989.

[**Bo1**] P. Bóna: The dynamics of a class of quantum meanfield theories, J. Math. Phys. 29 (10), 2223 - 2235, 1988.

[**BO2**] P. Bóna: Quantum mechanics with mean - field backgrounds, Comenius University, Fac. of Math. a. Physics, Physics Preprint No. PH10–91, Oct. 1991.

[**BR1**] O. Bratteli, D. W. Robinson: Operator algebras and quantum statistical mechanics I, Springer Verlag Berlin etc., 1979.

[**DR**] E. Duffner, A. Rieckers: On the global dynamics of multi - lattice systems with non–linear classical effects, Z. Naturforsch. 43a, 521 - 532, 1988.

[**DW1**] N. G. Duffield, R. F. Werner: Mean–field dynamical semigroups on C^*–algebras, Rev. Math. Phys. 4, 383 –424, 1992.

[**DW2**] N. G. Duffield, R. F. Werner: Local dynamics of mean–field quantum systems, Helv. Phys. Acta 65, 1016 - 1054, 1992.

[**FAT**] H. O. Fattorini: The Cauchy problem, Encyclop. of Mathematics and its Appl., Addison–Wesley Publ. C., London etc. 1983.

[**GVV**] D. Goderis, A. Verbeure, P. Vets: Non–commutative central limits, Probability Theory, Rel. Fields 82, 527 - 544, 1989.

[**HL**] K. Hepp, E. Lieb: Phase transitions in reservoir–driven open systems with applications to lasers and superconductors, Helv. Phys. Acta 46, 573 - 603, 1973.

[**HM**] C. W. Henson, L. C. Moore: Nonstandard analysis and the theory of Banach spaces, in: A. E. Hurd (ed.): Nonstandard analysis — recent developments, Lecture Notes in Mathematics Vol. 983, Springer Verlag Berlin etc. 1983, 27 - 112.

[**HULO**] A. E. Hurd, P. A. Loeb: An introduction to nonstandard real analysis, Academic Press, Orlando etc. 1985.

[**MS**] G. Morchio, F. Strocchi: Mathematical structures for long–range dynamics and symmetry breaking, J. Math. Phys. 28, 622–635 (1987).

[**MSTV**] J. Manuceau, M. Sirugue, D. Testard, A. Verbeure: The smallest C^*–algebra for canonical commutation relations, Commun. Math. Phys. 32, 231, 1973.

[**N**] E. Nelson: Internal set theory, Bull. Amer. Math. Soc. 83, 1165 - 1198, 1977.

[**OO**] I. Ojima, M. Ozawa: Unitary representations of the hyperfinite Heisenberg group and the logical extension methods in physics, Research Institute for Mathem. Sciences, Kyoto University Kyoto, Japan, RIMS–899, Sept. 1992.

[**P**] G. K. Pedersen: C^*-algebras and their automorphism groups, Academic Press, London etc. 1979.

[**RI**] A. Rieckers: On the classical part of the meanfield dynamics for quantum lattice systems in grand canonical representations, J. Math. Phys. 25, 2593 - 2601, 1984.

[**RW**] G. A. Raggio, R. F. Werner, Quantum statistical mechanics of general meanfield systems, Helv. Phys. Acta 62, 980 - 1003, 1989.

[**SL**] K. D. Stroyan, W. A. J. Luxemburg: Introduction to the theory of infinitesimals, Academic Press, New York etc., 1976.

[**ST**] E. Størmer: Symmetric states of infinite tensor products of C^*–algebras, J. Funct. Anal. 3, 44 - 51, 1969.

[**TA**] M. Takesaki: Theory of operator algebras I, Springer Verlag New York etc. 1979

[**UN**] T. Unnerstall: The dynamics of infinite open quantum systems, Lett. Math. Phys. 20, 183 - 187, 1990.

[**WH**] M. P. H. Wolff, R. Honegger: On the algebra of fluctuation operators of a quantum meanfield system, Preprint Univ. Tübingen 1993.

Index